Two Manuals Combined into One Book

AF816539

Ranger Survival

And

Marine Scouting and Patrolling

Department of Defense

Pentagon
Publishing

RANGER CREED

Recognizing that I volunteered as a Ranger, fully knowing the hazards of my chosen profession, I will always endeavor to uphold the prestige, honor, and high esprit de corps of the Rangers.

Acknowledging the fact that a Ranger is a more elite Soldier who arrives at the cutting edge of battle by land, sea, or air, I accept the fact that as a Ranger my country expects me to move further, faster, and fight harder than any other soldier.

Never shall I fail my comrades I will always keep myself mentally alert, physically strong, and morally straight and I will shoulder more than my share of the task whatever it may be, one hundred percent and then some.

Gallantly will I show the world that I am a specially selected and well trained Soldier. My courtesy to superior officers, neatness of dress, and care of equipment shall set the example for others to follow.

Energetically will I meet the enemies of my country. I shall defeat them on the field of battle for I am better trained and will fight with all my might. Surrender is not a Ranger word. I will never leave a fallen comrade to fall into the hands of the enemy and under no circumstances will I ever embarrass my country.

Readily will I display the intestinal fortitude required to fight on to the Ranger objective and complete the mission, though I be the lone survivor.

STANDING ORDERS, ROGERS' RANGERS

1. Don't forget nothing.
2. Have your musket clean as a whistle, hatchet scoured, sixty rounds powder and ball, and be ready to march at a minute's warning.
3. When you're on the march, act the way you would if you was sneaking up on a deer. See the enemy first.
4. Tell the truth about what you see and what you do. There is an army depending on us for correct information. You can lie all you please when you tell other folks about the Rangers, but don't never lie to a Ranger or officer.
5. Don't never take a chance you don't have to.
6. When we're on the march we march single file, far enough apart so one shot can't go through two men.
7. If we strike swamps, or soft ground, we spread out abreast, so it's hard to track us.
8. When we march, we keep moving till dark, so as to give the enemy the least possible chance at us.
9. When we camp, half the party stays awake while the other half sleeps.
10. If we take prisoners, we keep' em separate till we have had time to examine them, so they can't cook up a story between' em.
11. Don't ever march home the same way. Take a different route so you won't be ambushed.
12. No matter whether we travel in big parties or little ones, each party has to keep a scout 20 yards ahead, 20 yards on each flank, and 20 yards in the rear so the main body can't be surprised and wiped out.
13. Every night you'll be told where to meet if surrounded by a superior force.
14. Don't sit down to eat without posting sentries.
15. Don't sleep beyond dawn. Dawn's when the French and Indians attack.
16. Don't cross a river by a regular ford.
17. If somebody's trailing you, make a circle, come back onto your own tracks, and ambush the folks that aim to ambush you.
18. Don't stand up when the enemy's coming against you. Kneel down, lie down, hide behind a tree.
19. Let the enemy come till he's almost close enough to touch, then let him have it and jump out and finish him up with your hatchet.

--MAJOR ROBERT ROGERS, 1759

RANGER HISTORY

The history of the American Ranger is a long and colorful saga of courage, daring and outstanding leadership. It is a story of men whose skills in the art of fighting have seldom been surpassed. Only the highlights of their numerous exploits are told here.

Rangers primarily performed defensive missions until Benjamin Church's Company of Independent Rangers from Plymouth Colony proved successful in raiding hostile Indians during King Phillip's War in 1675. In 1756, Major Robert Rogers, a native of New Hampshire, recruited nine companies of American colonists to fight for the British during the French and Indian War. Ranger techniques and methods of operation were an inherent characteristic of the American frontiersmen; however, Major Rogers was the first to capitalize on them and incorporate them into the fighting doctrine of a permanently organized fighting force.

The method of fighting used by the first Rangers was further developed during the Revolutionary War by Colonel Daniel Morgan, who organized a unit known as "Morgan's Riflemen". According to General Burgoyne, Morgan's men were "....the most famous corps of the Continental Army, all of them crack shots."

Francis Marion, the "Swamp Fox," organized another famous Revolutionary War Ranger element known as "Marion's Partisans." Marion's Partisans, numbering anywhere from a handful to several hundred, operated both with and independent of other elements of General Washington's Army. Operating out of the Carolina swamps, they disrupted British communications and prevented the organization of loyalists to support the British cause, substantially contributing to the American victory.

The American Civil War was again the occasion for the creation of special units such as Rangers. John S. Mosby, a master of the prompt and skillful use of cavalry, was one of the most outstanding Confederate Rangers. He believed that by resorting to aggressive action he could compel his enemies to guard a hundred points. He would then attack one of the weakest points and be assured numerical superiority.

With America's entry into the Second World War, Rangers came forth to add to the pages of history. Major William O. Darby organized and activated the 1st Ranger Battalion on June19, 1942 at Carrickfergus, North Ireland. The members were all hand-picked volunteers; 50 participated in the gallant Dieppe Raid on the northern coast of France with British and Canadian commandos. The 1st, 3rd, and 4th Ranger Battalions participated with distinction in the North African, Sicilian and Italian campaigns. Darby's Ranger Battalions spearheaded the Seventh Army landing at Gela and Licata during the Sicilian invasion and played a key role in the subsequent campaign which culminated in the capture of Messina. They infiltrated German lines and mounted an attack against Cisterna, where they virtually annihilated an entire German parachute regiment during close in, night, bayonet, and hand-to-hand fighting.

The 2nd and 5th Ranger Battalions participated in the D-Day landings at Omaha Beach, Normandy. It was during the bitter fighting along the beach that the Rangers gained their official motto. As the situation became critical on Omaha Beach, the division commander of the 29th Infantry Division stated that the entire force must clear the beach and advance inland. He then turned to Lieutenant Colonel Max Schneider, Commander of the 5th Ranger Battalion, and said, "Rangers, lead the way." The 5th Ranger Battalion spearheaded the breakthrough and thus enabled the allies to drive inland away from the invasion beaches.

The 6th Ranger Battalion, operating in the Pacific, conducted Ranger type missions behind enemy lines which involved reconnaissance and hard-hitting, long-range raids. They were the first American contingent to return to the Philippines, destroying key coastal installations prior to the invasion. A reinforced company from the 6th Ranger Battalion formed the rescue force which liberated American and allied prisoners of war from the Japanese prison camp at Cabanatuan.

Another Ranger-type unit was the 5307th Composite Unit (Provisional), organized and trained as a long-range penetration unit for employment behind enemy lines in Japanese occupied Burma. The unit commander was Brigadier General (later Major General) Frank D. Merrill. Its 2,997 officers and men became popularly known as "Merrill's Marauders."

The men composing Merrill's Marauders were volunteers from the 5th, 154th, and 33rd Infantry Regiments and from other Infantry regiments engaged in combat in the southwest and South Pacific. These men responded to a call from then Chief of Staff, General George C. Marshall, for volunteers for a hazardous mission. These volunteers were to have a high state of physical ruggedness and stamina and were to come from jungle-trained and jungle-tested units.

Prior to their entry into the Northern Burma Campaign, Merrill's Marauders trained in India under the overall supervision of Major General Orde C. Wingate, British Army. There, they were trained from February to June 1943 in long-range penetration tactics and techniques of the type developed and first employed by General Wingate. The operations of the Marauders were closely coordinated with those of the Chinese 22nd and 38th Divisions in a drive to recover northern Burma and clear the way for the construction of Ledo Road, which was to link the Indian railhead at Ledo with the old Burma Road to China. The Marauders marched and fought through jungle and over mountains from Hukwang Valley in northwest Burma to Myitkyina and the Irrawaddy River. In 5 major and 30 minor engagements, they met and defeated the veteran soldiers of the Japanese 18th Division. Operating in the rear of the main force of the Japanese, they prepared the way for the southward advances of the Chinese by disorganizing supply lines and communications. The climax of the Marauder's operations was the capture of Myitkyina Airfield, the only all-weather strip in northern Burma. This was the final victory of "Merrill's Marauders," which disbanded in August 1944. Remaining personnel were consolidated into the 475th Infantry Regiment, which fought its last battle on February 3, and 4,1945, at Loi-Kang Ridge, China. This Infantry Regiment would serve as the forefather of today's 75th Ranger Regiment.

Shortly after the outbreak of the Korean War in June 1950, the 8th Army Ranger Company was formed of volunteers from American units in Japan. The Company was trained in Korea and distinguished itself in combat during the drive to the Yalu River, performing task force and spearhead operations. In November 1950 during the massive Chinese intervention, this small unit, though vastly outnumbered, withstood five enemy assaults on its position.

In September 1950, a Department of the Army message called for volunteers to be trained as Airborne Rangers. In the 82nd Airborne Division, five thousand regular Army paratroopers volunteered, and from that number nine hundred men were selected to form the initial eight Airborne Ranger Companies. An additional nine companies were formed from volunteers of regular Army and National Guard Infantry Divisions. These seventeen Airborne Ranger companies were activated and trained at Fort Benning, Georgia, with most receiving additional training in the mountains of Colorado.

In 1950 and 1951, some 700 men of the 1st, 2nd, 3rd, 4th, 5th, and 8th Airborne Ranger companies fought to the front of every American Infantry Division in Korea. Attacking by land, water, and air, these six Ranger companies conducted raids, deep penetrations, and ambush operations against North Korean and Chinese forces. They were the first Rangers in history to make a combat jump. After the Chinese intervention, these Rangers were the first Americans to re-cross the 38th parallel. The 2nd Airborne Ranger Company was the only African American Ranger unit in the history of the American Army. The men of the six Ranger companies who fought in Korea paid the bloody price of freedom. One in nine of this gallant brotherhood died on the battlefields of Korea.

Other Airborne Ranger companies led the way while serving with infantry divisions in the United States, Germany, and Japan. Men of these companies volunteered and fought as members of line infantry units in Korea. One Ranger, Donn Porter, would be posthumously awarded the Medal of Honor. Fourteen Korean War Rangers became general officers and dozens became colonels, senior noncommissioned officers, and leaders in civilian life. They volunteered for the Army, the Airborne, the Rangers, and for combat. The first men to earn and wear the coveted Ranger Tab, these men are the original Airborne Rangers.

In October 1951, the Army Chief of Staff, General J. Lawton Collins directed, "Ranger training be extended to all combat units in the Army." The Commandant of the Infantry School was directed to establish a Ranger Department. This new department would develop and conduct a Ranger course of instruction. The objective was to raise the standard of training in all combat units. This program was built upon what had been learned from the Ranger Battalions of World War II and the Airborne Ranger companies of the Korean conflict.

During the Vietnam Conflict, fourteen Ranger companies consisting of highly motivated volunteers served with distinction from the Mekong Delta to the DMZ. Assigned to separate brigade, division, and field force units, they conducted long-range reconnaissance and exploitation operations into enemy-held areas providing valuable combat intelligence. Initially designated at LRRP, then LRP companies, these units were later designated as C, D,E,F,G,H,I,K,L,M,N,O and P (Ranger) 75th Infantry.

Following Vietnam, recognizing the need for a highly trained and highly mobile reaction force, the Army Chief of Staff, General Abrams directed the activation of the first battalion-sized Ranger units since World War II, the 1st and 2nd Battalions (Ranger), 75th Infantry. The 1st Battalion was trained at Fort Benning, Georgia and was activated February 8, 1974 at Fort Stewart, Georgia with the 2nd Battalion being activated on October 3, 1974. The 1st Battalion is now located at Hunter Army Airfield, Georgia and the 2nd Battalion at Fort Lewis, Washington.

The farsightedness of General Abrams' decision, as well as the combat effectiveness of the Ranger battalions, was proven during the United States' invasion of the island of Grenada in October 1983 to protect American citizens there, and to restore democracy. As expected, Rangers led the way! During this operation, code named "Urgent Fury," the Ranger battalions conducted a daring, low level airborne assault (from 500 feet) to seize the airfield at Point Salines, and then continued operations for several days to eliminate pockets of resistance, and rescue American medical students.

As a result of the demonstrated effectiveness of the Ranger battalions, the Department of the Army announced in 1984 that it was increasing the strength of Ranger units to its highest level in 40 years. To do this, it activated another Ranger battalion as well as a Ranger Regimental Headquarters. These new units, the 3rd Battalion (Ranger), 75th Infantry, and Headquarters Company (Ranger), 75th Infantry, have increased the Ranger strength of the Army to over 2,000 soldiers actually assigned to Ranger units. On February 3, 1986, the 75th Infantry was renamed the 75th Ranger Regiment.

On December 20, 1989, the 75th Ranger Regiment was once again called upon to demonstrate its effectiveness in combat. For the first time since its reorganization in 1984, the Regimental Headquarters and all three Ranger battalions were deployed on Operation "Just Cause" in Panama. During this operation, the 75th Ranger Regiment spearheaded the assault into Panama by conducting airborne assaults onto Torrijos/Tocumen Airport and Rio Hato Airfield to facilitate the restoration of democracy in Panama, and protect the lives of American citizens. Between December 20, 1989 and January 7, 1990, numerous follow-on missions were performed in Panama by the Regiment.

Early in 1991, elements of the 75th Ranger Regiment deployed to Saudi Arabia in support of Operation Desert Storm.

In August 1993, elements of the 75th Ranger Regiment deployed to Somalia in support of Operation Restore Hope, and returned November 1993.

In 1994, elements of the 75th Ranger Regiment deployed to Haiti in support of Operation Uphold Democracy.

In 2000 – 2001, elements of the 75th Ranger Regiment deployed to Kosovo in support of Operation Joint Guardian.

Since September 11, 2001, the 75th Ranger Regiment has led the way in the Global War on Terrorism. In October 2001, elements of the 75th Ranger Regiment deployed to Afghanistan in support of Operation Enduring Freedom. In March 2003, elements of the Regiment deployed in support of Operation Iraqi Freedom.

The performance of these Rangers significantly contributed to the overall success of these operations and upheld the Ranger tradition of the past. As in the past, the Regiment stands ready to execute its mission to conduct special operations in support of the United States' policies and objectives.

RANGER MEDAL OF HONOR WINNERS

Millett, Lewis L. Sr	Captain	Feb 7 1951	Co. E 2/27th Infantry
Porter, Donn F.*	Sergeant	Sept 7 1952	Co. G 2/14th Infantry
Mize, Ola L.	Sergeant	June 10-11 1953	Co. K 3/15th Infantry
Dolby, David C.	Staff Sergeant	May 21 1966	Co. B 1/8th (ABN) Calvary
Foley, Robert F.	Captain	Nov 5 1966	Co. A 2/27th Infantry
Zabitosky, Fred M.	Staff Sergeant	Feb 19 1968	5th Special Forces
Bucha, Paul W.	Captain	May 16-19 1968	Co. D 3/187 Infantry
Rabel, Laszlo*	Staff Sergeant	Nov 13 1968	74th Infantry (LRRP)
Howard, Robert L.	Sergeant First Class	Dec 30 1968	5th Special Forces
Law, Robert D. *	Specialist 4	Feb 22 1969	Co. I 75th Infantry (Ranger)
Kerrey, J. Robert	Lieutenant	Mar 14 1969	Seal Team 1
Doane, Stephen H.*	1st Lieutenant	Mar 25 1969	Co. B 1/5th Infantry
Pruden, Robert J.*	Staff Sergeant	Nov 22 1969	Co. G 75th Infantry (Ranger)
Littrell, Gary L.	Sergeant First Class	April 4-8 1970	Advisory Team 21 (Ranger)
Lucas, Andre C.*	Lt Colonel	Jul 1-23 1970	HHC 2/506 Infantry
Gordon, Gary I. *	Master Sergeant	Oct 3 1993	Task Force Ranger
Shughart, Randall D. *	Sergeant First Class	Oct 3 1993	Task Force Ranger

*Awarded posthumously

SUMMARY OF CHANGES

THROUGHOUT	UPDATED doctrine, technology, terms; clarified verbiage
CHAPTER 1	UPDATED duties and responsibilities ADDED FO
CHAPTER 2	UPDATED WARNO ADDED sample WARNO UPDATED OPORD ADDED terrain model
CHAPTER 3	ADDED PLOT-CR ADDED risk estimate distances
CHAPTER 4	IDENTIFIED TL positions during movement
CHAPTER 5	REORGANIZED combat operations
CHAPTER 6	UPDATED battle drills IAW FM 3-21.8 CONSOLIDATED building clearing techniques under Battle Drill 5 (previously in Chapter 14)
CHAPTER 7	ADDED section on antennas
CHAPTER 8	UPDATED aircraft
CHAPTER 9	UPDATED rope bridge equipment UPDATED sequence and responsibilities
CHAPTER 10	UPDATED equipment and descriptions UPDATED climbing commands
CHAPTER 11	NA
CHAPTER 12	ADDED first aid techniques
CHAPTER 13	ADDED figure 'Lighting a time fuse with a match' ADDED expedient explosives ADDED table of US demolitions ADDED breaching and timber cutting charges
CHAPTER 14	MOVED building clearing techniques to Battle Drill 5 in Chapter 6 ADDED marking SOP
CHAPTER 15	NEW chapter covering vehicle convoy operations
APPENDIX A	NEW - Resources
APPENDIX B	NEW - Quick Reference Cards

TABLE OF CONTENTS

Chapter 1
LEADERSHIP

Leadership, the most essential element of combat power, gives purpose, direction, and motivation in combat. The leader balances and maximizes maneuver, firepower, and protection against the enemy. This chapter discusses how he does this by exploring the principles of leadership (BE, KNOW, DO); the duties, responsibilities, and actions of an effective leader; and the leader's assumption of command.

1-1. **PRINCIPLES**. The principles of leadership are BE, KNOW, and DO.

BE

- Technically and tactically proficient
- Able to accomplish to standard all tasks required for the wartime mission.
- Courageous, committed, and candid.
- A leader with integrity.

KNOW

- The four major factors of leadership and how they affect each other are--
  ---Led
  ---Leader
  ---Situation
  ---Communications
- Yourself, and the strengths and weaknesses in your character, knowledge, and skills. Seek continual self-improvement, that is, develop your strengths and work to overcome your weaknesses.
- Your Rangers, and look out for their well-being by training them for the rigors of combat, taking care of their physical and safety needs, and disciplining and rewarding them.

DO--

- Seek responsibility and take responsibility for your actions; exercise initiative; demonstrate resourcefulness; and take advantage of opportunities on the battlefield that will lead to you to victory; accept fair criticism, and take corrective actions for your mistakes.
- Assess situations rapidly, make sound and timely decisions, gather essential information, announce decisions in time for Rangers to react, and consider the short- and long-term effects of your decision.
- Set the example by serving as a role model for your Rangers. Set high but attainable standards; be willing do what you require of your Rangers; and share dangers and hardships with them.
- Keep your subordinates informed to help them make decisions and execute plans within your intent; encourage initiative, improve teamwork, and enhance morale.
- Develop a sense of responsibility in subordinates by teaching, challenging, and developing them. Delegate to show you trust them. This makes them want more responsibility.
- Ensure the Rangers understand the task; supervise them, and ensure they accomplish it. Rangers need to know what you expect: when and what you want them to do, and to what standard.
- Build the team by training and cross-training your Rangers until they are confident in their technical and tactical abilities. Develop a team spirit that motivates them to go willingly and confidently into combat.
- Know your unit's capabilities and limitations, and employ them accordingly.

1-2. **DUTIES, RESPONSIBILITIES, AND ACTIONS**. To complete all assigned tasks, every Ranger in the patrol must do his job. Each must accomplish his specific duties and responsibilities and be a part of the team.

PLATOON LEADER (PL)

- Is responsible for what the patrol does or fails to do. This includes tactical employment, training, administration, personnel management, and logistics. He does this by planning, making timely decisions, issuing orders, assigning tasks, and supervising patrol activities. He must know his Rangers and how to employ the patrol's weapons. He is responsible for positioning and employing all assigned or attached crew-served weapons and employment of supporting weapons.
- Establishes time schedule using backwards planning. Considers time for execution, movement to the objective, and the planning and preparation phase of the operation.
- Takes the initiative to accomplish the mission in the absence of orders. Keeps higher informed by using periodic situation reports (SITREP).
- Plans with the help of the platoon sergeant (PSG), squad leaders, and other key personnel (team leaders, FO, attachment leaders).
- Stays abreast of the situation through coordination with adjacent patrols and higher HQ; supervises, issues FRAGOs, and accomplishes the mission.
- If needed to perform the mission, requests more support for his patrol from higher headquarters.
- Directs and assists the platoon sergeant in planning and coordinating the patrol's sustainment effort and casualty evacuation (CASEVAC) plan.
- During planning, receives on-hand status reports from the platoon sergeant and squad leaders.
- Reviews patrol requirements based on the tactical plan.
- Ensures that all-round security is maintained at all times.
- Supervises and spot-checks all assigned tasks, and corrects unsatisfactory actions.
- During execution, positions himself where he can influence the most critical task for mission accomplishment; usually with the main effort, to ensure that his platoon achieves its decisive point
- Is responsible for positioning and employing all assigned and attached crew-served weapons.
- Commands through his squad leaders IAW the intent of the two levels higher commanders.
- Conducts rehearsals.

PLATOON SERGEANT (PSG)

The PSG is the senior NCO in the patrol and second in succession of command. He helps and advises the patrol leader, and leads the patrol in the leader's absence. He supervises the patrol's administration, logistics, and maintenance, and he prepares and issues paragraph 4 of the patrol OPORD.

DUTIES

- Organizes and controls the patrol CP IAW the unit SOP, patrol leader's guidance, and METT-TC factors.
- Receives squad leader's requests for rations, water, and ammunition. Work with the company first sergeant or XO to request resupply. Directs the routing of supplies and mail.
- Supervises and directs the patrol medic and patrol aid-litter teams in moving casualties to the rear.
- Maintains patrol status of personnel, weapons, and equipment; consolidates and forwards the patrol's casualty reports (DA Forms 1155 and 1156); and receives and orients replacements.
- Monitors the morale, discipline, and health of patrol members.
- Supervises task-organized elements of patrol:
  - — Quartering parties.
  - — Security forces during withdrawals.
  - — Support elements during raids or attacks.
  - — Security patrols during night attacks.
- Coordinates and supervises patrol resupply operations.
- Ensures that supplies are distributed IAW the patrol leader's guidance and direction.
- Ensures that ammunition, supplies, and loads are properly and evenly distributed (a critical task during consolidation and reorganization).
- Ensures the casualty evacuation plan is complete and executed properly.

- Ensures that the patrol adheres to the platoon leader's time schedule.
- Assists the platoon leader in supervising and spot-checking all assigned tasks, and corrects unsatisfactory actions.

ACTIONS DURING MOVEMENT AND HALTS

- Takes actions necessary to facilitate movement.
- Supervises rear security during movement.
- Establishes, supervises, and maintains security during halts.
- Knows unit location.
- Performs additional tasks as required by the patrol leader and assists in every way possible. Focuses on security and control of patrol.

ACTIONS AT DANGER AREAS

- Directs positioning of near-side security (usually conducted by the trail squad or team).
- Maintains accountability of personnel.

ACTIONS ON THE OBJECTIVE AREA

- Assists with ORP occupation.
- Supervises, establishes, and maintains security at the ORP.
- Supervises the final preparation of men, weapons, and equipment in the ORP IAW the patrol leader's guidance.
- Assists the patrol leader in control and security.
- Supervises the consolidation and reorganization of ammunition and equipment.
- Establishes, marks, and supervises the planned CCP, and ensures that the personnel status (to include WIA/KIA) is accurately reported to higher.
- Performs additional tasks assigned by the patrol leader and reports status to platoon leader.

ACTIONS IN THE PATROL BASE

- Assists in patrol base occupation.
- Assist in establishing and adjusting perimeter.
- Enforces security in the patrol base.
- Keeps movement and noise to a minimum.
- Supervises and enforces camouflage.
- Assigns sectors of fire.
- Ensures designated personnel remain alert and equipment is maintained to a high state of readiness.
- Requisitions supplies, water, and ammunition, and supervises their distribution.
- Supervises the priority of work and ensures its accomplishment.
  - Security plan.
    - Ensures crew-served weapons have interlocking sectors of fire.
    - Ensures Claymores are emplaced to cover dead space.
    - Ensures range cards and sector sketch are complete.
  - Alert plan.
  - Evacuation plan.
  - Withdrawal plan.
  - Alternate patrol base.
  - Maintenance plan.
  - Hygiene plan.
  - Messing plan.
  - Water plan.
  - Rest plan.
- Performs additional tasks assigned by the patrol leader and assists him in every way possible.

SQUAD LEADER (SL)

Is responsible for what the squad does or fails to do. He is a tactical leader who leads by example.

DUTIES

- Completes casualty feeder reports and reviews the casualty reports completed by squad members.
- Directs the maintenance of the squad's weapons and equipment.

- Inspects the condition of Rangers' weapons, clothing, and equipment.
- Keeps the PL and PSG informed on status of squad.
- Submits ACE report to PSG.

ACTIONS THROUGHOUT THE MISSION

- Obtains status report from team leaders and submits reports to the PL and PSG.
- Makes a recommendation to the PL/PSG when problems are observed.
- Delegates priority task to team leaders, and supervises their accomplishment IAW squad leader's guidance.
- Uses initiative in the absence of orders.
- Follows the PL's plan and makes recommendations.

ACTIONS DURING MOVEMENT AND HALTS

- Ensures heavy equipment is rotated among members and difficult duties are shared.
- Notifies PL of the status of the squad.
- Maintains proper movement techniques while monitoring route, pace, and azimuth.
- Ensures the squad maintains security throughout the movement and at halts.
- Prevents breaks in contact.
- Ensures subordinate leaders are disseminating information, assigning sectors of fire, and checking personnel.

ACTION IN THE OBJECTIVE AREA

- Ensures special equipment has been prepared for actions at the objective.
- Maintains positive control of squad during the execution of the mission.
- Positions key weapons systems during and after assault on the objective.
- Obtains status reports from team leaders and ensures ammunition is redistributed and reports status to PL.

ACTIONS IN THE PATROL BASE

- Ensures patrol base is occupied according to the plan.
- Ensures that his sector of the patrol base is covered by interlocking fires; makes final adjustments, if necessary.
- Sends out LP or OPs in front of assigned sector. (METT-TC dependent).
- Ensures priorities of work are being accomplished, and reports accomplished priorities to the PL and PSG.
- Adheres to time schedule.
- Ensures personnel know the alert and evacuation plans, and the locations of key leaders, OPs, and the alternate patrol base.

WEAPONS SQUAD LEADER

Is responsible for all that the weapons squad does or fails to do. His duties are the same as those of the squad leader. Also, he controls the machine guns in support of the patrol's mission. He advises the PL on employment of his squad.

DUTIES

- Supervises machine gun teams to ensure they follow priorities of work.
- Inspects machine gun teams for correct range cards, fighting positions, and understanding of fire plan.
- Supervises maintenance of machine guns, that is, ensures that maintenance is performed correctly, that deficiencies are corrected and reported, and that the performance of maintenance does not violate the security plan.
- Assists PL in planning.
- Positions at halts and danger areas and according to the patrol SOP any machine guns not attached to squads.
- Rotates loads. Machine gunners normally get tired first.
- Submits ACE report to PSG.
- Designates sectors of fire, principal direction of fire (PDF), and secondary sectors of fire for all guns.
- Gives fire commands to achieve maximum effectiveness of firepower:
  - — Shifts fires.
  - — Corrects windage or elevation to increase accuracy.
  - — Alternates firing guns.
  - — Controls rates of fire and fire distribution.
- Knows locations of assault and security elements, and prevents fratricide.
- Reports to the PL.

TEAM LEADER (TL)

Controls the movement of his fire team and the rate and placement of fire. To do this, leads from the front and uses the proper commands and signals. Maintains accountability of his Rangers, weapons, and equipment. Ensures his Rangers maintain unit standards in all areas, and are knowledgeable of their tasks and the operation. The following checklist outlines specific duties and responsibilities of team leaders during mission planning and execution. The TL leads by example:

ACTIONS DURING PLANNING AND PREPARATION

- Warning Order.
  - — Assists in control of the squad.
  - — Monitors squad during issuance of the order.
- OPORD Preparation.
  - — Posts changes to schedule.
  - — Posts and updates team duties on warning order board.
  - — Submits ammunition and supply requests.
  - — Picks up ammunition and supplies.
  - — Distributes ammunition and special equipment.
  - — Performs all tasks given in the SL's special instructions paragraph.
- OPORD Issuance and Rehearsal.
  - — Monitors squad during issuance of the order.
  - — Assists SL during rehearsals.
- Takes actions necessary to facilitate movement.
- Enforces rear security.
- Establishes, supervises, and maintains security at all times.
- Performs additional tasks as required by the SL, and assists him in every way possible, particularly in control and security.

ACTION IN THE ORP

- Assists in the occupation of the ORP.
- Helps supervise, establish, and maintain security.
- Supervises the final preparation of Rangers, weapons, and equipment in the ORP IAW the SL's guidance.
- Assists in control of personnel departing and entering the ORP.
- Reorganizes perimeter after the reconnaissance party departs.
- Maintains communication with higher headquarters.
- Upon return of reconnaissance party, helps reorganize personnel and redistribute ammunition and equipment; ensures accountability of all personnel and equipment is maintained.
- Disseminates PIR to his team.
- Performs additional tasks assigned by the SL.

ACTIONS IN THE PATROL BASE

- Inspects the perimeter to ensure team has interlocking sectors of fire; prepares team sector sketch.
- Enforces the priority of work and ensures it is properly accomplished.
- Performs additional tasks assigned by the SL and assists him in every way possible.

MEDIC

Assists the PSG in directing aid and litter teams; monitors the health and hygiene of the platoon.

DUTIES

- Treats casualties, conducts triage, and assists in CASEVAC under the control of the PSG.
- Aids the PL or PSG in field hygiene matters. Personally checks the health and physical condition of platoon members.
- Requests Class VIII (medical) supplies through the PSG.
- Provides technical expertise to and supervision of combat lifesavers.
- Ensures casualty feeder reports are correct and attached to each evacuated casualty.
- Carries out other tasks assigned by the PL or PSG.

RADIO OPERATOR Is responsible for establishing and maintaining communications with higher headquarters and within the patrol. *DUTIES DURING PLANNING* • Enters the net at the specified time. • Ensures that all frequencies, COMSEC fills, and net IDs, are preset in squad/platoon radios. • Informs SL and PL of changes to call signs, frequencies, challenge and password, and number combination based on the appropriate time in the ANCD. • Ensures the proper function of all radios and troubleshoots and reports deficiencies to higher. • Weatherproofs all communications equipment. *DUTIES DURING EXECUTION* • Serves as en route recorder during all phases of the mission. • Records all enemy contact and reports it to higher in a SALUTE format. • Reports all OPSKEDs to higher. • Consolidates and records all PIR.
FORWARD OBSERVER (FO) Works for the PL. Serves as the eyes of the FA and mortars. Is mainly responsible for locating targets, and for calling for and adjusting indirect fire support. Knows the terrain where the platoon is operating; knows the tactical situation. Knows the mission, the concept, and the unit's scheme of maneuver and priority of fires. *DUTIES DURING PLANNING* • Selects targets to support the platoon's mission based on the company OPORD, platoon leader's guidance, and analysis of METT-TC factors. • Prepares and uses situation maps, overlays, and terrain sketches. *DUTIES DURING EXECUTION* • Informs the FIST headquarters of platoon activities and of the fire support situation • Selects new targets to support the platoon's mission based on the company OPORD, the platoon leader's guidance, and an analysis of METT-TC factors. • Calls for and adjusts fire support. • Operates as a team with the radio operator. • Selects OPs. • Maintains communications as prescribed by the FSO. • Maintains the current 8-digit coordinate of his location at all times.

1-3. **ASSUMPTION OF COMMAND**. Any platoon/squad member might have to take command of his element in an emergency, so every Ranger must be prepared to do so. During an assumption of command, situation permitting, the Ranger assuming command accomplishes the following tasks (not necessarily in this order) based on METT-TC:

- *INFORMS* the unit's subordinate leaders of the command and notifies higher.
- *CHECKS* security.
- *CHECKS* crew-served weapons.
- *PINPOINTS* location.
- *COORDINATES* and *CHECKS* equipment.
- *CHECKS* personnel status.
- *ISSUES* FRAGO (if required).
- *REORGANIZES* as needed, maintaining unit integrity when possible.
- *MAINTAINS* noise and light discipline.
- *CONTINUES* patrol base activities, especially security, if assuming command in a patrol base.
- *RECONNOITERS* or, at the least, conducts a map reconnaissance.
- *FINALIZES* plan.
- *EXECUTES* the mission.

NOTES

Chapter 2
OPERATIONS

This chapter provides techniques and procedures used by Infantry platoons and squads throughout the planning and execution phases of tactical operations. Specifically, it discusses the troop-leading procedures, combat intelligence, combat orders, and planning techniques and tools needed to prepare a platoon to fight. These topics are time sensitive and apply to all combat operations. When they have time, leaders can plan and prepare in depth. If they have less time, they must rely on previously rehearsed actions, battle drills, and standing operating procedures (SOPs).

2-1. TROOP-LEADING PROCEDURES

Troop leading procedures comprise the following steps. They are what a leader does to prepare his unit to accomplish a tactical mission. The TLP starts when the leader is alerted for a mission or receives a change or new mission. He can perform Steps 3 through 8 in any order, or at the same time. (He can also use the tools of the tactician shown in Figure 2-1):

1. Receive the mission.	5. Reconnoiter.
2. Issue a warning order.	6. Complete the plan.
3. Make a tentative plan.	7. Issue the complete order.
4. Initiate movement.	8. Supervise.

Figure 2-1. TOOLS OF THE TACTICIAN RELATIONSHIP

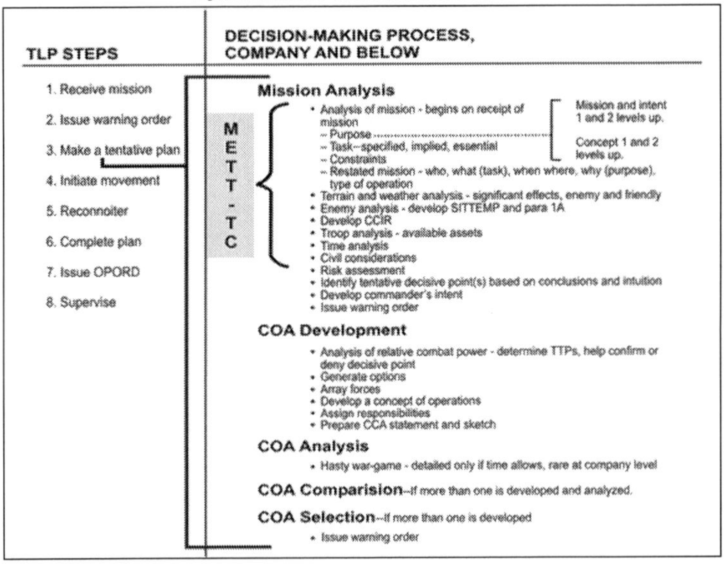

a. **STEP 1--RECEIVE THE MISSION**. The leader may receive the mission in a warning order, an operation order (OPORD), or a fragmentary order (FRAGO). He should use no more than one third of the available time for his own planning and for issuing his OPORD. The remaining two thirds is for subordinates to plan and prepare for the operation. Leaders should also consider other factors such as available daylight and travel time to and from orders and rehearsals.

b. **STEP 2--ISSUE A WARNING ORDER**. The leader provides initial instructions in a warning order. The warning order contains enough information to begin preparation as soon as possible. The warning order mirrors the five-paragraph OPORD format. A warning order may include--

- The mission or nature of the operation (mission statement).
- Time and place for issuance of the operation (coordinating instructions).
- Who is participating in the operation (coordinating instructions).
- Time of the operation (timeline).

c. **STEP 3--MAKE A TENTATIVE PLAN**. The leader uses METT-TC to develop an estimate of the situation, which he will use as the basis for his tentative plan. This set of actions is referred to as the leader's mission analysis:

(1) **Conduct a detailed mission analysis**.

(a) *Concept and Intent*. Higher commanders' concept and intent two levels up. This information is found in paragraph 1b for two levels up and in paragraphs 2 and 3 for one higher.

(b) *Unit Tasks*. Tasks that are clearly stated in the order (Specified Tasks) or tasks that become apparent as the OPORD is analyzed (Implied Tasks).

EXAMPLE SPECIFIED TASKS

- Retain hill 545 to prevent envelopment of B Co.
- Provide one squad to the 81-mm platoon to carry ammo.
- Establish an OP vic GA124325 NLT 301500 NOV 89.

EXAMPLE IMPLIED TASKS

- Provide security during movement.
- Conduct resupply operations.
- Coordinate with adjacent units.

(c) *Unit Limitations*. The leader next determines all control measures or instructions in the OPORD that restrict his freedom of action; these are called limitations. In every operation, there are some limitations on the unit.

COMMON LIMITATIONS

Graphic control measures.
Cross the LD at 100030 OCT 94 (time).
MOPP4 in effect (uniform and environmental considerations).
ADA weapons status, tight; warning status, yellow (rules of engagement).

(d) *Mission-Essential Task(s)*. After reviewing all the above factors, the leader identifies his mission-essential task(s). Failure to accomplish a mission-essential task results in the unit's failure to accomplish its primary purpose for that operation. The mission essential task should be found in the maneuver paragraph.

(e) *Restated Mission*. The restated mission statement becomes the focus for the remainder of the estimate process. This is a clear, concise statement of the mission essential task(s) to be accomplished by the unit and the purpose to be achieved. The mission statement will state WHO, WHAT (the task), WHEN (the critical time), WHERE (usually a grid coordinate), and WHY (the purpose the unit must achieve). Some examples of restated missions follow:

EXAMPLE RESTATED MISSIONS
(WHO) 1st Platoon attacks (WHAT) to seize (WHERE) HILL 482 vic NB 457371 (OBJ BLUE) (WHEN) NLT 090500Z Dec 92 L 482 (WHY) to enable the company's main effort to destroy enemy command bunker.
(WHO) 1st Platoon, C Company defends (WHAT) to destroy from (WHERE) AB163456 to AB163486 to AB123486 to AB123456 (WHEN) NLT 281530Z Oct 97 (WHY) to prevent enemy forces from enveloping B Company, 1-66 Infantry (L) from the south.

(2) Analyze the situation and develop a course of action. Each COA must be--suitable, acceptable, feasible, distinguishable, and complete.

(3) Upon developing a COA, the unit leader will assign C2 headquarters, complete generic task organization assigning all organic and attached elements, and prepare COA statement and sketch.

(4) With the restated mission from Step 1 to provide focus, the leader continues the estimate process using the remaining factors of METT-TC.

(a) What is known about the **ENEMY**?

Composition	This is an analysis of the forces and weapons that the enemy can bring to bear. Determine what weapons systems they have available, and what additional weapons and units are supporting him.
Disposition	The enemy's disposition is how he is arrayed on the terrain, such as in defensive positions, in an assembly area, or moving in march formation.
Strength	Percentage strength.
Recent Activities	Identify recent and significant enemy activities that may indicate future intentions.
Reinforcement Capabilities	Determine positions for reserves and estimated time to counterattack or reinforce.
Possible Courses of Action	Determine the enemy's possible COAs. This will help the leader determine how to best array his own forces against the enemy and fight the battle.

(b) How will **TERRAIN** and **WEATHER** affect the operation? Analyze terrain using OAKOC.

- **Observation and Fields of Fire**. Determine locations that provide the best observation and fields of fire along friendly and enemy avenues of approach, near the objective, and on key terrain. The analysis of fields of fire is mainly concerned with the ability to cover the terrain with direct fire.
- **Avenues of Approach**. Analyze and identify friendly and enemy avenues of approach. Identify avenues of approach en route to the objective, on and around the objective, and for possible enemy counterattack. Also, consider aerial and subterranean avenues.

Offensive Considerations (Friendly)	• How can these avenues support my movement? • What are the advantages/disadvantages of each? (Consider enemy, speed, cover, and concealment.) • What are the likely enemy counterattack routes?

Offensive Considerations (Enemy)	• How can the enemy use these approaches? • Which avenue is most dangerous? Least? (Prioritize each approach.) • Which avenues would support a counterattack?

- **Cover and Concealment**. The analysis of cover and concealment is often inseparable from the fields of fires and observation. Weapon positions must have both to be effective and to be survivable. Infantry units are capable of improving poor cover and concealment by digging in and camouflaging their positions. When moving, the terrain is used to provide cover and concealment.
- **Obstacles**. Identify the existing and reinforcing obstacles and hindering terrain that will affect mobility.
- **Key Terrain**. Key terrain is any location or area that the seizure, retention, or control of affords a marked advantage to either combatant. Using the map, aerial photos, and information already gathered, look for key terrain that dominates avenues of approach or the objective area. Next, look for decisive terrain that if held or controlled will have an extraordinary impact on the mission.

(5) Analyze courses of action (war-game). This analysis is conducted by war-gaming friendly courses of action against the enemy's most probable courses of action. The leader can--and should--war-game with his subordinates.

(6) Compare courses of action. The leader compares the COAs and selects the one that is most likely to accomplish the assigned mission. He considers the advantages and disadvantages for each COA. He also considers how the critical events impact on COAs.

(7) Make a decision. The leader selects the COA that he believes has the best chance of accomplishing the mission.

d. **STEP 4--START NECESSARY MOVEMENT**. The unit may need to begin movement while the leader is still planning or forward reconnoitering. This step may occur anytime during the TLP.

e. **STEP 5--RECONNOITER**. If time allows, the leader makes a personal reconnaissance. When time does not allow, the leader must make a map or aerial photo reconnaissance. Sometimes the leader must rely on others, such as scouts, to conduct the reconnaissance.

f. ***STEP 6--COMPLETE THE PLAN***. The leader completes his plan based on the reconnaissance and any changes in the situation.

g. **STEP 7--ISSUE THE COMPLETE ORDER**. Platoon and squad leaders normally issue oral operations orders to aid subordinates in understanding the concept of the mission. Leaders may require subordinates to repeat part of the order, demonstrate it on a terrain model, or sketch their understanding of the operation. Leaders should also quiz their Rangers to ensure that all Rangers understand the mission.

h. **STEP 8--SUPERVISE AND REFINE**. The leader supervises the unit's preparation for combat by conducting rehearsals and inspections.

(1) ***Rehearsals***. Rehearsals include the practice of having squad leaders brief their planned actions in execution sequence to the platoon leader. The leader should conduct rehearsals on terrain that resembles the actual ground, and in similar light conditions. At a minimum, the leader should conduct a backbrief (least preferred), full force rehearsal (most preferred), or reduced force rehearsal (key leaders), and a rock drill.

(a) *Purpose*. The leader uses rehearsals to--

- Practice essential tasks (improve performance).
- Reveal weaknesses or problems in the plan.
- Coordinate the actions of subordinate elements.
- Improve Ranger understanding of the concept of the operation (foster confidence in Rangers).

(b) *Times and Tasks*. The platoon may begin rehearsals of battle drills and other SOP items before the receipt of the operation order. Once the order has been issued, it can rehearse mission specific tasks. Some important tasks to rehearse include--

- Actions on the objective.
- Assaulting a trench, bunker, or building.
- Actions at the assault position.
- Breaching obstacles (mine and wire).
- Using special weapons or demolitions.
- Actions on unexpected enemy contact.

(c) *Types.*

- Backbrief.
  - -- Key leaders sequentially brief the actions required during operation.
  - -- Patrol leader controls.
  - -- Conducted twice: right after FRAGO (confirmation brief), and again after subordinates develop own plan.
- Reduced force.
  - -- Conducted when time is key constraint.
  - -- Conducted when security must be maintained.
  - -- Key leaders normally attend.
  - -- Mock-ups, sand tables, and small scale replicas used.
- Full force.
  - -- Most effective type.
  - -- First executed in daylight and open terrain.
  - -- Secondly conduct in same conditions as operation.
  - -- All Rangers participate.
  - -- May use force on force.
- Techniques.
  - -- Force on force.
  - -- Map (limited value and limited number of attendees).
  - -- Radio (cannot mass leaders; confirms communications).
  - -- Sand table or terrain model (key leaders; includes all control measures).
  - -- Rock drill (similar to sand table/terrain model; subordinates actually move themselves). Very effective for coordinating timing and sequence of events--especially when air assets are used.

(d) *Inspections.* Squad leaders should conduct initial inspections shortly after receipt of the warning order. The platoon sergeant spot-checks throughout the unit's preparation for combat. The platoon leader and platoon sergeant make a final inspection of--

- Weapons and ammunition.
- Uniforms and equipment.
- Mission-essential equipment.
- Soldier's understanding of the mission and their specific responsibilities.
- Communications.
- Rations and water.
- Camouflage.
- Deficiencies noted during earlier inspections.

2-2. **COMBAT INTELLIGENCE**. Gathering information is one of the most important aspects of conducting patrolling operations. The following details what information to collect and how to report it:

a. **Reports**. All information must be quickly, completely, and accurately reported. Use the SALUTE report format for reporting and recording information.

SIZE	–	Seven enemy personnel
ACTIVITY	–	Traveling SW
LOCATION	–	GA123456
UNIT/UNIFORM	–	OD uniforms with red six-point star on left shoulder
TIME	–	210200JAN99
EQUIPMENT	–	Carry one machine gun and one rocket launcher

b. **Field Sketches**. Try to include a sketch with each report. Include only any aspects of military importance such as targets, objectives, obstacles, sector limits, or troop dispositions and locations (use symbols from FM 1-02). Use notes to explain the drawing, but they should not clutter the sketch.

c. **Captured Documents**. The leader collects documents and turns them in with his reports. He marks each document with the time and place of capture.

d. **Prisoners**. If prisoners are captured during a patrolling operation, they should be treated IAW the Geneva Convention and handled by the 5-S rule:

(1) Search
(2) Silence
(3) Segregate
(4) Safeguard
(5) Speed to rear

e. **Debriefs**. Immediately upon return from a mission, the unit is debriefed. The intelligence officer will generally have a unit-specific format for debriefing a patrol.

2-3. **WARNING ORDER**. Warning orders give subordinates advance notice of upcoming operations. This gives them time to prepare. A warning order must be brief, but complete. *A warning order only authorizes execution when it clearly says so.* (Figure 2-2 shows an example format; Figure 2-3 shows an example warning order.)

Figure 2-2. WARNING ORDER FORMAT

WARNING ORDER ________

*Roll call, pencil/pen/paper, RHB, map, protractor, leader's monitor, hold all questions till the end
References: Refer to higher headquarters' OPORD, and identify map sheet for operation.
Time Zone Used throughout the Order: (Optional)
Task Organization: (Optional; see paragraph 1c.)

1. **SITUATION** (Higher's OPORD para 1a[1]--[3])

a. **Enemy Forces**. Include significant changes in enemy composition, dispositions, and courses of action. Information not available for inclusion in the initial WARNO can be included in subsequent warning orders. (who, what, where)

b. **Friendly Forces**. (Optional) only address if essential to the WARNO.

(1) Higher commander's mission (who, what, where, why).
(2) Higher commander's intent. (Higher's [go to mapboard] OPORD para 1b[2], give task and purpose)

c. **Attachments and Detachments**. Initial task organization, only address major unit changes.

(1) Orient relative to each point on the compass.
(2) Box in the entire AO with grid lines.
(3) Trace each Zone using boundaries.
(4) Familiarize by identifying 3 natural (terrain) and 3 man-made features in each zone.
(5) Point out the enemy and friendly locations on the map board.

2. **MISSION**. State mission twice (who, what, when, where, and why) and concisely state task and purpose.

3. **EXECUTION**.

a. **Concept of Operation**. Provide as much information as available. The concept should describe the form of maneuver or defensive technique, critical events, decisive point of the operation and why it is decisive, task and purpose for the main and supporting efforts, purposes of the warfighting systems, such as engineers, fire support, intelligence etc), acceptable risk, and the desired endstate. After the desired endstate is stated, leaders should talk through the concept of the operation like a story. Use a simple sketch to assist in relaying the information.

b. **Tasks to Maneuver Units**. Provide specified tasks to subordinate units. These are tactical instructions on how to execute the mission for each element in task organization. Planning guidance consists of tasks assigned to subordinate units, special teams, and key individuals.

c. **Tasks to Combat Support Units**. See paragraph 3b.

d. **Coordinating Instructions**. Include any information available at that time. If you know it, then at least cover--

- Uniform and equipment common to all. Consider the factors of METT-TC and tailor the load for each Ranger.
- Time line. (State 4 W's and specified and implied times. Reverse plan. Use 1/3-2/3 rule.)
- CCIR.
- Risk guidance.
- Deception guidance.
- Specific priorities, in order of completion.
- Guidance on orders and rehearsals.
- Orders group meeting (attendees, location, and time).
- Earliest movement time and degree of notice.

4. **SERVICE SUPPORT**. Include any known logistics preparation for the operation.

a. Special equipment. Identify requirements and any coordination measures the unit needs to take to transfer the special equipment. (State the equipment you will use, need, or want for the mission.)

b. Transportation. Identify requirements and any coordination needed to pre-position assets. (State unit's means of infil/exfil.)

5. **COMMAND AND SIGNAL**.

a. **Command**. State the chain of command if different from unit SOP.

b. **Signal**. Identify current SOI edition, and pre-position signal assets to support operation.

- Give subordinates guidance on tasks to complete for preparation of the OPORD and the mission.
- Give time, place, and uniform of the OPORD.
- Give a time hack and ask for questions.

NOTES

Figure 2-3. EXAMPLE WARNING ORDER

WARNING ORDER (SQUAD)

ROLL CALL, CAMP DARBY SPECIAL (CDS) 1:50,000 MAP, PEN, PAPER, PENCIL, PROTRACTOR, RANGER HANDBOOK (RHB), HOLD ALL QUESTIONS, AND TLs MONITOR AND TASK-ORGANIZE

1. **SITUATION**.

a. **Enemy**.

WHO: The Aragon Liberation Front (ALF).
WHAT: Has been conducting aggressive offensive operations.
WHERE: In Zones A, B, C.

b. **Friendly**:

(1) **Higher's Mission**.

WHO: 2nd PLT, C CO.
WHAT: Conducts area ambushes to destroy enemy personnel and equipment.
WHERE: Zone D.
WHY: To prevent enemy logistical resupply.

(2) **Higher Commander's Intent**.

- Rapid infiltration into Zone D.
- Find, fix, and finish enemy forces in Zone D.
- Enemy personnel and equipment located and targeted.
- Squads ready to conduct future combat operations.

c. **Attachments/Detachments**. MG Tm #1/2 PLT 160630-190630.

2. **MISSION**.

WHO: 1st SQD, 2nd PLT
WHAT: Conducts a point ambush to destroy enemy personnel and equipment.
WHERE: GA 1805 7878.
WHEN: NLT 25 1830 DEC 2007.
WHY: To prevent enemy logistical resupply and gather PIR.

3. **EXECUTION**.

a. **Concept of Operation**. (Orient Rangers to sketch or terrain model). We are currently located at Camp Darby, GA 1962 7902, which is on a ridge. Our ground tactical plan will begin when we will move generally southwest to our ORP at Grid GA 187778. Terrain feature there is a draw for 1,000 meters, by foot and it will take 1 hour. Here will we finalize the preparing of M, W, and E. We will then move generally southwest to our objective at Grid GA 184774. Terrain feature there is a ridge for 200 meters, by foot, and it will take us 30 minutes to an hour. The reason for this is stealth on the objective while we occupy our positions.

b. **Tasks to Maneuver Units**.

(1) HQs: HQ is responsible for command and control and communications. Commo/HQ is 2nd in the OOM. M240 will be placed in a position to provide supporting fires in the kill zone. RTO will be the recorder en route and during actions on the objective. SL will be located where he can best control his squad and observe the kill zone. SL is responsible for dissemination of information.

(2) ATM: ATM will be 1st OOM/responsible for navigation, and for clearing and securing the far side at all linear danger areas. ATM is flank security for AOO, will provide early warning and seal off objective/1-2 man EPW, and search TM/1-2 man aid and litter TM/1-2 man demo TM/1 man (P), 1 man (A) ORP clearing TM/ 2-2 man flank security. TM for AOO/ 1 compassman, 1 pace man, or ATL will be the sec. TM LDR along the most likely avenue of approach.

(3) BTM: BTM 3RD OOM/responsible for securing the near side of all linear danger areas/BTM. Is assault for AOO. Must provide suppressive fire and assault thru kill zone/1-2 man EPW and search team/1-2 man aid and litter team/1-2 man demolitions team; 1 man (P); 1 man (A) ORP clearing team; 1-2 man S/O team; 1-2 man security team for reentry; 1 compassman; 1 pace man; and BTL. You will be the assault element leader for AOO.

c. **Tasks to Combat Support Units**. None.

d. **Coordinating Instructions**.

(1) Uniform and equipment common to all. At a minimum, each Ranger wears or carries--

(a) A ballistic helmet w/camouflaged band and luminous tape; clean and serviceable ACUs, camelback, black belt w/buckle, blackened combat boots, serviceable OD or black socks, serviceable OD/brown T-shirt, ID tags w/breakaway lanyard around neck (silenced). Right cargo pocket; (RHB, CDS [map], protractor, pencil, pen, and paper) waterproofed, left cargo pocket; patrol cap w/ luminous tape, ID card in right breast pocket, SL will carry sterile fire support overlay in left breast pocket.

(b) LCE, consisting of pistol belt, suspenders, 2 ammo pouches, seven 30-round magazines, one first aid dressing with pouch, one compass and pouch, two 1-quart canteens w/covers and cups, one whistle, and four M67 fragmentation grenades. All are tied down IAW 4th Bn SOP, except that the pace cord is tied to suspenders.

(c) Rucksack, complete; one set serviceable ACUs, two 2-quart canteens w straps; E-tool with cover; two pair socks; two brown T-shirts; weapons cleaning kit; one waterproof bag; a sewing kit; a patrol cap; MREs; poncho and liner; wet weather top; personal hygiene kit; one brown towel; and camouflage stick. RTO also packs and ties off the ANCD.

(d) Optional bug juice, 550 cord, foot powder, black tape, knife, all other items will be approved by SL 15 minutes prior to completion time of OPORD.

(2) Time schedule.

WHEN	WHAT	WHERE	WHO	REF
*0630	WARNING ORDER	BAY AREA	All	3D1
0700	INITIAL INSPECTION	BAY AREA	All	
*0730	Required Ammo/Supplies	CO TOC	BTL	4B1
*0745	P/U-Ammo/Supplies	CO TOC	BTL/Detail	4B1
0800	Test Fire	T/F AREA	All	
*0830	S-2/S-3 Coord	CO BAY AO	SL/RTO	3D1
*0830	Fire Spt Coord	BN TOC	SL/ RTO	3A2
0850	Adj Unit Coord	BN TOC	ATL/compass man	3D1
*0900	Enter Net	BAY AREA	RTO	5B1
0930	Complete Plan	BAY AREA	All	
*1000	Issue OPORD	BAY AREA	All	3D1
1300	Rehearsal	BAY AREA	All	
1400	Final Inspection	BAY AREA	All	
1430	Conduct Move	BAY AREA	All	
1800	In ORP	GA	All	
2000	In Position	GA 184774	All	
*2300	Mission Complete	GA 186874	All	3D1
*0200	Reenter FFU	GA	All	
0500	S-2 Debrief	BN TOC	All	
* Specified time				

4. **SERVICE AND SUPPORT**.

a. **Special Equipment**.

(1) HQs: One hand-held radio and extra batteries; one AN/PRC-119 radio and extra batteries, one ANCD, one AN/PSN-11 and extra batteries, one M-22 binos, one AN/PVS-14 and extra batteries; one sheets overlay material, three sheets carbon paper; two HE smoke, two red smoke, one M240-B MG complete w/tripod, flex mount, T&E, and Pintle, one AN/PVS-4 and extra batteries, one AN/PVS-7D, and extra batteries.

(2) ATM: One hand-held radio and extra batteries, one M-24 binos, two AN/PVS-14 and extra batteries, two HE smoke, two yellow smoke, one thermite grenade, two gags, two flex cuffs, two sandbags, one CLS bag, one pole-less litter, two M18A1 claymore mines, one AT-4, one LAW, one M-203 vest, two M-249 SAW pouches, one small metal flashlight, and extra batteries secured to rifle.

(3) BTM: One hand-held radio and extra batteries, one M-24 binos, two AN/PVS 14 and extra batteries, two HE smoke, two yellow smoke, one thermite grenade; two each gags, flex cuffs, and sandbags; one CLS bag, one pole-less Litter, two M18A1 claymore mines, one AT-4, one LAW, one M-203 vest, two M249 SAW pouches, one small metal flashlight, and extra batteries secured to rifle.

b. **Ammo**. (per weapon system).

M16/M4	M249	M203	M240	SL AND TLs
210 Rounds 5.56-MM Ball	625 Rounds 5.56-MM (4:1 Mix)	16 Rounds 40-MM HEDP 5 Rounds 40-MM Shot 4 Rounds 40-MM Illumination 2 Rounds 40-MM TP-T	825 Rounds 7.62-MM (4:1 Mix)	42 Rounds 5.56-MM Tracer

c. **Transportation**. NA

5. **COMMAND AND SIGNAL**.

a. **Command**.

PERSONNEL	POS	SOC
HQ/SUPPORT		
BARNETT	SL	1
MARENGO	RTO	4
CASELY	MG	11
WADE	AG	12
ATM/SECURITY		
TRUITT	ATL	3
SILCOX	GRN	5
WISE	R	6
EDMUNDS	R	7
BTM/ASSAULT		
SPENCER	BTL	2
CHRISTENSEN	GRN	8
RUFUS	R	9
ROBINSON	R	10

SUBORDINATE GUIDANCE:	
ATL:	Write paragraph 1 of squad OPORD, and for drawing all sketches, terrain models, routes, and nonsterile fire support overlays.
BTL:	You are second in the chain of command and in charge at all times during my absence. Write paragraph 4 of squad OPORD Write the supply, DX, and ammo lists, draw and issue all items, ensure that everyone conducts a test fire, and all equipment is tied down IAW 4th RTBn SOP, and update the squad status card and hand receipt.
RTO:	Write paragraph 5 of squad OPORD, ensure all radios are operational with proper freqs loaded, and ensure we enter the net on time.
TL:	TLs: You will update the Warning Order board with all the correct information. As a task is accomplished, you will line it out. Also, post your succession of command (SOC), duty positions, and job descriptions (specialty teams and key individuals). Come and see me for further guidance at the conclusion of this WARNO.

b. **Signal**.

TACTICAL FREQUENCY	62,000
MEDEVAC FREQUENCY	35,000
CHALLENGE	BAYONET
PASSWORD	MILLETT
RUNNING PASSWORD	RANGER
NUMBER COMBO	7
TIME, PLACE, AND UNIFORM FOR OPORD. TIME HACK. ANY QUESTIONS?	

NOTES

2-4. **OPERATIONS ORDER**. An operations order (OPORD) is a directive issued by a leader to his subordinates in order to effect the coordinated execution of a specific operation. A five-paragraph format (shown below) is used to organize the briefing, to ensure completeness, and to help subordinate leaders understand and follow the order. Use a terrain model or sketch along with a map to explain the order. When possible, such as in the defense, give the order while observing the objective. The platoon/squad leader briefs his OPORD orally off notes that follow the five-paragraph format. Before the issuance of the OPORD, the leader ensures that the following resources are in place: pencil, pen, paper; RHB; map; protractor; leader's monitor. Then he calls roll and says "*Please hold all questions till the end.*" Figure 2-4 shows an example OPORD format. Figure 2-5 shows an OPORD shell.

Figure 2-4. EXAMPLE OPORD FORMAT

OPERATIONS ORDER

[Plans and orders normally contain a code name and are numbered consecutively within a calendar year.]

References: [The heading of the plan or order includes a list of maps, charts, datum, or other related documents the unit will need to understand the plan or order. The user need not reference the SOP, but may refer to it in the body of the plan or order. He refers to a map by map series number (and country or geographic area, if required), sheet number and name, edition, and scale, if required. "*Datum*" refers to the mathematical model of the earth that applies to the coordinates on a particular map. It is used to determine coordinates. Different nations use different datum for printing coordinates on their maps. The datum is usually referenced in the marginal information of each map.]

Time zone used throughout the order: The time zone used throughout the order (including annexes and appendixes) is the time zone applicable to the operation. Operations across several time zones use Zulu time.
Task organization: Describe the allocation of forces to support the commander's concept. Task organization may be shown in one of two places: preceding paragraph one, or in an annex, if the task organization is long and complicated.

**Weather and light data, and general forecast (Only discuss what will affect your patrol. Note effects on friendly and enemy personnel and equipment.):*

High
Moonrise
Sunrise
Low
Moonset
Sunset
Wind Speed
Moon Phase
BMNT
Wind Direction
Percent Illumination
EENT

**This is the information the leader received from higher's OPORD.*

Terrain: Make a statement concerning the effects terrain will have on both friendly and enemy forces in the area of operation using the OAKOC format.

1. **SITUATION**.

a. **Enemy Forces**. The enemy situation in higher headquarters' OPORD (paragraph 1a) is the basis for this, but the leader refines this to provide the detail required by his subordinates.

(1) Enemy's composition, disposition, strength.
(2) Recent activities.
(3) Known/suspected locations and capabilities.
(4) Describe the enemy's most likely and most dangerous course of action on a map.

b. **Friendly Forces**. This information is in paragraphs 1b, 2, and 3 of higher headquarters' OPORD.

(1) State the mission of the Higher Unit (2 levels up).

(2) State the mission of the Higher Unit (1 level up).

(3) State intent 2 levels up.

(4) State locations of units to the left, right, front, and rear. State those units' tasks and purposes; and say how those units will influence your unit's patrol.

(a) Show other units locations on map board.

(b) Include statements about the influence each of the above patrols will have on your mission, if any.

- Obtain this information from higher's OPORD. It gives each leader an idea of what other units are doing and where they are going. This information is in paragraph 3a (1) (execution, concept of the operation, maneuver).
- Also include any information obtained when the leader conducts adjacent unit coordination.

c. **Attachments and Detachments**. Do not repeat information already listed under Task Organization. Try to put all information in the Task Organization. However, when not in the Task Organization, list units that are attached or detached to the headquarters that issues the order. State when attachment or detachment is to be effective if different from when the OPORD is effective, such as on order, on commitment of the reserve). Use the term "*remains attached*" when units will be or have been attached for some time.

2. **MISSION**. State the mission derived during the planning process. There are no subparagraphs in a mission statement. Include the 5 W's: Who, What (task), Where, When, and Why (purpose).

- Make it a clear and concise statement and read it twice.
- Go to map and point out the exact location of the OBJ and point out the unit's present location.

3. **EXECUTION**. State the commander's intent.

a. **Concept of Operation**. The concept statement should be concise and clear. It describes how the unit will accomplish its mission from start to finish. The number of subparagraphs, if any, is based on what the leader considers appropriate for the level of leadership and complexity of the operation. Generally, the concept should describe the form of maneuver or defensive technique and critical events; identify what the decisive point of the operation is and why; describe the task and purpose for the main effort and supporting efforts; state the purpose of the warfighting functions, such as engineers, fire support, or intelligence; define acceptable risk; and identify the desired endstate. After stating the desired endstate, leaders should talk through the concept of the operation like a story, using a simple sketch to help them convey the information. Figure 1 shows the six warfighting functions (WFF):

Figure 1. Warfighting functions.

- Fire support
- Movement and Maneuver
- Protection
- Command and Control
- Intelligence
- Sustainment (formerly called "CSS")

(1) ***Maneuver***. The maneuver paragraph details the mechanics of the operations. This subparagraph specifically addresses all subordinate units and attachments by name, giving each its mission in the form of a task and purpose. The main effort's mission must be supported by the subordinate unit's missions. Actions on the objective will comprise the majority of this paragraph and should include a detailed plan for engagement/disengagement criteria, an alternate plan in the event of compromise or unplanned movement of enemy forces, and a withdrawal plan.

(2) ***Fires***. State scheme of fires to support the overall concept. This paragraph should state which maneuver unit has priority of fires. Fires should be planned using the PLOT-CR format (purpose, location, observer, trigger, communication method, resources). A target list worksheet and overlay are referenced here, if applicable. Specific targets are discussed and pointed out on the terrain model (see Chapter 3, Fire Support).

(3) ***Casualty Evacuation***.

(a) *Plan*. State a detailed CASEVAC plan during each phase of the operation. Include CCP locations, tentative extraction points, and methods of extraction.

b. **Tasks to Maneuver Units**. Clearly state the missions or tasks for each maneuver unit that reports directly to the headquarters issuing the order. List the units in the task organization, including reserves. Use a separate subparagraph for each maneuver unit. Only state the tasks that are necessary for comprehension, clarity, and emphasis. Place tactical tasks that affect two or more units in Coordinating Instructions (subparagraph 3d). Platoon leaders task their subordinate squads. Those squads may be tasked to provide any of the following special teams: reconnaissance and security, assault, support, aid and litter, EPW and search, clearing, and demolitions. Detailed instructions may also be given to the platoon sergeant, RTO, compassman, and paceman.

(1) ***Tasks to Combat Support Units***. Add subparagraphs here only as necessary. List CS units in subparagraphs from the task organization. Use CS subparagraphs to list only those specific tasks that CS units must accomplish and that are not specified or implied elsewhere. Include organization for combat if not clear from task organization.

c. **Coordinating Instructions**. This is always the last subparagraph under paragraph 3. List only the instructions that apply to two or more units, and which are seldom covered in unit SOPs. Refer the user to an annex for more complex instructions. The following information is required:

1. Time schedule (state time, place, uniform, and priority of rehearsals, backbriefs, inspections, and movement).

2. Give the commander's critical information requirements (CCIR), which includes priority intelligence requirements (PIR), essential elements of friendly information (EEFI), and friendly force information requirements (FFIR):

a. PIR is intelligence that the commander must have for planning and decision making.

b. EEFI are critical aspects of friendly operations that, if known by the enemy, would compromise, lead to failure, or limit success of the operation.

c. FFIR include what the commander needs to know about friendly forces available for the operation. It can include personnel status, ammunition status, and leadership capabilities.

3. Risk reduction control measures. These are measures unique to the operation. They supplement the unit SOP and can include mission-oriented protective posture, operational exposure guidance, vehicle recognition signals, and fratricide prevention measures.

4. Rules of engagement (ROE).

5. Environmental considerations.

6. Force protection

7. Movement plan. Use terrain model, sketch, or both. State azimuths, directions, and grid coordinates. Use names of real Rangers on the sketches.

a. Sketch out the order of movement, the formation, and the movement technique.

b. Sketch actions at halts (long and short).

c. Routes (use terrain model):

1) Orient the terrain model (North, South, East, and West).

2) Identify the grid lines, both North-South and East-West.

3) Brief the legend.

4) Box in the AO with distinguishable natural or man-made terrain features such as roads, ridgelines, rivers, and streams.

5) Brief all the routes depicted on the terrain model from start to finish using specific azimuths and distance.

a) The primary route from the start point to the ORP, from the ORP to the objective, and from the objective to the patrol base while describing the terrain over which the squad/platoon will travel. (Include near/far side rally points for danger areas.)

b) The alternate route from the start point to the ORP, from the ORP to the objective, and from the objective to the patrol base while describing the terrain over which the squad/platoon will travel. (Include near/far side rally points for danger areas.)

6) Brief the fire support plan (if not given in fires paragraph). When covering fires the leader should cover the following (PLOT-CR):

- *Purpose.*
- *Location and terrain feature to the targets.*
- *Observer.*
- *Trigger.*
- *Communication method.*
- *Resources.*

7) State where the sterile fire support overlay is being carried.

a) Rally points and actions at all rally points (include grid location and terrain reference, and use sketch).

b) Actions at danger areas. This is a general plan for unknown linear, small open areas and large open areas; specific plan for all known danger areas that unit will encounter along the route (use sketch).

c) Actions on enemy contact; battle drills (use sketch).

4. **SERVICE SUPPORT**. Address service support in the areas shown below as needed to clarify the service support concept. Subparagraphs can include:

a. **General**. Refer to the SOPs that govern the sustainment operations of the unit. Provide current and proposed company trains locations, casualty, and damaged equipment collection points and routes.

b. **Materiel and Services**.

(1) ***Supply***.

Class I--Rations plan.
Class III--Petroleum.
Class V--Ammunition.
Class VII--Major end items.
Class VIII--Medical.
Class IX--Repair parts.
Distribution Methods.

(2) ***Transportation***.

(3) ***Services***. (Laundry and showers.)

(4) ***Maintenance***. (Weapons and equipment.)

(a) *Medical Evacuation and Hospitalization*. Method of evacuating dead and wounded, friendly and enemy personnel. Include priorities and location of CCP.

(b) *Personnel Support*. Method of handling EPWs and designation of the EPW collection point.

5. **COMMAND AND SIGNAL**.

This paragraph states where command and control facilities and key leaders are located during the operation. Use the OPORD shell to save time and organize your thoughts while preparing an OPORD.

a. **Command**.

- Location of the higher unit commander and CP.
- Location of key personnel and CP during each phase of the operation.

- Succession of command.
- Adjustments to the patrol SOP.

b. **Signal**.
   - SOI index in effect.
   - Methods of communication in priority.
   - Pyrotechnics and signals, to include arm and hand signals (demonstrate).
   - Code words such as OPSKEDs.
   - Challenge and password (use behind friendly lines).
   - Number combination (use forward of friendly lines).
   - Running password.
   - Recognition signals (near/far and day/night).

c. **Special Instructions to RTOs**.

d. **Actions after Issuance of OPORD--**
   (1) Issue annexes.
   (2) Give time hack.
   (3) ASK for questions.

Figure 2-5. OPORD SHELL

OPORD

DTG:

REFERENCE:

TASK ORGANIZATION:

1ST PLT	2ND PLT	TM CONTROL:

1. **SITUATION**.

BMNT:	SR:	SS:	EENT:	MR:	MS:	HI:	LO:	ILLUM:

a. **Enemy**.

Terrain | Terrain

(1) Higher mission:

Cdr's intent:

End state | End state

(2) Units to N, S, E, W.

2. **MISSION**:

3. **EXECUTION**:

a. **Concept of Operations**.

Cdr's intent:

(ME): T: P:	(SE1): T: P:	(SE2): T: P:	(SE3): T: P:	(SE4): T: P:	(SE5): T: P:

b. **Maneuver**

PHASE 1:
ME:
SE1:
SE2:
SE3:
SE4:
SE5:
FIRES:
MTRS:
C2:

PHASE 2:
ME:
SE1:
SE2:
SE3:
SE4:
SE5:
FIRES:
MTRS:
C2:

PHASE 3:
ME:
SE1:
SE2:
SE3:
SE4:
SE5:
FIRES:
MTRS:
C2:

PHASE 4:
ME:
SE1:
SE2:
SE3:
SE4:
SE5:
FIRES:
MTRS:
C2:

c. **Tasks to Maneuver Units.**

(1) Risk-reduction control measures.
Rules of engagement.
(2) Environmental considerations.
(3) Non-Airborne personnel.

(ME): (SE1): (SE2):

(SE3): (SE4): (SE5):

d. **Tasks to Subordinate Units**

Mortars

Coordinating instructions:

PIR:

EEFI:

FFIR:

Timeline.

Coordinating Instructions.

WHEN	WHO	WHAT	WHERE

4. **SERVICE SUPPORT**.

a. **General**.

b. **Material and Services**.

(1) Supply:

Class I.

Class III:

Class V:

Class VIII:

Class IX:

(2) LOGSTAT.

(3) The chain of command is responsible to ensure that ammo lasts between resupplies.

c. **Medical Evacuation**.

(1) Air MEDEVAC is on call through [this] frequency. JP capable; use standard 9-line, 24-hr coverage.

(2) Establish friendly CCPs in AAs. Adjust CCPs as mission progresses.

(3) Troop medical clinic located vic [this location].

(4) Precedence for evacuation of WIA is US military, host nation military, host nation civilian, and enemy military.

(5) Precedence for evacuation of KIA is same. Evacuation of KIA will be as per mission constraints.

d. **Personnel**.

(1) Method of handling EPWs:

ANNEXES:

5. **COMMAND AND SIGNAL**.

a. **Command**.

(1) TF CP:

(2) Succession of Command:

b. **Signal**.

(1) Challenge and Password:

(2) Number Combination:

(3) Running Password:

(4) Recognition:

(a) Far:

(b) Near:

(7) Code word:

(8) OPSKED:

(9) Reports. SALUTE to higher on enemy contact.

2-5. **FRAGMENTARY ORDER**. A FRAGO is an abbreviated form of an operations order, usually issued on a day-to-day basis that eliminates the need for restating information contained in a basic operations order. It is issued after an OPORD to change or modify that order or to execute a branch or sequel to that order. Figure 2-6 shows an annotated FRAGO format:

Figure 2-6. ANNOTATED FRAGO FORMAT

FRAGMENTARY ORDER____________

Time Zone referenced throughout order:

Task Organization:

Weather and Light Data:

High:	BMNT:	Moonrise:
Low:	Sunrise:	Moonset:
Wind Speed:	Sunset:	% Illum:
Wind Direction:	EENT:	
Forecast:		

Terrain (changes that will effect operation in new area of operations):

OAKOC: Observation and fields of fire, avenues of approach, key terrain, observation, and cover and concealment.

1. **SITUATION**. (Brief changes from base OPORD specific to this day's operation.)

a. **Enemy Situation**.

(1) Composition, disposition and strength.

(2) Capabilities.

(3) Recent activities.
(4) Most likely COA.

b. **Friendly Situation**.
(1) Higher mission.
(2) Adjacent patrols task/purpose.
(3) Adjacent patrol objective/route (if known).

2. **MISSION** (Who, what [task], when, where, why [purpose]—from higher HQ's maneuver paragraph).

3. **EXECUTION**

a. **Concept of Operation**. (Explain how unit will accomplish mission in general terms. Identify mission essential task, designate the main effort and how the supporting efforts support the main effort.)

(1) ***Maneuver***. (Assign task/purpose and discuss actions on the objective in detail. Use a sketch or terrain model to brief.)

(2) ***Fires***. (In support of today's mission, portions that do not change are not briefed.) PLOT-CR:
(a) Purpose.
(b) Location.
(c) Observer.
(d) Trigger.
(e) Communication method.
(f) Resource.

(3) ***CASEVAC***.

b. **Tasks to Maneuver Units**. (List unique tasks that apply to each squad.)

c. **Coordinating Instructions**.

(1) ***Timeline***.
(a) Hit time:
(b) ORP time:
(c) Movement time from PB:
(d) Final inspection:
(e) Rehearsals:
(f) FRAGO complete:

(2) ***Movement Plan***. (Use a sketch to brief.)
(a) Routes (primary and alternate):
(b) OOM:
(c) Formations:
(d) Movement technique:

(3) ***PIR***. (Specific to this mission):

(4) ***Rehearsal Plan***.

(5) ***Patrol Base Plan***. (If not IAW SOP):
(a) Teams:
(b) Occupation plan:
(c) Operations plan (security plan, alert plan, Black and Gold):
(d) Priorities of work:

(6) ***Air Assault Plan***. (Provided with higher's FRAGO, if applicable):
(a) Number, type, and ACL aircraft:
(b) PZ grid, DOL, and PZ posture time:
(c) Load time/lift time/flight time:
(d) Number of lifts and composition:
(e) Air checkpoints en route:
(f) LZ grid/DOL/LZ time:
(g) Actions after getting off aircraft:
(h) Actions on contact on LZ:

(7) ***Linkup Plan***. (If applicable):

(a) Time of linkup:
(b) Location of linkup site:
(c) Stationary element:
(d) Moving element:
(e) Rally points:
(f) Actions at linkup point:
(g) Near/far recognition signals (day and night):

4. **SUPPORT**. (Only cover changes from base order that apply for today's mission).

a. **Material and Services**.

(1) Any changes in classes of supply:
(2) Resupply plan:
(3) Water resupply plan:
(4) Aerial resupply plan (if applicable) :
(5) Truck plan:
(6) Maintenance issues specific to plan:

b. **Medical Evacuation Plan**. (Specific to mission):

(1) CCP point and markings:
(2) Aid and litter duties (if not SOP):

5. **COMMAND AND SIGNAL**.

a. **Command**.

(1) Location of Company CP:
(2) Location of PL:
(3) Location of key leaders:
(4) Succession of command:

b. **Signal**.

(1) Location of radios:

(a) During movement:
(b) During actions on the objective:

(2) SOI in effect:

(a) Bn command freq:	Co CP call sign:
(b) Bn MEDEVAC freq:	PL call sign:
(c) Retrans freq:	PSG call sign:
(d) Company freq:	1 SL call sign:
(e) Platoon freq:	2 SL call sign:
	3 SL call sign:
	WSL call sign:

(3) Pyro signals in use:
(4) Running password/number combination:
(5) OPSKEDs in effect:

(a) The field FRAGO should take no more than 40 minutes to issue, with 30 minutes for the target. The proposed planning guide is as follows:

Paragraphs 1 and 2:	5 minutes
Paragraph 3:	20 to 30 minutes
Paragraphs 4 and 5:	5 minutes

(b) The FRAGO should focus on actions on the objective. The PL may use subordinates to prepare para 1, 4, 5 and routes and fires for the FRAGO. It is acceptable for subordinates to brief the portions of the FRAGO they prepare.

(c) Use of sketches and a terrain model are critical to allow rapid understanding of the operation/FRAGO.

(d) Rehearsals are critical as elements of the constrained planning model. The FRAGO used in conjunction with effective rehearsals reduces preparation time and allows the PL more time for movement and recon.

(e) Planning in a field environment will necessarily reduce the amount of time leaders have for in-depth mission planning. The TLP gives leaders a framework to plan missions and produce orders when time is short.

2-6. **ANNEXES**. Operation order annexes are necessary to complete the plan and to provide greater clarity and understanding during complex or critical aspects of the operation. Information issued in annex form include the: aerial re-supply, truck movement, air assault, patrol base, small boat, linkup, and stream crossing annexes. Annexes are prepared only if the subject is not addressed thoroughly enough in the OPORD; brevity remains the standard. Annexes are always issued after the operation order.

AIR MOVEMENT ANNEX

1. **SITUATION**.
   a. **Enemy Situation**.
      (1) Enemy air capability.
      (2) Enemy ADA capability.
      (3) Include in weather: % Illum, illum angle, NVG window, ceiling, and visibility.
   b. **Friendly Situation**.
      (1) Unit(s) supporting operation.
      (2) Friendly ADA status.
2. **MISSION**.
3. **EXECUTION**.
   a. **Concept of Operation**.
   b. **Subunit Missions**.
   c. **Coordinating instructions**.
      (1) ***Pickup Zone***.
         (a) Name/Number/
         (b) Coordinates/
         (c) Load Time/
         (d) Takeoff Time/
         (e) Markings/
         (f) Control/
         (g) Landing Formation/
         (h) Approach/Departure Direction.
         (i) Alternate PZ Name/Number.
         (j) Penetration Points.
         (k) Extraction Points.
      (2) ***Landing Zone***.
         (a) Name/Number.
         (b) Coordinates.
         (c) H-Hour.
         (d) Markings.
         (e) Control.
         (f) Landing Formation/Direction.
         (g) Alt LZ Name/Number.
         (h) Deception Plan.
         (i) Extraction LZ.
      (3) ***Laager Site***.
         (a) Communications.
         (b) Security Force.
      (4) ***Flight Routes and Alternates***.
      (5) ***Abort Criteria***.
      (6) ***Down Aircraft/Crew*** (designated area of recovery [DAR]).
      (7) ***Special Instructions***.
      (8) ***Cross-FLOT Considerations***.
      (9) ***Aircraft Speed***.

(10) ***Aircraft Altitude***.
(11) ***Aircraft Crank Time***.
(12) ***Rehearsal Schedule/Plan***.
(13) ***Actions on Enemy Contact*** (en route and on the ground).

4. **SERVICE SUPPORT**.
   a. **Forward Area Refuel/Rearm Points**.
   b. **Classes I, III, and V** (specific).

5. **COMMAND AND SIGNAL**.
   a. **Command**.
   b. **Signal**.
      (1) Air/ground call signs and frequencies.
      (2) Air/ground emergency code.
      (3) Passwords/number combinations.
      (4) Fire net/Quickfire net.
      (5) Time zone.
      (6) Time hack.

AERIAL RESUPPLY ANNEX

1. **SITUATION**.
   a. **Enemy Forces** (include weather).
   b. **Friendly Forces**.
   c. **Attachments and Detachments**.

2. **MISSION**.

3. **EXECUTION**.
   a. **Concept of Operation**.
      (1) ***Maneuver***.
      (2) ***Fires***.
   b. **Tasks to Combat Units**.
      (1) ***Command and Control***.
      (2) ***Security***.
      (3) ***Marking***.
      (4) ***Recovery/Transport***.
   c. **Tasks to Combat Support Units**.
   d. **Coordinating Instructions**.
      (1) Flight Route.
         (a) General.
         (b) Checkpoints.
         (c) Communication checkpoint (CCP).
            - Marking of CCP.
            - Report time.
         (d) Heading from CCP.
      (2) Landing/Drop Zone.
         (a) Location.
            - Primary.
            - Alternate.
         (b) Marking.
            - Near.
            - Far.

(3) Drop Information.
    (a) Date/time of resupply (and alternates).
    (b) Code letter at DZ/LZ.
    (c) Length of DZ in seconds or dimensions of LZ.
    (d) Procedures for turning off DZ/LZ.
    (e) Formation, altitude, and air speed.
    - En route.
    - At DZ/LZ.

(4) Actions on enemy contact during resupply.
(5) Abort Criteria: En route and at DZ/LZ.
(6) Actions at DZ/LZ.
- Rehearsals.

4. **SERVICE SUPPORT**.

5. **COMMAND AND SIGNAL**.
   a. **Command**.
      (1) Location of patrol leader.
      (2) Location of assistant patrol leader.
      (3) Location of members not involved in resupply.
   b. **Signal**.
      (1) Air to ground call-signs and frequencies (primary and alternate).
      (2) Long range visual signals.
      (3) Short range visual signals.
      (4) Emergency procedures and signals.
      (5) Air drop communication procedures.
      (6) Code words.

PATROL BASE ANNEX

1. **SITUATION**
   a. Enemy Forces
   b. Friendly Forces
   c. Attachments and Detachments

2. **MISSION**.

3. **EXECUTION**.
   a. **Concept of Operation**.
      (1) ***Maneuver***.
      (2) ***Fires***.
   b. **Tasks to Combat Units**.
      (1) ***Teams***.
      - Security
      - Recon
      - Surveillance
      - LP/OPs

      (2) ***Individuals***.
   c. **Tasks to Combat Support Units**.
   d. **Coordinating instructions**.
      (1) Occupation plan.
      (2) Operations plan.
      - Security Plan.
      - Alert Plan.

- Priority of work.
- Evacuation plan.
- Alternate patrol base (used when primary is unsuitable or compromised).

4. **SERVICE SUPPORT**.
   a. Water plan.
   b. Maintenance plan.
   c. Hygiene plan.
   d. Messing plan.
   e. Rest plan.

5. **Command and Signal**.
   a. **Command**.
      (1) Location of patrol leader.
      (2) Location of assistant patrol leader.
      (3) Location of patrol CP.
   b. **Signal**.
      (1) Call signs and frequencies.
      (2) Code words.
      (3) Emergency signals.

SMALL BOAT ANNEX

1. **SITUATION**.
   a. **Enemy Forces**.
      (1) **Weather**.
         (a) Tide.
         (b) Surf.
         (c) Wind.
      (2) **Terrain**.
         (a) River width.
         (b) River depth and water temperature.
         (c) Current.
         (d) Vegetation.
      (3) **Identification, Location, Activity, and Strength**.
   b. **Friendly Forces** (unit furnishing support).
   c. **Attachments and Detachments**.
   d. **Organization for Movement**.

2. **MISSION**.

3. **EXECUTION**.
   a. **Concept of Operation**.
      (1) ***Maneuver***.
      (2) ***Fires***.
   b. **Tasks to Combat Units**.
      (1) Security.
      (2) Tie-down teams.
         (a) Load equipment.
         (b) Secure equipment.
      (3) Designation of coxswains and boat commanders.
      (4) Selection of navigator(s) and observer(s).

c. **Coordinating Instructions**.
(1) Formations and order of movement.
(2) Route and alternate route.
(3) Method of navigation.
(4) Actions on enemy contact.
(5) Rally points.
(6) Embarkation plan.
(7) Debarkation plan.
(8) Rehearsals.
(9) Time schedule.

4. **SERVICE SUPPORT**.
a. **Ration Plan**.
b. **Weapons and Ammunition**.
c. **Uniform and Equipment**.
(1) Method of distribution of paddles and life jackets.
(2) Disposition of boat, paddles and life jackets upon debarkation.

5. **COMMAND AND SIGNAL**.
a. **Command**.
(1) Location of patrol leader.
(2) Location of assistant patrol leader.
b. **Signal**.
(1) Signals used between boats and in boats.
(2) Code words.

STREAM CROSSING ANNEX

1. **SITUATION**.
a. **Enemy Forces**.
(1) ***Weather***.
(2) ***Terrain***.
(a) River width.
(b) River depth and water temperature.
(c) Current.
(d) Vegetation.
(e) Obstacles.
(3) ***Enemy*** (location, identification, activity).
b. **Friendly Forces**.
c. **Attachments and Detachments**.

2. **MISSION**.

3. **EXECUTION**.
a. **Concept of Operation**.
(1) ***Maneuver***.
(2) ***Fires***.
b. **Tasks to Combat Units**.
(1) ***Elements***.
(2) ***Teams***.
(3) ***Individuals***.

c. **Tasks to Combat Support Units**.

d. **Coordinating Instructions**.

(1) Crossing procedure/techniques.
(2) Security.
(3) Order of crossing.
(4) Actions on enemy contact.
(5) Alternate plan.
(6) Rallying points.
(7) Rehearsal plan.
(8) Time schedule.

4. **SERVICE SUPPORT**.

5. **COMMAND AND SIGNAL**.

a. **Command**.

(1) Location of patrol leader.
(2) Location of assistant patrol leader.
(3) Location of CP.

b. **Signal**.

(1) Emergency signals.
(2) Signals.

TRUCK ANNEX

1. **SITUATION**.

a. **Enemy Forces**.

b. **Friendly Forces**.

c. **Attachments and Detachments**.

2. **MISSION**.

3. **EXECUTION**.

a. **Concept of Operation**.

(1) ***Maneuver***.
(2) ***Fires***.

b. **Tasks to Combat Units**.

c. **Tasks to Combat Support Units**.

d. **Coordinating Instructions**.

(1) Time of departure and return.
(2) Loading plan and order of movement.
(3) Route (primary and alternate).
(4) Air Guards.
(5) Actions on enemy contact (vehicle ambush) during movement, loading, and downloading.
(6) Actions at the detrucking point.
(7) Rehearsals.
(8) Vehicle speed, separation, and recovery plan.
(9) Broken vehicle instructions.

4. **SERVICE SUPPORT**.

5. **COMMAND AND SIGNAL**.
   a. **Command**. Location of PL and PSG
   b. **Signal**.
      (1) Radio call signs and frequencies
      (2) Code words

2-7. **COORDINATION CHECKLISTS**. The following checklists are items that a platoon/squad leader must check when planning for a combat operation. In some cases, he will coordinate directly with the appropriate staff section, in most cases this information will be provided by the company commander or platoon leader. The platoon/squad leader, to keep him from overlooking anything that may be vital to his mission, may carry copies of these checklists:

Figure 2-7. Intelligence coordination checklist
Figure 2-8. Operations coordination checklist
Figure 2-9. Fire support coordination checklist
Figure 2-10. Coordination with forward unit checklist
Figure 2-11. Adjacent unit coordination checklist
Figure 2-12. Rehearsal area coordination checklist
Figure 2-13. Army aviation coordination checklist
Figure 2-14. Vehicular movement coordination checklist

Figure 2-7. INTELLIGENCE COORDINATION CHECKLIST

INTELLIGENCE COORDINATION CHECKLIST

In this coordination, the leader is informed of any changes in the situation as given in the operation order of mission briefing. He must keep himself constantly updated to ensure the plan is sound:

1. Identification of enemy unit.
2. Weather and light data.
3. Terrain update.
   a. Aerial photos.
   b. Trails and obstacles not on map.
4. Known or suspected enemy locations.
5. Weapons.
6. Probable course of action.
7. Recent enemy activities.
8. Reaction time of reaction forces.
9. Civilians on the battlefield.
10. Update to CCIR.

Figure 2-8. OPERATIONS COORDINATION CHECKLIST

OPERATIONS COORDINATION CHECKLIST

This coordination occurs with the platoon leader/company commander so that the platoon/squad leader can confirm his mission and operational plan, receive any last-minute changes to his mission or plan, and to update his subordinates or issue a FRAGO, if required:

1. Mission backbrief.
2. Identification of friendly units.
3. Changes in the friendly situation.
4. Route selection, LZ/PZ/DZ selection.
5. Linkup procedures.
6. Transportation/movement plan.

7. Resupply (in conjunction with S4).
8. Signal plan.
9. Departure and re-entry of forward units.
10. Special equipment requirements.
11. Adjacent units operating in the area of operations.
12. Rehearsal areas.
13. Method of insertion/extraction.

Figure 2-9. FIRE SUPPORT COORDINATION CHECKLIST

FIRE SUPPORT COORDINATION CHECKLIST

The platoon/squad leader will normally coordinate the following with the platoon forward observer (FO):

1. Mission backbrief.
2. Identification of supporting unit.
3. Mission and objective.
4. Route to and from the objective (include alternate routes).
5. Time of departure and expected time of return.
6. Unit target list (from fire plan).
7. Type of available support (artillery, mortar, naval gunfire, and aerial support to include Army, Navy, and Air Force) and their locations.
8. Ammunition available (to include different fuses).
9. Priority of fires.
10. Control measures.
    a. Checkpoints.
    b. Boundaries.
    c. Phase lines.
    d. Fire support coordination measures.
    e. Priority targets (target list).
    f. RFA (restrictive fire area).
    g. RFL (restrictive fire line).
    h. NFA (no-fire areas).
    i. Precoordinated authentication.
11. Communication (include primary and alternate means, emergency signals and code words).

Figure 2-10. COORDINATION WITH FORWARD UNIT CHECKLIST

COORDINATION WITH FORWARD UNIT CHECKLIST

A platoon/squad that requires foot movement through a friendly forward unit must coordinate with that unit's commander for a safe and orderly passage. If no time and place has been designated for coordination with the forward unit, the platoon/squad leader should set a time and place when he coordinates with the S-3. He must talk with someone at the forward unit who has the authority to commit that unit in assisting the platoon/squad during departure. Coordination entails a two-way exchange of information:

1. Identification (yourself and your unit).
2. Size of platoon/squad.
3. Time(s) and place(s) of departure and return, location(s) of departure point(s), ERRP, and detrucking points.
4. General area of operations.
5. Information on terrain and vegetation.
6. Know or suspected enemy positions or obstacles.
7. Possible enemy ambush sites.
8. Latest enemy activity.
9. Detailed information on friendly positions such as crew-served weapons, FPF.

10. Fire and barrier plan.
    a. Support the unit can furnish. How long and what can they do?
        (1) Fire support.
        (2) Litter teams.
        (3) Navigational signals and aids.
        (4) Guides.
        (5) Communications.
        (6) Reaction units.
        (7) Other.
    b. Call signs and frequencies.
    c. Pyrotechnic plan.
    d. Challenge and password, running password, number combination.
    e. Emergency signals and code words.
    f. If the unit is relieved, pass the information to the relieving unit.
    g. Recognition signals.

Figure 2-11. ADJACENT UNIT COORDINATION CHECKLIST

ADJACENT UNIT COORDINATION CHECKLIST

Immediately after the operation order of mission briefing, the platoon/squad leader should check with other platoon/squad leaders who will be operating in the same areas. If the leader is not aware of any other units operating is his area, he should check with the S3 during the operations coordination. The S3 can help arrange this coordination if necessary. The platoon/squad leaders should exchange the following information with other units operating in the same area:

1. Identification of the unit.
2. Mission and size of unit.
3. Planned times and points of departure and reentry.
4. Route(s).
5. Fire support and control measures.
6. Frequencies and call signs.
7. Challenge and password, running password, number combination.
8. Pyrotechnic plan.
9. Any information that the unit may have about the enemy.
10. Recognition signals.

Figure 2-12. REHEARSAL AREA COORDINATION CHECKLIST

REHEARSAL AREA COORDINATION CHECKLIST

This coordination is conducted with the platoon leader/company commander to facilitate the unit's safe, efficient and effective use of rehearsal area prior to its mission:

1. Identification of your unit.
2. Mission.
3. Terrain similar to objective site.
4. Security of the area.
5. Availability of aggressors.
6. Use of blanks, pyrotechnics, and ammunition.
7. Mock-ups available.
8. Time the area is available (preferably when light conditions approximate light conditions of patrol).
9. Transportation.
10. Coordination with other units using area.

Figure 2-13. ARMY AVIATION COORDINATION CHECKLIST

ARMY AVIATION COORDINATION CHECKLIST

This coordination is conducted with the platoon leader/company commander and/or S3 Air to facilitate the time, detailed and effective use of aviation assets as they apply to your tactical mission:

1. **SITUATION**.
   a. **Enemy Situation**.
      (1) Enemy air capability.
      (2) Enemy ADA capability.
      (3) Include in Weather: Percent illum, illum angle, NVG window, ceiling, and visibility.
   b. **Friendly Situation**.
      (1) Unit(s) supporting operation, Axis of movement/corridor/routes.
      (2) Friendly ADA status.

2. **MISSION**.

3. **EXECUTION**.
   a. **Concept of the Operation**. Overview of what requesting unit wants to accomplish with the air assault/air movement.
   b. **Tasks to Combat Units**.
      (1) Infantry.
      (2) Attack aviation.
   c. **Tasks to Combat Support Units**.
      (1) Artillery.
      (2) Aviation (lift).
   d. **Coordinating Instructions**.
      (1) ***Pickup Zone***.
      - Direction of landing.
      - Time of landing/flight direction.
      - Location of PZ/alternate PZ.
      - Loading procedures.
      - Marking of PZ (panel, smoke, SM, lights).
      - Flight route planned (SP, ACP, RP).
      - Formations: PZ, en route, LZ.
      - Code words:
        -- PZ secure (prior to landing), PZ clear (lead bird and last bird).
        -- Alternate PZ (at PZ, en route, LZ), names of PZ/alt PZ.
      - TAC air/artillery.
      - Number of pax per bird and for entire lift.
      - Equipment carried by individuals.
      - Marking of key leaders.
      - Abort criteria (PZ, en route, LZ).

      (2) ***Landing Zone***.
      - Direction of landing.
      - False insertion plans.
      - Time of landing (LZ time).
      - Location of LZ and Alternate LZ.
      - Marking of LZ (panel, smoke, SM, lights).
      - Formation of landing.
      - Code words, LZ name, alternate LZ name.
      - TAC air/artillery preparation, fire support coordination.
      - Secure LZ or not?

4. **SERVICE AND SUPPORT**.
   a. Number of aircraft per lift and number of lifts.
   b. Refuel/rearm during mission or not?
   c. Special equipment/aircraft configuration for weapons carried by unit personnel.
   d. Bump plan.
5. **COMMAND AND SIGNAL**.
   a. Frequencies, call signs and code words.
   b. Locations of air missions commander ground tactical commander, and air assault task force commander.

Figure 2-14. VEHICULAR MOVEMENT COORDINATION CHECKLIST

VEHICULAR MOVEMENT COORDINATION CHECKLIST

This is coordinated with the supporting unit through the platoon sergeant/first sergeant to facilitate the effective, detailed, and efficient use of vehicular support and/or assets:

1. Identification of the unit.
2. Supporting unit identification.
3. Number and type of vehicles and tactical preparation.
4. Entrucking point.
5. Departure/loading time.
6. Preparation of vehicles for movement.
   a. Driver responsibilities.
   b. Platoon/squad responsibilities.
   c. Special supplies/equipment required.
7. Availability of vehicles for preparation/rehearsals/inspection (time and location).
8. Routes.
   a. Primary.
   b. Alternate.
   c. Checkpoints.
9. Detrucking points.
   a. Primary.
   b. Alternate.
10. March internal/speed.
11. Communications (frequencies, call signs, codes).
12. Emergency procedures and signals.

2-8. DOCTRINAL TERMS

TACTICAL TASK--A clearly defined, measurable activity accomplished by individuals and organizations. Tasks are specific activities that contribute to the accomplishment of encompassing missions or other requirements. A task should be definable, measurable, and decisive (achieve the purpose).

Actions by Friendly Force		Effects on Enemy Force
Assault	Follow and Assume	Block
Attack-by-Fire	Follow and Support	Canalize
Breach	Linkup	Contain
Bypass	Occupy	Defeat
Clear	Reconstitution	Destroy
Combat Search and Rescue	Reduce	Disrupt
Consolidation and Reorganization	Retain	Fix
Control	Secure	Interdict
Counterreconnaissance	Seize	Isolate
Disengagement	Support-by-Fire	Neutralize
Exfiltrate	Suppress	Penetrate
		Turn

PURPOSE (in order to)--The desired or intended result of the tactical operation stated in terms related to the enemy or the desired situation. The why of the mission statement. The most important component of the mission statement.

Allow	Divert	Prevent
Cause	Enable	Protect
Create	Envelop	Support
Deceive	Influence	Surprise
Deny	Open	

OPERATION--A military action or the carrying out of a military action to gain the objectives of any battle or campaign.

Types of operations include:	Forms of Offensive Maneuver:	Types of Defensive Techniques:	Other Operations:
Movement to Contact	Envelopment	Area Defense	Reconnaissance Ops
Search and Attack	Frontal Attack	Mobile Defense	Security Ops
Attack	Infiltration	Retrograde Ops	Information Ops
Ambush	Penetration	Delay	Combined Arms
Demonstration	Turning Movement	Withdrawal	Breach
Feint		Retirement	Passage of Lines
Raid			Relief in Place
Spoiling Attack			River Crossing Ops
Exploitation			Troop Movement
Pursuit			Admin
			Approach March
			Road March

2-9. **TERRAIN MODEL**. The terrain model (Figure 2-15) is an outstanding visual means used during the planning process to communicate the patrol routes and also detailed actions on the objective. At a minimum, the model is required to during planning to display routes to the objective and to highlight prominent terrain features the patrol will encounter during movement. A second terrain model is usually constructed of the objective area, enlarged to a sufficient size and detail to brief the patrol's actions on the objective.

a. **Checklist**.

(1) North-seeking arrow.
(2) Scale.
(3) Grid lines.
(4) Objective location.
(5) Exaggerated terrain relief, water obstacles.
(6) Friendly patrol locations.
(7) Targets (indirect fires, including grid and type of round).
(8) Routes, primary and alternate.
(9) Planned RPs (ORP, L/URP, RP).
(10) Danger areas (roads, trails, open areas).
(11) Legend.
(12) Blowup of objective area.

b. **Construction**. The following are some field-expedient techniques to aid in terrain model construction:

(1) Use a 3 x 5 card, MRE box, or piece of paper to label the objective or key sites.

(2) Grid lines can be made using string from the guts of 550 cord or colored tape. (Grid lines are identified by writing numbers on small pieces of paper.)

(3) Trees and vegetation are replicated by using moss, green or brown spray paint, pine needles, crushed leaves, or cut up grass.

(4) Water is designated by blue chalk, blue spray paint, blue yarn, tin foil, or MRE creamer.

(5) North seeking arrows are made from sharpened twigs, pencils, or colored yarn.

(6) Enemy positions are designated using red yarn, M-16 rounds, toy Rangers, or poker chips.

(7) Friendly positions such as security elements, support by fire, and assault elements are made using M-16 rounds, toy Rangers, poker chips, small MRE packets of sugar and coffee, or pre-printed acetate cards.

(8) Small pieces of cardboard or paper can identify target reference points and indirect fire targets. Ensure grids are shown for each point.

(9) Breach, support by fire, and assault positions are made using colored yarn or string so that these positions can be easily identified.

(10) Bunkers and buildings are constructed using MRE boxes or tongue depressors/sticks.

(11) Perimeter wire is constructed from a spiral notebook.

(12) Key phase lines are constructed with colored string or yarn.

(13) Trench lines are replicated by colored tape or yarn, by digging a furrow and coloring it with colored chalk or spray paint.

Note: All symbols used on the terrain model must be clearly identified in the legend.

Figure 2-15. TERRAIN MODEL

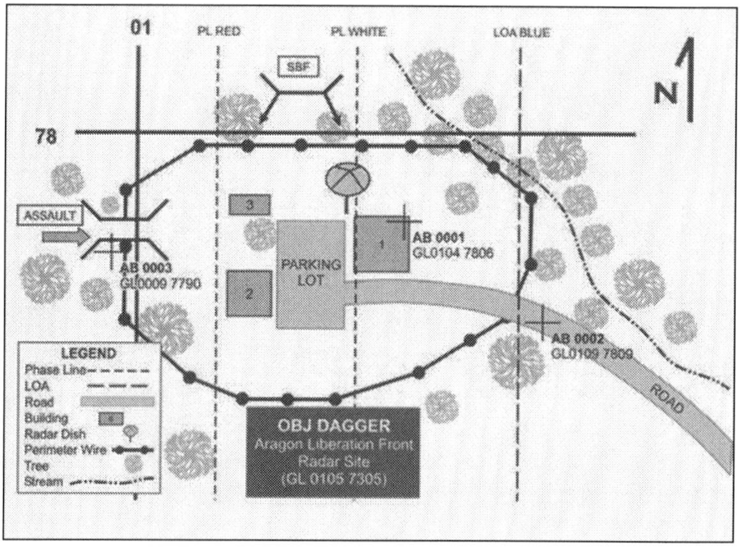

NOTES

Chapter 3
FIRE SUPPORT

Indirect fire support can greatly increase the combat effectiveness and survivability of any Infantry unit. The ability to plan for and effectively use this asset is a task that every Ranger and small unit leader should master. Fire support assets can assist a unit suppress, fix, destroy, or neutralize the enemy. Employment of indirect fire support should be considered throughout every offensive and defensive operation. This chapter discusses planning, tasks, capabilities, risk estimate distances, target overlays, close air support, elements and sequence of call for fire, and example call-for-fire transmissions.

3-1. **PLANNING**. Planning is the continual process of selecting targets on which fires are prearranged to support a phase of the commander's plan. The principles of planning follow:

- Consider what the commander wants to do.
- Plan early and continuously.
- Exploit all available targeting assets.
- Use all available lethal and nonlethal fire support means.
- Use the lowest echelon able to furnish effective support.
- Observe all fires.
- Use the most effective fire support asset available.
- Provide adequate fire support.
- Avoid unnecessary duplication.
- Provide for safety of friendly forces and installations.
- Provide for flexibility.
- Furnish the type of fire support requested.
- Consider the airspace.
- Provide rapid and effective coordination.
- Keep all fire support informed.

3-2. **TASKS**.

a. **All Operations**.
   (1) Locate targets.
   (2) Integrate all available assets.
   (3) Destroy, neutralize, or suppress all enemy direct and indirect fire systems.
   (4) Provide illumination and smoke.
   (5) Provide fires in support of JAAT and SEAD missions.
   (6) Deliver scatterable mines.
   (7) Prepare for future operations.
   (8) Provide positive clearance of fires.

b. **Offensive Operations**.
   (1) Support the movement to contact, chance contact.
   (2) Soften enemy defenses before the attack by arranging short, violent preparations, where required.
   (3) Provide support during the attack by attacking high payoff targets.
   (4) Plan for deep and flanking fires.
   (5) Plan fires during consolidation.
   (6) Provide counterfires.

c. **Defensive Operations**.
   (1) Execute all plans as the commander intends.
   (2) Disorganize, delay, and weaken the enemy before the attack.
   (3) Provide counterfire.
   (4) Provide fires in support of planned engagement areas (EAs).
   (5) Attack high payoff targets.
   (6) Plan fires in support of barriers and obstacle plans.
   (7) Plan for deep, flanking, and rear fires.

(8) Provide fires in support of counterattacks.
(9) Plan final protective fire (FPFs).

3-3. **CAPABILITIES**. Tables 3-1 and 3-2 show capabilities of field artillery and mortars.

Table 3-1. CAPABILITIES OF FIELD ARTILLERY

WEAPON	MAX RANGE (meters)	MIN RANGE (meters)	MAX RATE Rds per Min	Burst Radius	SUSTAINED RATE Rds per Min
105-mm Howitzer M119, Towed	14,000 m	0	6 for 2 min	35 m	3 Rds for 30 min, then 1 rd per min
155-mm Howitzer M198, Towed	18 m,100 m 30,000 m (RAP)	0	4 for 3 min 2 for 30 min	50 m	1 rd per min Temp Dependent
155-mm Howitzer M109A6 SP	18 m, 100 m 30,000 m (RAP)	0	4 for 3 min	50 m	1 for 60 min 0.5

Table 3-2. CAPABILITIES OF MORTARS

WPN	MUNITIONS AVAILABLE	MAX HE RANGE	MIN RANGE (meters)	MAX RATE (Rds per Min)	BURST RADIUS	SUSTAINED RATE (Rds per Min)
60mm	HE, WP, Illum	3,500 m	70 m (HE)	30 for 4 min	30 m	20
81mm	HE, WP, Illum	5,600 m	70 m (HE)	25 for 2 min	38 m	8
120mm	HE, Smoke, Illum	7,200 m	180 m (HE)	15 for 1 min	60 m	5

DANGER

DANGER CLOSE

1. **When the target is within 600 meters of any friendly troops (for mortars and field artillery), announce DANGER CLOSE in the Method of Engagement portion of the call for fire.**
2. **When adjusting 5-inch or smaller naval guns on targets within 750 meters, announce DANGER CLOSE. For larger naval guns, announce DANGER CLOSE for targets within 1,000 meters. Failure to adhere to this guidance can result in fratricide.**
3. **Avoid making corrections using the bracketing method of adjustment, because doing so can cause serious injury or death. Use only the creeping method of adjustment during DANGER CLOSE missions. Make corrections of no more than 100 meters by creeping the rounds to the target.**

3-4. **RISK ESTIMATE DISTANCES**. RED applies to combat only. Minimum safe distances (Table 3-3) apply to training IAW AR 350-1. RED takes into account the bursting radius of particular munitions and the characteristics of the delivery system and associates this combination with a percentage representing the likelihood of becoming a casualty, that is, the percentage of risk. RED is defined as the minimum distance friendly troops can approach the effects of friendly fires without suffering appreciable casualties of 0.1% PI or higher.

WARNING

[Commanders] use risk-estimate distance formulas to determine acceptable risk levels in combat only. Specifically, use them to identify the risk to your Rangers at various distances from their targets. Risk estimate distances apply only in combat. In training, use minimum safe distances (MSD).

a. **Casualty Criterion**. The casualty criterion is the 5-minute assault criterion for a prone soldier in winter clothing and helmet. Physical incapacitation means that a soldier is physically unable to function in an assault within a 5-minute period after an attack. A PI value of less than 0.1 percent can be interpreted as being less than or equal to one chance in one thousand.

Table 3-3. RISK ESTIMATE DISTANCES FOR MORTARS AND CANNON ARTILLERY

	Risk Estimate Distances (Meters)					
	10 % PI			0.1 % PI		
System	1/3 range	2/3 range	Max range	1/3 range	2/3 range	Max range
60-mm mortar	60	65	65	100	150	175
81-mm mortar	75	80	80	165	185	230
120-mm mortar	100	100	100	150	300	400
105-mm howitzer	85	85	90	175	200	275
155-mm howitzer	100	100	125	200	280	450
155-mm DPICM	150	180	200	280	300	475

b. **Risk**. Using echelonment of fires within the specified RED for a delivery system requires the unit to assume some risks. The maneuver commander determines by delivery system how close he will allow fires to fall in proximity to his forces. The maneuver commander makes the decision for this risk level, but he relies heavily on the FSO's expertise.

3-5. **TARGET OVERLAYS**.

a. Figure 3-1 shows contents of fire support overlay.

Figure 3-1. CONTENTS OF FIRE SUPPORT OVERLAY

- Unit and official capacity of person making overlay.
- Date the overlay was prepared.
- Map Sheet Number.
- Effective Period of Overlay (DTG).
- Priority target.
- ORP Location.
- Call signs and frequencies. (PRI/ALT)
- Routes--Primary/Alternate.
- Phase Lines/Checkpoints used by the patrol.
- Spares.
- Index marks to position overlay on map.
- Objective.
- Target Symbols.
- Description, location and remarks column, complete.

b. Using a PLOT-CR checklist helps ensure the leader's fire support plan is complete:
   (1) Purpose--Purpose of the planned indirect fires.
   (2) Location--Planned targets with an 8-digit grid (minimum).
   (3) Observer--Planned observer to see the impact of the rounds and adjust.
   (4) Trigger--Method to initiate fires.
   (5) Communication--Method to communicate between the observer and the supporting unit.
   (6) Resources--Planned allocated resource for each target.

c. Sterile overlay includes---
   (1) Index marks to position overlay on map.
   (2) Target symbols.

d. Target overlay symbols (Figure 3-2) include--
   (1) Point target:
   (2) Linear target:
   (3) Circular target:

Figure 3-2. TARGET OVERLAY SYMBOLS--POINT (TOP), LINEAR (CENTER), AND CIRCULAR (BOTTOM)

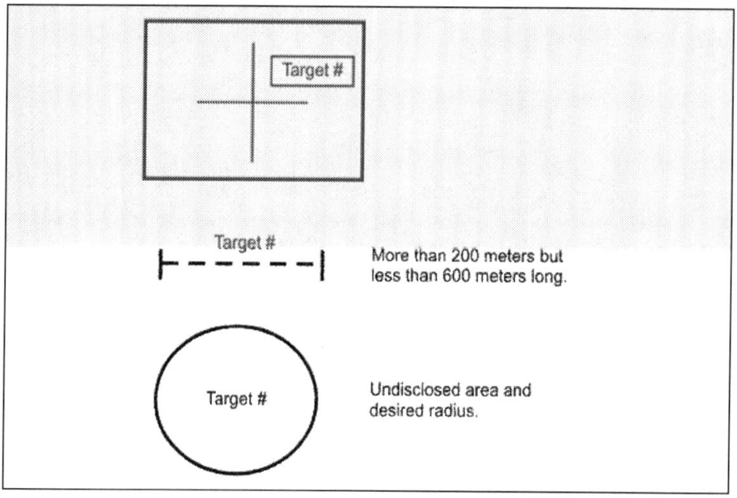

3-6. **CLOSE AIR SUPPORT**. The two types of close air support requests are planned and immediate. Planned requests are processed by the Army chain to corps for approval. Immediate requests are initiated at any level and processed by the battalion S-3 FSO, and air liaison officer.

a. Format for requesting immediate CAS:
   (1) Observer identification.
   (2) Warning order (request close air).
   (3) Target location (grid).

(4) Target description. (Description must include, as a minimum: type and number of targets, activity or movement; point or area targets, include desired results on target and time on target.

(5) Location of friendly forces.

(6) NAV details (elevation).

(7) Threats--ADA, small arms, and so on.

(8) Hazards--friendly aircraft in area.

(9) Wind direction.

b. Close air support capabilities are shown in Table 3-4.

Table 3-4. CLOSE AIR SUPPORT

AIRCRAFT	SERVICE	CHARACTERISTICS
A-10 *	AF	Specialized CAS aircraft; subsonic; typical load 6,000 lbs, max load 16,000 lbs; 30-mm gun.
F-16 *	AF	A multirole aircraft; complements the F-4 and F-15 in an air-to-air role. Most accurate ground delivery system in the inventory; supersonic; typical load 6,000 lbs, max load 10,600 lbs.
F-18 *	N/MC	A multirole fighter scheduled to replace the F-4. A wide variety of air-to-surface weapons. Typical load 7,000 lbs, max load 17,000 lbs, 20-mm gun mounted in the nose; air-to-air missiles.
AC-130	AF/R	A specialized CAS/RACO aircraft, propeller driven. Two models: The "A" model has two 40-mm guns, two 20-mm guns and two 7.62-mm miniguns. The similar "H" model has no 7.62-mm miniguns, and has a 105-mm howitzer in place of one of the 40-mm guns. Both models have advanced sensors and a target-acquisition system, including FLIR and low light TV. Very accurate. Vulnerable to enemy air defense systems, so must operate in a low-threat environment.
*FM capability.		

3-7. **CALL FOR FIRE**.

a. **Observer's Identification - Call Signs**.

b. **Warning Order**.

(1) ***Type of mission***.

(a) Adjust fire.

(b) Fire for effect.

(c) Suppress.

(d) Immediate suppression/immediate smoke.

(2) ***Size of element to Fire for Effect***. When the observer does not specify what size element to fire, the battalion FDC will decide.

(3) ***Method of Target Location***.

(a) Polar plot.

(b) Shift from a known point.

(c) Grid.

(4) **Location of Target**.

(a) *Grid Coordinate - 6 Digit*. 8 digit if greater accuracy is required.

(b) *Shift from a Known Point*. Send observer-target (OT) direction:

- Mils (nearest 10).
- Degrees.
- Cardinal direction.
- Send lateral shift (right/left, nearest 10 mils).
- Send range shift (add/drop, nearest 100 mils).
- Send vertical shift (up/down). Use only if it exceeds 35 mils (nearest 5 mils).

(c) *Polar Plot*. Send direction. (Nearest 10 mils)

- Send distance (nearest 100 mils).
- Send vertical shift (nearest 5 mils).

(5) ***Description of Target***.
   (a) Type.
   (b) Activity.
   (c) Number.
   (d) Degree of protection.
   (e) Size and shape (length/width or radius).

(6) ***Method of Engagement, Type of Adjustment***. When the observer does not request a specific type of fire control adjustment, area fire is issued.
   (a) Area fire--moving target.
   (b) Precision fire--point target.

(7) ***Danger Close***. Use when friendly troops are within--
   (a) 600 mils for mortars.
   (b) 600 mils for artillery.
   (c) 750 mils for naval guns 5 inches or smaller.

(8) ***Mark***. Use to orient observer or to indicate targets.

(9) ***Trajectory***.
   (a) Low angle (standard).
   (b) High angle (mortars or if requested).

(10) ***Ammunition***. Use HE quick used unless observer specifies otherwise.
   (a) Projectile (HE, illum, ICM, smoke, and so on).
   (b) Fuse (quick, time, and so on).
   (c) Volume of fire (observer may request the number of rounds to be fired).

(11) ***Distribution***.
   (a) 100 mil sheaf (standard).
   (b) Converged sheaf (used for small hard targets).
   (c) Special sheaf (any length, width, and attitude).
   (d) Open sheaf (separate bursts).
   (e) Parallel sheaf (linear target).

(12) ***Method of Fire and Control***.
   (a) *Method of Fire*. Specific guns and a specific interval between rounds. Normally adjust fire, one gun is used with a 5-second interval between rounds.
   (b) *Method of Control*.
   - AT MY COMMAND - FIRE--Remains in effect until observer announces, "*Cancel at my command*."
   - CANNOT OBSERVE--Observer can't see the target.
   - TIME ON TARGET--Observer tells FDC when he wants the rounds to impact.
   - CONTINUOUS ILLUMINATION. Calculated by the FDC, otherwise observer indicates interval between rounds in seconds.
   - COORDINATED ILLUMINATION. Observer may order the interval between illum and HE shells.
   - CEASE LOADING--Stop loading rounds.
   - CHECK FIRING--Stop firing at once.
   - CONTINUOUS FIRE--Load and fire as fast as possible.
   - REPEAT--Fire another round(s) with or without adjustments.

(13) ***Authentication***. Challenge and reply.

(14) ***Message to Observer***.
   (a) Battery(ies) to fire for effect.
   (b) Adjusting battery.
   (c) Changes to the initial call for fire.
   (d) Number of rounds (per tube) to be fired for effect.
   (e) Target numbers.

(f) Additional information:

- Time of flight. Moving target mission.
- Probable error in range. 38 meters or greater (normal mission).
- Angle T. 500 mils or greater.

c. **Correction of Errors**. When FDC has made an error when reading back he fire support data, the observer announces "*CORRECTION*" and transmits the correct data in its entirety.

d. **Call-for-Fire Transmissions**. Example call-for-fire transmissions are shown in Table 3-5.

Table 3-5. EXAMPLE CALL-FOR-FIRE TRANSMISSIONS.

GRID MISSION	
OBSERVER	**FIRING UNIT**
F24, this is J42, ADJUST FIRE, OVER.	J42, this is F24, AJUST FIRE, OUT.
GRID WM180513, DIRECTION 0530, OVER.	GRID WM180513, DIRECTION 0530, OUT.
Infantry platoon dug in, OVER.	Infantry platoon dug in, OUT.
	SHOT OVER.
SHOT OUT.	
	SPLASH, OVER.
SPLASH OUT.	
End of mission, 15 casualties, platoon dispersed, OVER.	End of mission, 15 casualties, platoon dispersed, OUT.
SHIFT FROM KNOWN POINT	
OBSERVER	**FIRING UNIT**
J42, THIS IS F24, ADJUST FIRE, SHIFT AB1001, OVER.	F24, THIS IS J42, ADJUST FIRE, SHIFT AB1001, OUT.
DIRECTION 2420, RIGHT 400, ADD 400, OVER.	DIRECTION 2420, RIGHT 400, ADD 400, OUT.
FIVE T-72 TANKS AT POL SITE, OVER.	FIVE T-72 TANKS AT POL SITE, AUTHENTICATE JULIET NOVEMBER, OVER.
I AUTHENTICATE TANGO, OVER.	
	SHOT, OVER.
SHOT OUT.	SPLASH, OVER.
SPLASH OUT.	
END OF MISSION, 2 TANKS DESTROYED, 3 IN WOODLINE, OVER	END OF MISSION, 2 TANKS DESTROYED, 3 IN WOODLINE, OUT

POLAR	
OBSERVER	**FIRING UNIT**
J42, this is F24, adjust fire, polar OVER.	F24, this is J42, adjust fire, polar, OUT
Direction 2300, distance 4,000, OVER.	Direction 2300, distance 4,000, OUT.
Infantry platoon dug in, OVER	Infantry platoon dug in, OUT
	Shot OVER.
Shot OUT.	Splash, OVER.
Splash OUT.	
End of mission, 15 casualties, platoon dispersed, OVER.	End of mission, 15 casualties, platoon dispersed, OUT.

NOTES

Chapter 4
MOVEMENT

To survive on the battlefield, stealth, dispersion, and security must be enforced in all tactical movements. The leader must be skilled in all movement techniques.

4-1. **FORMATIONS**. Movement formations are comprised of elements and Soldiers in relation to each other. Fire teams, squads and platoons use several movement formations. Formations provide the leader control based on a METT-TC analysis. Leaders position themselves where they can best command and control formations. The formations below allow the fire team leader to lead by example, "*Follow me, and do as I do.*" All Soldiers in the team must be able to see their leader. The formations in Figure 4-1 reflect fire team formations. Squad formations are very similar with more Soldiers. Squads can operate in lines and files similar to fire teams. When squads operate in wedges or in echelon, the fire teams use those formations and simply arrange themselves in column or one team behind the other. Squads may also use the vee, with one team forming the lines of the vee, and the SL at the apex for command and control. Platoons work on the same basic formations as the Squads. When operating as a platoon, the platoon leader must carefully select the location for his machine guns in the movement formation.

Figure 4-1. FORMATIONS

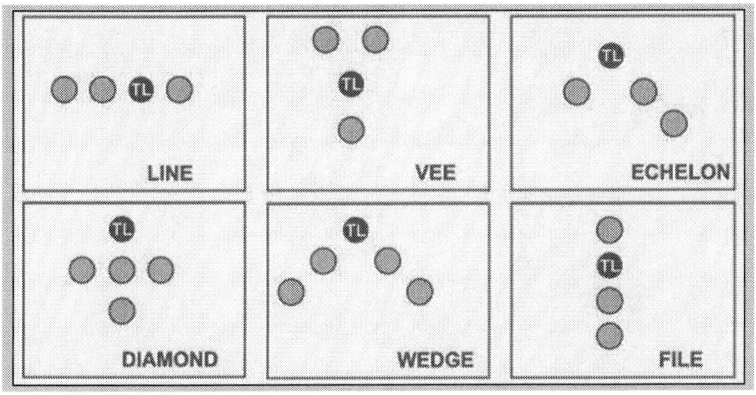

a. **Techniques**. A movement technique is the method a unit uses to traverse terrain. There are three movement techniques: traveling, traveling overwatch, and bounding overwatch. The selection of a movement technique is based on the likelihood of enemy contact and the need for speed. Factors to consider for each technique are control, dispersion, speed, and security. Movement techniques are not fixed formations. They refer to the distances between Soldiers, teams, and squads that vary based on METT-TC. Soldiers must be able to see their fire team leaders. The platoon leader should be able to see his lead squad leader. Leaders control movement with hand-and-arm signals and use radios only when needed.

b. **Standards**.

(1) Unit moves on designated route or arrives at specified location IAW OPORD maintaining accountability of all assigned/attached personnel.

(2) Unit uses movement formation and technique ordered by the leader based on METT-TC.

(3) Leaders remain oriented (within 200 meters) and follow planned route unless METT-TC dictates otherwise.

(4) Unit maintains 360 degree security and remains 100% alert during movement.

(5) Unit maintains 360 degree security and a minimum of 75% security during halts.

(6) If contact with the enemy is made, it is made with the smallest element possible.

(7) Control measures are used during movement such as head counts, rally points, or phase lines.

c. **Fundamentals**.

(1) ***Land Navigation***. Mission accomplishment depends on successful land navigation. The patrol should use stealth and vigilance to avoid chance contact. Designate a primary and alternate compass and pace man per patrol.

Note: The point man will not be tasked to perform compass or pace duties. The point man's sole responsibility is forward security for the element.

(2) ***Avoidance of Detection***. Patrols must use stealth, and use the cover and concealment of the terrain to its maximum advantage. Whenever possible, move during limited visibility in order to maximize technological advantages gained by night vision devices and to hinder the enemy's ability to detect the patrol. Exploit the enemy's weaknesses, and attempt to time movements to coincide with other operations that distract the enemy. The enemy threat and terrain determines which of the three movement techniques will be used:

(a) Fire teams maintain visual contact, but the distance between them is such that the entire patrol does not become engaged if contact is made. Fire teams can spread their formations as necessary to gain better observation to the flanks. Although widely spaced, men retain their relative position in their wedge and follow their team leader. Only in extreme situations should the file be used.

(b) The lead squad must secure the front along with assuming responsibility for navigation. For a long movement, the PL may rotate the lead squad responsibilities. The fire team/squad in the rear is charged with rear security.

(c) Vary movement techniques to meet the changing situation.

(d) With the exception of fire team leaders, leaders move inside their formations where they can maintain the best control.

(3) ***Security***. The patrol must use both active and passive security measures constantly. Assign subunits responsibility for security at danger areas, patrol bases, and most importantly in the objective area.

(4) ***Fire Support***. Plan fire support (mortars, artillery, tactical air, attack helicopter, naval gunfire).

(5) ***Three-Dimensional Battlefield***. 360 Degree Security is achieved through high and low security. Within a fire team, squad, and so on, the leader must assign appropriate sectors of fire to their subordinate in order to ensure all aspects of the battlefield are covered. This includes trees, multiple storied structures, tunnels, sewers, ditches.

d. **Movement Techniques**.

(1) The traveling technique is used when enemy contact is not likely but speed is necessary.

(2) The traveling overwatch technique is used when enemy contact is possible.

(3) The bounding overwatch technique is used when enemy contact is likely, or when crossing a danger area.

e. **Traveling**. In the traveling technique, the distance between individuals is about 10 meters with 20 meters between squads. It has the following characteristics:

(1) More control than traveling overwatch but less than bounding overwatch.

(2) Minimum dispersion.

(3) Maximum speed.

(4) Minimum security.

f. **Traveling Overwatch**. The traveling overwatch technique is the basic movement technique. The distance between individuals is about 20 meters, between teams about 50 meters.

(1) In platoon traveling overwatch, the lead squad must be far enough ahead of the rest of the platoon to detect or engage any enemy before the enemy observes or fires on the main body. However, it must be close enough to be supported by the platoon's small arm's fires. This is normally between 50 to 100 meters, depending on terrain, vegetation, and light and weather conditions.

(2) In a column formation, only the lead squad should use the traveling overwatch; however, if greater dispersion is desired, all squads may use it.

(3) In other formations, all squads use traveling overwatch unless the platoon leader specifies otherwise.

(4) Traveling overwatch has the following characteristics:

- Good control.
- Good dispersion.
- Good speed.
- Good security forward.

g. **Bounding Overwatch**. In the bounding overwatch technique (Figure 4-2), the distance between men remains about 20 meters. The distance between teams and squads varies.

(1) The squad or platoon has a bounding element and an overwatch element. The bounding element moves while the overwatch element occupies an overwatch position that can cover the route of the bounding element by fire. Each bound is within supporting range of the overwatch element.

(2) There are two types of bounding, successive and alternating (Figure 4-2). Successive is nothing more than one squad moving to a position, then the overwatching squad moving to a position generally online with the first squad. Alternating bounding is when one squad moves into position, then the overwatching squad moves to a position in front of the first squad.

Figure 4-2. SQUAD BOUNDING OVERWATCH

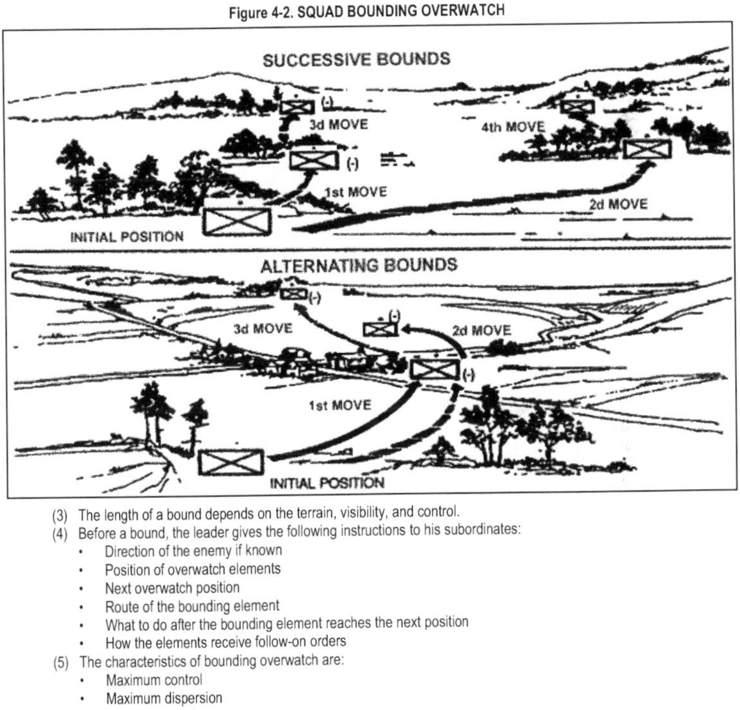

(3) The length of a bound depends on the terrain, visibility, and control.

(4) Before a bound, the leader gives the following instructions to his subordinates:

- Direction of the enemy if known
- Position of overwatch elements
- Next overwatch position
- Route of the bounding element
- What to do after the bounding element reaches the next position
- How the elements receive follow-on orders

(5) The characteristics of bounding overwatch are:

- Maximum control
- Maximum dispersion
- Minimum speed
- Maximum security

h. **Platoon Bounding Overwatch** (Figure 4-3).

(1) ***Method One***. When platoons use bounding overwatch, one squad bounds and one squad overwatches; the third squad awaits orders. Forward observers stay with the overwatching squad to call for fire. Platoon leaders normally stay with the overwatching squad who use machine guns and attached weapons to support the bounding squad.

(2) ***Method Two***. Another way is to have one squad use bounding overwatch and have the other two squads use traveling or traveling overwatch technique.

Figure 4-3. PLATOON BOUNDING OVERWATCH

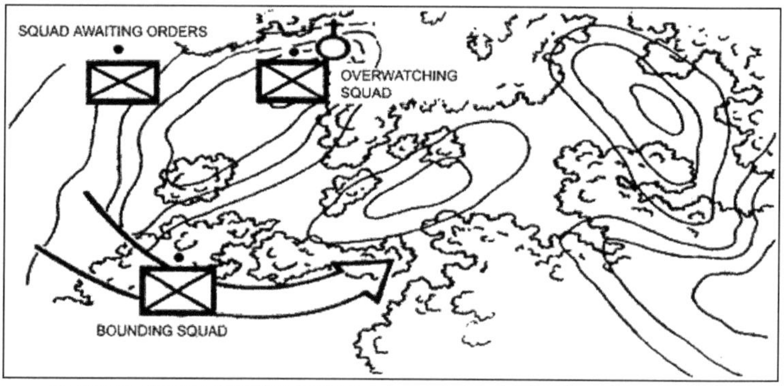

(3) ***Movement Considerations***. When deciding where to move the bounding element, consider--

- Where the enemy is likely to be.
- The mission.
- The routes to the next overwatch position.
- The weapons ranges of the overwatching unit.
- The responsiveness of the rest of the unit.
- The fields of fire at the next overwatch position.

4-2. **TACTICAL MARCHES**. Platoons conduct two types of marches with the company: foot marches and motor (road) marches.

a. **Purpose/General**. A successful foot march is when troops arrive at their destination at the prescribed time, physically able to execute their tactical mission.

b. **Standard**.

(1) The unit crosses the start point and release point at the time specified in the order.

(2) The unit follows the prescribed route, rate of march, and interval without deviation unless required otherwise by enemy action or higher headquarters action.

c. **Fundamentals**.

(1) Effective control.

(2) Detailed planning.

(3) Rehearsals.

d. **Considerations**.

(1) ***METT-TC***.

- Mission ....Task and purpose.
- Enemy ....Intentions, capabilities, and course of action.
- Terrain and weather.................Road condition/trafficability, and visibility.

- Troops/equipment....................Condition of Soldiers and their loads, numbers and types of weapons and radios.
- Time...Start time, release time, rate of march, time available.
- Civilians ....Movement through populated areas, refugees, OPSEC.

(2) ***Task Organization***:
- Security....................................Advance and trail teams.
- Main body................................Two remaining line squads and weapons squad.
- Headquarters...........................Command and control.
- Control measures.

(3) ***Start Point and Release Point*** (given by higher):
(a) Checkpoints.............................At checkpoints report to higher and use checkpoint to remain oriented.
(b) Rally or rendezvous points ......Used when elements become separated.
(c) Location of leaders ....Where they can best control their elements.
(d) Commo plan ....Locations of radios, frequencies, call signs, and OPSKEDs.
(e) Dispersion between Soldiers--
- 3 to 5 meters / day.
- 1 to 3 meters / night.

(4) ***March Order***. May be issued as an OPORD, FRAGO, or annex to either (must use operational overlay or strip map)"
(a) Formations and order of movement
(b) Route of march........................Assembly area, start point, release point, rally points, checkpoints, break/halt points.
(c) Start point time, release point time, and rate of march.
(d) March interval for squads, teams, and individuals.
(e) Actions on enemy contact........Air and ground.
(f) Actions at halts.
(g) Fires ....Detailed plan of fire support for the march.
(h) Water supply plan.
(i) MEDEVAC plan.

e. **Duties and Responsibilities**.

(1) ***Platoon Leader***:
- Before ....Issues warning order and FRAGO; inspects and supervises.
- During ....Ensures unit makes movement time; ensures interval is maintained and that unit remains oriented; maintains security; checks condition of men; and enforces water discipline and field sanitation.
- After...Ensures men are prepared to accomplish their mission, supervises SLs, and ensures medical coverage is provided to men as needed.

(2) ***Platoon Sergeant***:
- Before.......................................Assists PL, makes recommendations, and enforces uniform and packing lists.
- During.......................................Controls stragglers, helps platoon leader maintain proper interval and security.
- At Halts.....................................Enforces security, ensures welfare of men, enforces field sanitation and litter discipline, and enforces use of preventive medicine.
- After..Coordinates for water, rations, and medical supplies. Recovers casualties.

(3) ***Squad Leaders***:
- Before.......................................Provides detailed instruction to TLs, inspects boots and socks for serviceability and proper fit, inspects adjustment of equipment, inspects to ensure canteens are full, and ensures equal distribution of loads.
- During.......................................Controls squad, maintains proper interval between men and equipment, enforces security, remains oriented.

- At Halts....................................Ensures security is maintained, provides men for water resupply as detailed. Physically checks the men in his squad, ensures they drink water, and that they change socks as necessary. Rotates heavy equipment. (Units should plan this in detail to avoid confusion before, during, and after halts.)
- After..Occupies squad sector of assembly area, conducts foot inspection, reports condition of men to PL, and prepares men to accomplish the mission.

(4) ***Security Squad***:
- Lead Team.
  - -- Serves as point element for platoon.
  - -- Recons route to SP.
  - -- Calls in checkpoints.
  - -- Provides early warning.
  - -- Maintains rate of march.
  - -- Moves 10 to 20 meters in front of main body.
- Trail Team.
  - -- Provides rear security.
  - -- Moves 10 to 20 meters behind main body.

(5) ***Medic***:
- Assesses and treats march casualties.
- Advise chain of command on evacuation and transportation requirements of casualties.

(6) ***Individual***:
- Maintains interval.
- Follows TL's examples.
- Relays hand and arm signals; remains alert during movement and at halts.
- Remains alert during movement and at halts.

4-3. **MOVEMENT IN LIMITED VISIBILITY CONDITIONS**. During hours of limited visibility, a platoon will use surveillance, target acquisition, and night observation (STANO) devices to enhance their effectiveness. Leaders must be able to control, navigate, maintain security, and move during limited visibility.

a. **Control**. When visibility is poor, the following methods aid in control.

(1) Use of night vision devices.
(2) Leaders move closer to the front.
(3) Platoon reduces speed.
(4) Use of luminescent tape on equipment.
(5) Reduce intervals between men and elements.
(6) Conducts headcount regularly.

b. **Navigation**. While navigating during limited visibility the same techniques are used as during the day; however, leaders exercise more care to keep the patrol oriented.

c. **Security**.

(1) Enforce strict noise and light discipline.
(2) Use radio-listening silence.
(3) Use camouflage.
(4) Use terrain to avoid detection by enemy surveillance or night vision devices.
(5) Make frequent listening halts; Conduct SLLS (Stop, Look, Listen, Smell).
(6) Mask the sounds of movement when possible. (Rain, wind, and flowing water will mask the sounds of movement).

d. **Rally Points**. Plan actions to be taken at rally points in detail. All elements must maintain communications at all time. There are two techniques for actions at rally points:

(1) ***Minimum Force***: Patrol members assemble at the rally point, and the senior leader assumes command. When the minimum force (designated in the OPORD) is assembled and organized the patrol will continue the mission.

(2) ***Time Available***: The senior leader determines if the patrol has enough time remaining to accomplish the mission.

e. **Actions at Halts**. During halts, post security and cover all approaches into the sector with key weapons.

(1) ***Short Halt***. Typically 1-2 minutes long. Soldiers seek immediate cover and concealment and take a knee. Leaders assign sectors of fire.

(2) ***Long Halt***. More than 2 minutes. Soldiers assume the prone position behind cover and concealment. (Ensure soldiers have clear fields of fire.) Leaders assign sectors of fire.

4-4. **DANGER AREAS**. A danger area is any place on a unit's route where the leader determines his unit may be exposed to enemy observation or fire. Some examples of danger areas are open areas, roads and trails, built-up areas, enemy positions, and natural and man-made obstacles. Bypass danger areas whenever possible.

a. **Standards**.

(1) The unit prevents the enemy from surprising the main body.
(2) The unit moves all personnel and equipment across the danger area.
(3) The unit prevents decisive engagement by the enemy.

b. **Fundamentals**.

(1) Designate near and far side rally points.
(2) Secure near side, left and right flank, and rear security.
(3) Recon and secure the far side.
(4) Cross the danger area.
(5) Plan for fires on all known danger areas.

c. **Technique for Crossing Danger Areas**.

(1) ***Linear Danger Area*** (LDA, Figure 4-4) ***Actions for a Squad***:

STEP 1: The alpha team leader (ATL) observes the linear danger area and sends the hand and arm signal to the SL who determines to bound across.

STEP 2: SL directs the ATL to move his team across the LDA far enough to fit the remainder of the squad on the far side of the LDA. Bravo team moves to the LDA to the right or left to provide an overwatch position prior to A team crossing.

STEP 3: SL receives the hand and arm signal that it is safe to move the rest of the squad across (B team is still providing overwatch).

STEP 4: SL moves himself, radio operator, and B team across the LDA. (A team provides overwatch for squad missions.)

STEP 5: A team assumes original azimuth at SL's command or hand and arm signal.

Figure 4-4. LINEAR DANGER AREA

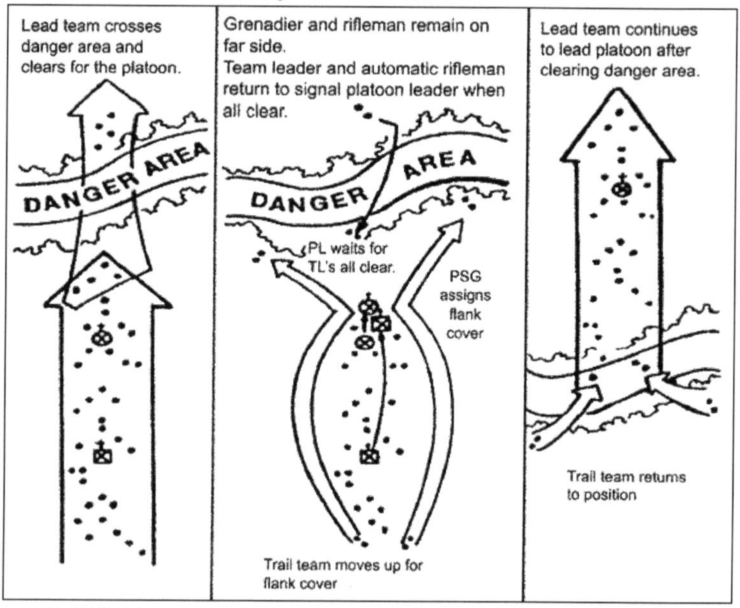

(2) ***LDA Crossing for a Platoon***:

(a) The lead squad halts the platoon, and signals danger area.

(b) The platoon leader moves forward to the lead squad to confirm the danger area and decides if current location is a suitable crossing site.

(c) The platoon leader confirms danger area/crossing site and establishes near and far side rally points.

(d) On the platoon leader's signal, the A team of the lead squad establishes an overwatch position to the left of the crossing site. Prior to crossing, the compassman with the lead two squads confirm azimuth and pace data.

(e) B team of the lead squad establishes an overwatch position to the right of the crossing site.

(f) Once overwatch positions are established, the platoon leader gives the second squad in movement the signal to bound across by fire team.

(g) Once across, the squad is now lead in movement and continues on azimuth

(h) Once Stop, Look, Listen, and Smell (SLLS) is conducted, squad leader signals platoon leader all clear.

- Day time—hand and arm signal such as thumbs up.
- Night time—Clandestine signal such as IR, red lens.

(i) Platoon leader receives all clear and crosses with radio operator, FO, WSL, and 2 gun teams.

(j) Once across, PL signals the 3rd squad in movement to cross.

(k) PSG with medic and one gun team crosses after 2nd squad is across (sterilizing central crossing site).

(l) PSG signals security squad to cross at their location.

Notes: Platoon leader will plan for fires at all known LDA crossing sites.
Nearside security in overwatch will sterilize signs of the patrol .

(3) ***Danger Area (Small/Open)***

(a) The lead squad halts the platoon and signals danger area.

(b) The PL moves forward to the lead squad to confirm the danger area.

(c) The platoon leader confirms danger area and establishes near and far side rally points.

(d) The PL designates lead squad to bypass danger area using the detour-bypass method.

(e) Paceman suspends current pace count and initiates an interim pace count. Alternate pace/compass man moves forward and offsets compass 90 degrees left or right as designated and moves in that direction until clear of danger area.

(f) After moving set distance (x-meters as instructed by PL). Lead squad assumes original azimuth, and primary pace man resumes original pace.

(g) After the open area, the alternate pace/compass man offsets his compass 90 degrees left or right and leads the platoon/squad the same distance (x-meters) back to the original azimuth.

(4) ***Danger Areas (Series)***: A series of danger areas is two or more danger areas within an area that can be either observed or covered by fire.

- Double linear danger area (use linear danger area technique and cross as one LDA).
- Linear/small open danger area (use by-pass/contour technique. Figure 4-5).
- Linear/large open danger area (use platoon wedge when crossing).

Note: A series of danger areas is crossed using the technique which provides the most security.

(5) ***Danger Area (Large)***:

(a) Lead squad halts the platoon, and signals danger area.

(b) The platoon leader moves forward with RTO and FO and confirms danger area.

(c) The platoon leader confirms danger area and establishes near and far side rally points.

(d) PL designates direction of movement.

(e) PL designates change of formation as necessary to ensure security.

Notes: Before point man steps into danger area, PL and FO adjust targets to cover movement.
If far side of danger area is within 250 meters, PL establishes overwatch, and designates lead squad to clear wood line on far side.

Figure 4-5. SMALL OPEN AREA

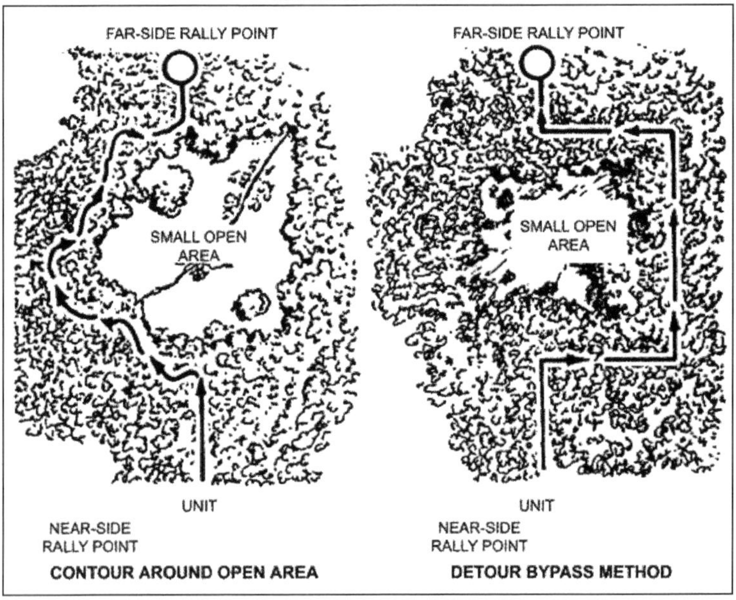

NOTES

Chapter 5
PATROLS

Infantry platoons and squads primarily conduct two types of patrols: reconnaissance and combat. This chapter describes the principles of patrolling, planning considerations, types of patrols, supporting tasks, patrol base, and movement to contact (FM 7-8, FM 3-0, and FM 101-5-1). In this chapter, the terms "*element*" and "*team*" refer to the squads, fire teams, or buddy teams that perform the tasks as described.

5-1. **PRINCIPLES**. All patrols are governed by five principles:

a. **Planning**. Quickly make a simple plan and effectively communicate it to the lowest level. A great plan that takes forever to complete and is poorly disseminated isn't a great plan. Plan and prepare to a realistic standard and rehearse everything.

b. **Reconnaissance**. Your responsibility as a Ranger leader is to confirm what you think you know, and to learn that which you do not already know.

c. **Security**. Preserve your force as a whole. Every Ranger and every rifle counts; anyone could be the difference between victory and defeat.

d. **Control**. Clear understanding of the concept of the operation and commander's intent, coupled with disciplined communications, to bring every man and weapon available to overwhelm the enemy at the decisive point.

e. **Common Sense**. Use all available information and good judgment to make sound, timely decisions.

5-2. **PLANNING**. This paragraph provides the planning considerations common to most patrols. It discusses task organization, initial planning and coordination, completion of the plan, and contingency planning.

a. **Task Organization**. A patrol is a mission, not an organization. To accomplish the patrolling mission, a platoon or squad must perform specific tasks, for example, secure itself, cross danger areas, recon the patrol objective, breach, support, or assault. As with other missions, the leader tasks elements of his unit in accordance with his estimate of the situation, identifying those tasks his unit must perform and designating which elements of his unit will perform which tasks. Where possible, in assigning tasks, the leader should maintain squad and fire team integrity. The chain of command continues to lead its elements during a patrol. Squads and fire teams may perform more than one task in an assigned sequence; others may perform only one task. The leader must plan carefully to ensure that he has identified and assigned all required tasks in the most efficient way. Elements and teams for platoons conducting patrols include--

(1) Elements common to all patrols:

(a) *Headquarters Element*. The headquarters consists of the platoon leader (PL), RTO, platoon sergeant (PSG), FO, RTO, and medic. It may include any attachments that the PL decides that he or the PSG must control directly.

(b) *Aid and Litter Team*. Aid and litter teams are responsible for buddy aid and evacuation of casualties.

(c) *Enemy Prisoner of War Team*. EPW teams control enemy prisoners using the five S's and the leader's guidance.

(d) *Surveillance Team*. The surveillance team keeps watch on the objective from the time that the leader's reconnaissance ends until the unit deploys for actions on the objective. They then rejoin their parent element.

(e) *En Route Recorder*. Part of the HQ element, maintains communications with higher and acts as the recorder for all CCIR collected during the mission.

(f) *Compass Man*. The compass man assists in navigation by ensuring the patrol remains on course at all times. Instructions to the compass man must include initial and subsequent azimuths. As a technique, the compass man should preset his compass on the initial azimuth before the unit moves out, especially if the move will be during limited visibility conditions. The platoon or squad leader should also designate an alternate compass man.

(g) *Point/Pace Man*. As required, the PL designates a primary and alternate point man and a pace man for the patrol. The pace man aids in navigation by keeping an accurate count of distance traveled. The point man selects the actual route through the terrain, guided by the compass man or team leader. In addition, the point man also provides frontal security.

(2) Elements common to all combat patrols:

(a) *Assault Element*. The assault element seizes and secures the objective and protects special teams as they complete their assigned actions on the objective.

(b) *Security Element*. The security element provides security at danger areas, secures the ORP, isolates the objective, and supports the withdrawal of the rest of the patrol once actions on the objective are complete. The security element may have separate security teams, each with an assigned task or sequence of tasks.

(c) *Support Element*. The support element provides direct and indirect fire support for the unit. Direct fires include machine guns, medium and light antiarmor weapons, small recoilless rifles. Indirect fires available may include mortars, artillery, CAS, and organic M203 weapon systems.

(d) *Demolition Team*. Demolition teams are responsible for preparing and detonating the charges to destroy designated equipment, vehicles, or facilities on the objective.

(e) *EPW and Search Teams*. The assault element may provide two-man (buddy teams) or four-man (fire team) search teams to search bunkers, buildings, or tunnels on the objective. These teams will search the objective or kill zone for any PIR that may give the PL an idea of the enemy concept for future operations. Primary and alternate teams may be assigned to ensure enough prepared personnel are available on the objective.

(f) *Breach Element*. The breach team conducts initial penetration of enemy obstacles to seize a foothold and allow the patrol to enter an objective. This is typically done IAW METT-TC and the steps outlined in the "*Conduct an Initial Breach of a Mined Wire Obstacle*" battle drill in Chapter 6 of this Handbook.

(3) Elements common to all reconnaissance patrols:

(a) *Reconnaissance Team*. Reconnaissance teams reconnoiter the objective area once the security teams are in position. Normally these are two-man teams (buddy teams) to reduce the possibility of detection.

(b) *Reconnaissance and Security Teams*. R&S teams are normally used in a zone reconnaissance, but may be useful in any situation when it is impractical to separate the responsibilities for reconnaissance and security.

(c) *Security Element*. When the responsibilities of reconnaissance and security are separate, the security element provides security at danger areas, secures the ORP, isolates the objective, and supports the withdrawal of the rest of the platoon once the recon is complete. The security element may have separate security teams, each with an assigned task or sequence of tasks.

b. **Initial Planning and Coordination**. Leaders plan and prepare for patrols using the troop-leading procedures and the estimate of the situation, as described in Chapter 2. Through an estimate of the situation, leaders identify required actions on the objective (mission analysis) and plan backward to departure from friendly lines and forward to reentry of friendly lines. Because patrolling units act independently, move beyond the direct-fire support of the parent unit, and operate forward of friendly units, coordination must be thorough and detailed. Coordination is continuous throughout planning and preparation. PLs use checklists to preclude omitting any items vital to the accomplishment of the mission.

(1) ***Coordination with Higher Headquarters***. This coordination includes Intelligence, Operations, and Fire Support IAW Chapter 2, *Coordination Checklists* (**page 2-34**). This initial coordination is an integral part of Step 3 of Troop-Leading Procedures, *Make a Tentative Plan*.

(2) ***Coordination with Adjacent Units***. The leader also coordinates his unit's patrol activities with the leaders of other units that will be patrolling in adjacent areas at the same time, IAW Chapter 2-7, *Coordination Checklists* (**page 2-34**).

c. **Completion of Plan**. As the PL completes his plan, he considers--

(1) ***Specified and Implied Tasks***. The PL ensures that he has assigned all specified tasks to be performed on the objective, at rally points, at danger areas, at security or surveillance locations, along the route(s), and at passage lanes. These make up the maneuver and tasks to maneuver units subparagraphs of the Execution paragraph.

(2) ***Key travel and Execution Times***. The leader estimates time requirements for movement to the objective, leader's reconnaissance of the objective, establishment of security and surveillance, completion of all assigned tasks on the objective, and passage through friendly lines. Some planning factors are--

- Movement: Average of 1 Kmph during daylight hours in Woodland Terrain; Average limited visibility ½ Kmph. Add additional time for restrictive, or severely restrictive terrain such as mountains, swamps, or thick vegetation.
- Leader's recon: NLT 1.5 hr.
- Establishment of security and surveillance: 0.5 hr.

(3) ***Primary and Alternate Routes***. The leader selects primary and alternate routes to and from the objective. The return routes should differ from the routes to the objective. The PL may delegate route selection to a subordinate, but is ultimately responsible for the routes selected.

(4) ***Signals***. The leader should consider the use of special signals. These include hand-and-arm signals, flares, voice, whistles, radios, and infrared equipment. Primary and alternate signals must be identified and rehearsed so that all Soldiers know their meaning.

(5) ***Challenge and Password Forward of Friendly Lines***. The challenge and password from the unit's ANCD must not be used beyond the FLOT.

(a) *Odd-Number System.* The leader specifies an odd number. The challenge can be any number less than the specified number. The password will be the number that must be added to it to equal the specified number, for example, the number is 7, the challenge is 3, and the password is 4.

(b) *Running Password.* ANCDs may also designate a running password. This code word alerts a unit that friendly Soldiers are approaching in a less than organized manner and possibly under pressure. The number of Soldiers approaching follows the running password. For example, if the running password is "*Ranger*," and five friendly Soldiers are approaching, they would say "*Ranger five.*"

(6) ***Location of Leaders***. The PL considers where he and the PSG and other key leaders are located during each phase of the mission. The PL positions himself where he can best control the actions of the patrol. The PSG is normally located with the assault element during a raid or attack to help the PL control the use of additional assaulting squads, and will assist with securing the OBJ. The PSG will locate himself at the CCP to facilitate casualty treatment and evacuation. During a reconnaissance mission, the PSG will stay behind in the ORP to facilitate the transfer of Intel to the higher headquarters, and control the recon elements movement into and out of the ORP.

(7) ***Actions on Enemy Contact***. Unless required by the mission, the unit avoids enemy contact. The leader's plan must address actions on chance contact at each phase of the patrol mission. The unit's ability to continue will depend on how early contact is made, whether the platoon is able to break contact successfully (so that its subsequent direction of movement is undetected), and whether the unit receives any casualties because of the contact. The plan must address the handling of seriously wounded Soldiers and KIAs. The plan must also address the handling of prisoners who are captured because of chance contact and are not part of the planned mission.

(8) ***Contingency Plans***. The leader leaves his unit for many reasons throughout the planning, coordination, preparation, and execution of his patrol mission. Each time the leader departs the patrol main body, he must issue a five-point contingency plan to the leader left in charge of the unit. The contingency plan is described by the acronym GOTWA, as follows. The patrol leader will additionally issue specific guidance stating what tasks are to be accomplished in the ORP in his absence:

- G: Where the leader is GOING.
- O: OTHERS he is taking with him.
- T: TIME he plans to be gone.
- W: WHAT to do if the leader does not return in time.
- A: The unit's and the leader's ACTIONS on chance contact while the leader is gone.

(9) ***Rally Points***. The leader considers the use and location of rally points. A rally point is a place designated by the leader where the unit moves to reassemble and reorganize if it becomes dispersed. Soldiers must know which rally point to move to at each phase of the patrol mission should they become separated from the unit. They must also know what actions are required there and how long they are to wait at each rally point before moving to another.

(a) *Criteria.* Rally points must be--

- Easily identifiable in daylight and limited visibility.
- Show no signs of recent enemy activity.
- Covered and concealed.
- Away from natural lines of drift and high-speed avenues of approach.
- Defendable for short periods of time.

(b) *Types.* The most common types of rally points are initial, en route, objective, and near-and-far-side rally points.

(10) ***Objective Rally Point***. The ORP is typically 200 to 400m from the objective, or at a minimum, one major terrain feature away. Actions at the ORP include--

- Conduct SLLS and pinpoint location.
- Leaders Recon of the Objective.
- Issuing a FRAGO, if needed.
- Making final preparations before continuing operations, for example, recamouflaging, preparing demolitions, lining up rucksacks for quick recovery. Preparing EPW bindings, first aid kits, litters, and inspecting weapons.
- Accounting for Soldiers and equipment after actions at the objective are complete.
- Reestablishing the chain of command after actions at the objective are complete.
- Disseminating information from reconnaissance, if contact was not made.

(11) ***Leader's reconnaissance of the objective***. The plan must include a leader's reconnaissance of the objective once the platoon or squad establishes the ORP. Before departing, the leader must issue a 5-point contingency plan. During his

reconnaissance, the leader pinpoints the objective, selects reconnaissance, security, support, and assault positions for his elements, and adjusts his plan based on his observation of the objective. Each type of patrol requires different tasks during the leader's reconnaissance. The platoon leader will bring different elements with him. (These are discussed separately under each type of patrol). The leader must plan time to return to the ORP, complete his plan, disseminate information, issue orders and instructions, and allow his squads to make any additional preparations. During the Leader's Reconnaissance for a Raid or Ambush, the PL will leave surveillance on the OBJ.

(12) ***Actions on the objective***. Each type of patrol requires different actions on the objective. Actions on the objective are discussed under each type of patrol.

5-3. **RECONNAISSANCE PATROLS**. Recon patrols are one of the two types of patrols. provide timely and accurate information on the enemy and terrain. They confirm the leader's plan before it is executed. Units on reconnaissance operations collect specific information (*priority intelligence requirements* [PIR]) or general information (*information requirements* [IR]) based on the instructions from their higher commander. The two types of recon patrols discussed here are area and zone. This section discusses the fundamentals of reconnaissance, task standards for the two most common types of recon, and actions on the objective for those types of recon.

a. **Fundamentals of Reconnaissance**. In order to have a successful area reconnaissance, the platoon leader must apply the fundamentals of the reconnaissance to his plan during the conduct of the operation.

(1) ***Gain all required information***. The parent unit tells the patrol leader (PL) what information is required. This is in the form of the IR and PIR. The platoon's mission is then tailored to what information is required. During the entire patrol, members must continuously gain and exchange all information gathered, but cannot consider the mission accomplished unless all PIR has been gathered.

(2) ***Avoid detection by the enemy***. A patrol must not let the enemy know that it is in the objective area. If the enemy knows he is being observed, he may move, change his plans, or increase his security measures. Methods of avoiding detection are--

(a) Minimize movement in the objective area (area reconnaissance).

(b) Move no closer to the enemy than necessary.

(c) If possible, use long-range surveillance or night vision devices.

(d) Camouflage, stealth, noise, and light discipline.

(e) Minimize radio traffic.

(3) ***Employ security measures***. A patrol must be able to break contact and return to the friendly unit with what information is gathered. If necessary, break contact and continue the mission. Security elements are emplaced so that they can overwatch the reconnaissance elements and suppress the enemy so the reconnaissance element can break contact.

(4) ***Task organization***. When the platoon leader receives the order, he analyzes his mission to ensure he understands what must be done. Then he task organizes his platoon to best accomplish the mission IAW METT-TC. Recons are typically squad-sized missions.

b. **Task Standards**.

(1) ***Area Reconnaissance***. The area recon patrol collects all available information on PIR and other intelligence not specified in the order for the area. The patrol completes the recon and reports all information by the time specified in the order. The patrol is not compromised.

(2) ***Zone Reconnaissance***. The zone recon patrol determines all PIR and other intelligence not specified in the order for its assigned zone. The patrol reconnoiters without detection by the enemy. The patrol completes the recon and reports all information by the time specified in the order.

c. **Actions on the Objective, Area Reconnaissance** (Figure 5-1).

(1) The element occupies the ORP as discussed in the section on occupation of the ORP. The RTO calls in spare for occupation of ORP. The leader confirms his location on map while subordinate leaders make necessary perimeter adjustments.

(2) The PL organizes the platoon in one of two ways: separate recon and security elements, or combined recon and security elements.

(3) The PL takes subordinates leaders and key personnel on a leader's recon to confirm the objective and plan.

(a) Issues a 5-point contingency plan before departure.

(b) Establishes a suitable release point that is beyond sight and sound of the objective if possible, but that is definitely out of sight. The RP should also have good rally point characteristics.

(c) Allow all personnel to become familiar with the release point and surrounding area.

(d) Identifies the objective and emplaces surveillance. Designates a surveillance team to keep the objective under surveillance. Issues a contingency plan to the senior man remaining with the surveillance team. The surveillance team is positioned with one man facing the objective, and one facing back in the direction of the release point.

(e) Takes subordinate leaders forward to pinpoint the objective, emplace surveillance, establish a limit of advance, and choose vantage points.

(f) Maintains commo with the platoon throughout the leader's recon.

(4) The PSG maintains security and supervises priorities of work in the ORP.

(a) Reestablishes security at the ORP.

(b) Disseminates the PLs contingency plan.

(c) Oversees preparation of recon personnel (personnel recamouflaged, NVDs and binos prepared, weapons on safe with a round in the chamber).

(5) The PL and his recon party return to the ORP.

(a) Confirms the plan or issues a FRAGO.

(b) Allows subordinate leaders time to disseminate the plan.

(6) The patrol conducts the recon by long-range observation and surveillance if possible.

(a) R&S elements move to observation points that offer cover and concealment and that are outside of small-arms range.

(b) Establishes a series of observation posts (OP) if information cannot be gathered from one location.

(c) Gathers all PIR using the SALUTE format.

(7) If necessary, the patrol conducts its recon by short-range observation and surveillance.

(a) Moves to an OP near the objective.

(b) Passes close enough to the objective to gain information.

(c) Gathers all PIR using the SALUTE format.

(8) R&S teams move using a technique such as the cloverleaf method to move to successive OP's. In this method, R&S teams avoid paralleling the objective site, maintain extreme stealth, do not cross the limit of advance, and maximize the use of available cover and concealment.

(9) During the conduct of the recon, each R&S team will return to the release point when any of the following occurs:

- They have gathered all their PIR.
- They have reached the limit of advance.
- The allocated time to conduct the recon has elapsed.
- Contact has been made.

(10) At the release point, the leader will analyze what information has been gathered and determine if he has met the PIR requirements.

(11) If the leader determines that he has not gathered sufficient information to meet the PIR requirements, or if the information he and the subordinate leader gathered differs drastically, he may have to send R&S teams back to the objective site. In this case, R&S teams will alternate areas of responsibilities. For example, if one team reconnoitered from the 6 – 3 – 12, then that team will now recon from the 6 – 9 – 12.

(12) The R&S element returns undetected to the ORP by the specified time.

(a) Disseminates information to all patrol members through key leaders at the ORP, or moves to a position at least one terrain feature or one kilometer away to disseminate. To disseminate, the leader has the RTO prepare three sketches of the objective site based on the leader's sketch and provides the copies to the subordinate leaders to assist in dissemination.

(b) Reports any information requirements and/or any information requiring immediate attention to higher headquarters, and departs for the designated area.

(13) If contact is made, move to the release point. The recon element tries to break contact and return to the ORP, secure rucksacks, and quickly move out of the area. Once they have moved a safe distance away, the leader will inform higher HQ of the situation and take further instructions from them.

(a) While emplacing surveillance, the recon element withdraws through the release point to the ORP, and follows the same procedures as above.

(b) While conducting the reconnaissance, the compromised element returns a sufficient volume of fire to allow them to break contact. Surveillance can fire an AT-4 at the largest weapon on the objective. All elements will pull off the objective and move to the release point. The senior man will quickly account for all personnel and return to the ORP. Once in the ORP, follow the procedures previously described.

Figure 5-1. ACTIONS ON THE OBJECTIVE, AREA RECONNAISSANCE

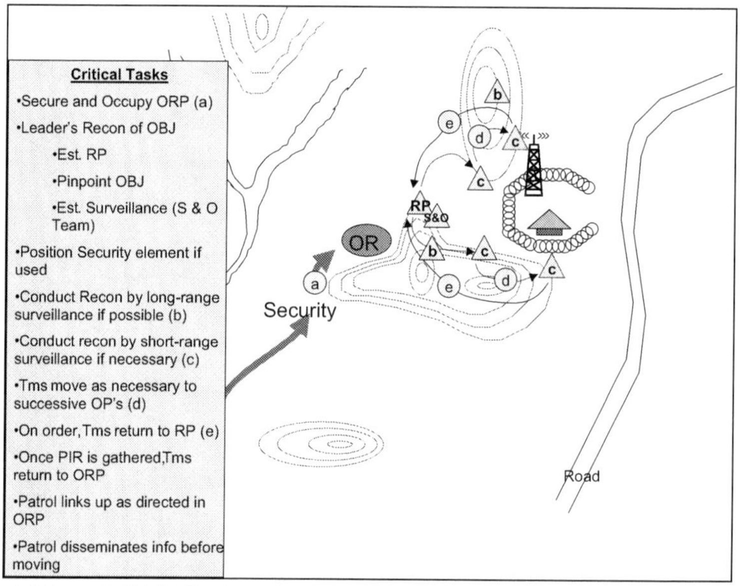

d. **Actions on the Objective, Zone Reconnaissance**.

(1) The element occupies the initial ORP as discussed in the section occupation of the ORP. The radio operator calls in spare for occupation of ORP. The leader confirms his location on map while subordinate leaders make necessary perimeter adjustments.

(2) The recon team leaders organize their recon elements.

(a) Designate security and recon elements.

(b) Assign responsibilities (point man, pace man, en route recorder, and rear security), if not already assigned.

(c) Designates easily recognizable rally points.

(d) Ensure local security at all halts.

(3) The patrol recons the zone.

(a) Moves tactically to the ORPs.

(b) Occupies designated ORPs.

(c) Follows the method designated by the PL (fan, converging routes, or box method, Figure 5-2).

Figure 5-2. COMPARISON OF ZONE RECONNAISSANCE METHODS

FAN METHOD	*CONVERGING ROUTES METHOD*	*BOX METHOD*
• Uses a series (fan) of ORPs. • Patrol establishes security at first ORP. • Each recon element moves from ORP along a different fan-shaped route. Route overlaps with that of other recon elements. This ensures recon of entire area. • Leader maintains reserve at ORP. • When all recon elements return to ORP, PL collects and disseminates all info before moving to next ORP.	• PL selects routes from ORP thru zone to a linkup point at the far side of the zone from the ORP. Each recon element moves and recons along a specified route. They converge (link up) at one time and place.	• PL sends recon elements from the first ORP along routes that form a box. He sends other elements along routes throughout the box. All teams link up at the far side of the box from the ORP.

(d) The recon teams reconnoiter.
- During movement, the squad will gather all PIR specified by the order.
- Recon team leaders will ensure sketches are drawn or digital photos are taken of all enemy hard sites, roads, and trails.
- Return to the ORP, or link up at the rendezvous point on time.
- When the squad arrives at new rendezvous point or ORP, the recon team leaders report to the PL with all information gathered.

(e) The PL continues to control the recon elements.
- PL moves with the recon element that establishes the linkup point.
- PL changes recon methods as required.
- PL designates times for the elements to return to the ORP or to linkup.
- PL collects all information and disseminates it to the entire patrol. PL will brief all key subordinate leaders on information gathered by other squads, establishing one consolidated sketch if possible, and allow team leaders time to brief their teams.
- PL and PSG account for all personnel.

(f) The patrol continues the recon until all designated areas have been reconned, and returns undetected to friendly lines.

5-4. **COMBAT PATROLS**. Combat patrols are the second type of patrol. Combat patrols are further divided into raids and ambushes Units conduct combat patrols to destroy or capture enemy Soldiers or equipment; destroy installations, facilities, or key points; or harass enemy forces. Combat patrols also provide security for larger units. This section describes overall combat patrol planning considerations, task considerations for each type of combat patrol, and finally actions on the objective for each type.

a. **Planning Considerations (General)**. In planning a combat patrol, the PL considers the following:

(1) ***Tasks to Maneuver Units***. Normally the platoon headquarters element controls the patrol on a combat patrol mission. The PL makes every try to maintain squad and fire team integrity as he assigns tasks to subordinates units.

(a) The PL must consider the requirements for assaulting the objective, supporting the assault by fire, and security of the entire unit throughout the mission.
- For the assault on the objective, the PL considers the required actions on the objective, the size of the objective, and the known or presumed strength and disposition of the enemy on and near the objective.
- The PL considers the weapons available, and the type and volume of fires required to provide fire support for the assault on the objective.
- The PL considers the requirement to secure the platoon at points along the route, at danger areas, at the ORP, along enemy avenues of approach into the objective, and elsewhere during the mission.
- The PL will also designate engagement/disengagement criteria.

(b) The PL assigns additional tasks to his squads for demolition, search of EPWs, guarding of EPWs, treatment and evacuation (litter teams) of friendly casualties, and other tasks required for successful completion of patrol mission (if not already in the SOP).

(c) The PL determines who will control any attachments of skilled personnel or special equipment.

(2) ***Leader's Reconnaissance of the Objective***. In a combat patrol, the PL has additional considerations for the conduct of his reconnaissance of the objective from the ORP.

(a) *Composition of the leader's reconnaissance party.* The platoon leader will normally bring the following personnel.

- Squad leaders to include the weapons squad leader.
- Surveillance team.
- Forward observer.
- Security element (dependent on time available).

(b) *Conduct of the leader's reconnaissance*. In a combat patrol, the PL considers the following additional actions in the conduct of the leader's reconnaissance of the objective.

- The PL designates a release point approximately half way between the ORP and this objective. The PL posts the surveillance team. Squads and fire teams separate at the release point, and then they move to their assigned positions.
- The PL confirms the location of the objective or kill zone. He notes the terrain and identifies where he can emplace claymores to cover dead space. Any change to his plan is issued to the squad leaders (while overlooking the objective if possible).
- If the objective is the kill zone for an ambush, the leader's reconnaissance party should not cross the objective; to do so will leave tracks that may compromise the mission.
- The PL confirms the suitability of the assault and support positions and routes from them back to the ORP.
- The PL issues a five-point contingency plan before returning to the ORP.

b. **Ambush**.

(1) ***Planning Considerations***. An ambush is a surprise attack from a concealed position on a moving or temporarily halted target. Ambushes are classified by category--hasty or deliberate; type--point or area; and formation--linear or L-shaped. The leader uses a combination of category, type, and formation in developing his ambush plan. The key planning considerations include--

(a) Coverage of entire kill zone by fire.

(b) METT-TC.

(c) Use of existing or reinforcing obstacles, including Claymores, to keep the enemy in the kill zone.

(d) Security teams (typically equipped with hand-held antitank weapons such as AT-4 or LAW; Claymores; and various means of communication.

(e) Protect the assault and support elements with claymores or explosives.

(f) Use security elements or teams to isolate the kill zone.

(g) Assault through the kill zone to the limit of advance (LOA). (The assault element must be able to move quickly through its own protective obstacles.)

(h) Time the actions of all elements of the platoon to preclude loss of surprise. In the event any member of the ambush is compromised, he may immediately initiate the ambush.

(i) When the ambush must be manned for a long time, use only one squad to conduct the entire ambush and determining movement time of rotating squads from the ORP to the ambush site.

(2) ***Categories***.

(a) *Hasty Ambush*. A unit conducts a hasty ambush when it makes visual contact with an enemy force and has time to establish an ambush without being detected. The actions for a hasty ambush must be well rehearsed so that Soldiers know what to do on the leader's signal. They must also know what action to take if the unit is detected before it is ready to initiate the ambush.

(b) *Deliberate Ambush*. A deliberate ambush is conducted at a predetermined location against any enemy element that meets the commander's engagement criteria. The leader requires the following detailed information in planning a deliberate ambush: size and composition of the targeted enemy, and weapons and equipment available to the enemy.

(3) ***Types***.

(a) Point ambush. In a point ambush, Soldiers deploy to attack an enemy in a single kill zone.

(b) Area ambush. In an area, Soldiers deploy in two or more related point ambushes.

(4) ***Formations***. (Figure 5-3).

(a) *Linear Ambush*. In an ambush using a linear formation, the assault and support elements deploy parallel to the enemy's route. This positions both elements on the long axis of the kill zone and subjects the enemy to flanking fire. This formation can be used in close terrain that restricts the enemy's ability to maneuver against the platoon, or in open terrain provided a means of keeping the enemy in the kill zone can be effected.

(b) *L-Shaped Ambush*. In an L-shaped ambush, the assault element forms the long leg parallel to the enemy's direction of movement along the kill zone. The support element forms the short leg at one end of and at right angles to the assault element. This provides both flanking (long leg) and enfilading fires (short leg) against the enemy. The L-shaped ambush can be used at a sharp bend in a trail, road, or stream. It should not be used where the short leg would have to cross a straight road or trail.

Figure 5-3. AMBUSH FORMATIONS

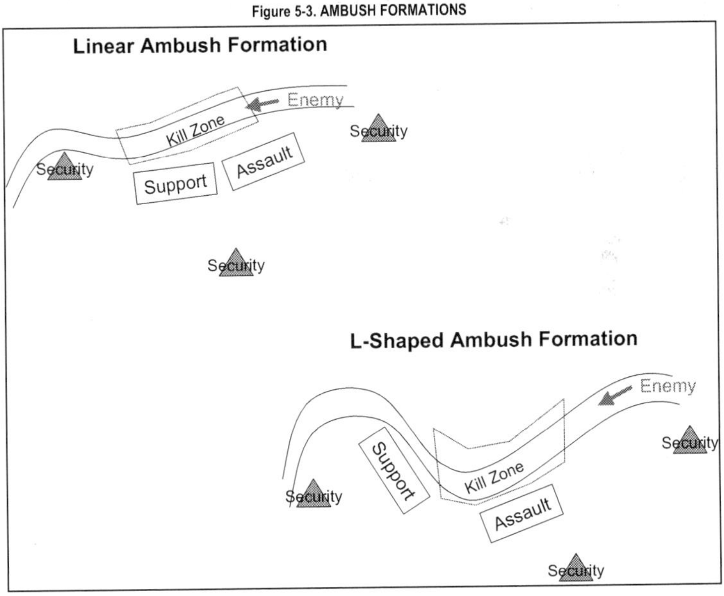

c. **Hasty Ambush**.

(1) ***Task Standards***. The platoon moves quickly to concealed positions. The ambush is not initiated until the majority of the enemy is in the kill zone. The unit does not become decisively engaged. The platoon surprises the enemy. The patrol captures, kills, or forces the withdrawal of all of the enemy within the kill zone. On order, the patrol withdraws all personnel and equipment in the kill zone from observation and direct fire. The unit does not become decisively engaged by follow-on elements. The platoon continues follow-on operations.

(2) ***Actions on the Objective (Hasty Ambush)***. See Figure 5-4.

(a) Using visual signals, any Soldier alerts the unit that an enemy force is in sight. The Soldier continues to monitor the location and activities of the enemy force until his team or squad leader relieves him, and gives the enemy location and direction of movement.

(b) The platoon or squad halts and remains motionless.

- The PL gives the signal to conduct a hasty ambush, taking care not to alert the enemy of the patrol's presence.
- The leader determines the best nearby location for a hasty ambush. He uses arm-and-hand signals to direct the unit members to covered and concealed positions.

(c) The leader designates the location and extent of the kill zone.

d) Teams and squads move silently to covered and concealed positions, ensuring positions are undetected and have good observation and fields of fire into the kill zone.

(e) Security elements move out to cover each flank and the rear of the unit. The leader directs the security elements to move a given distance, set up, and then rejoin the unit on order or, after the ambush (the sound of firing ceases). At squad level, the two outside buddy teams normally provide flank security as well as fires into the kill zone. At platoon level, fire teams make up the security elements.

(f) The PL assigns sectors of fire and issues any other commands necessary such as control measures.

(g) The PL initiates the ambush, using the greatest casualty-producing weapon available, when the largest percentage of enemy is in the kill zone. The PL--

- Controls the rate and distribution of fire.
- Employs indirect fire to support the ambush.
- Orders cease fire.
- (If the situation dictates) Orders the patrol to assault through the kill zone.

(h) The PL designates personnel to conduct a hasty search of enemy personnel and process enemy prisoners and equipment.

(i) The PL orders the platoon to withdraw from the ambush site along a covered and concealed route.

(j) The PL gains accountability, reorganizes as necessary, disseminates information, reports the situation, and continues the mission as directed.

d. **Deliberate (Point/Area) Ambush**.

(1) ***Task Standards***. The ambush is emplaced NLT the time specified in the order. The patrol surprises the enemy and engages the enemy main body. The patrol kills or captures all enemy in the kill zone and destroys equipment based on the commander's intent. The patrol withdraws all personnel and equipment from the objective, on order, within the time specified in the order. The patrol obtains all available PIR from the ambush and continues follow-on operations.

(2) ***Actions on the Objective (Deliberate Ambush).*** See Figure 5-5.

(a) The PL prepares the patrol for the ambush in the ORP.

(b) The PL prepares to conduct a leader's reconnaissance. He--

- Designates the members of the leader's recon party (typically includes squad leaders, surveillance team, FO, and possibly the security element.
- Issues a contingency plan to the PSG.

Figure 5-4. ACTIONS ON THE OBJECTIVE -- HASTY AMBUSH

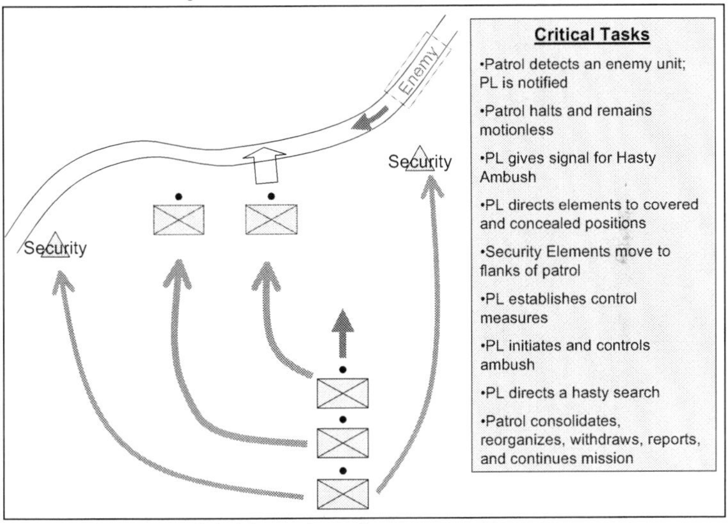

(c) The PL conducts his leader's reconnaissance. He--
- Ensures the leader's recon party moves undetected.
- Confirms the objective location and suitability for the ambush.
- Selects a kill zone.
- Posts the surveillance team at the site and issues a contingency plan.
- Confirms suitability of assault and support positions, and routes from them to the ORP.
- Selects the position of each weapon system in the support-by-fire position, and then designates sectors of fire.
- Identifies all offensive control measures to be used. Identifies the PLD, the assault position, LOA, any boundaries or other control measures. If available, the PL can use infrared aiming devices to identify these positions on the ground.

(d) The PL adjusts his plan based on info from the reconnaissance. He--
- Assigns positions.
- Designates withdrawal routes.

(e) The PL confirms the ambush formation.

(f) The security team(s) occupy first, securing the flanks of the ambush site, and providing early warning. The security element must be in position before the support and assault elements move forward of the release point. A security team remains in the ORP if the patrol plans to return to the ORP after actions on the objective. If the ORP is abandoned, a rear security team should be emplaced.

(g) Support element leader assigns sectors of fire. He--
- Emplaces claymores and obstacles as designated.
- Identifies sectors of fire and emplaces limiting stakes to prevent friendly fires from hitting other elements.
- Overwatches the movement of the assault element into position.

(h) Once the support element is in position, or on the PLs order, the assault element--

- Departs the ORP and moves into position.
- Upon reaching the PLD, the assault element transitions from the movement formation to the battle formation.
- Identifies individual sectors of fire as assigned by the PL. Emplaces aiming stakes or uses metal-to-metal contact with the machine gun tripods to prevent fratricide on the objective.
- Emplaces claymores to help destroy the enemy in the kill zone.
- Camouflages positions.

(i) The security element spots the enemy and notifies the PL, and reports the direction of movement, size of the target, and any special weapons or equipment carried. The security element also keeps the platoon leader informed if any enemy forces are following the lead force.

(j) The PL alerts other elements, and determines if the enemy force is too large, or if the ambush can engage the enemy successfully.

(k) The PL initiates the ambush using the highest casualty-producing device. He may use a command-detonated claymore. He must also plan a backup method for initiating the ambush, in case his primary means fails. This should also be a casualty-producing device such as his individual weapon. He passes this information to all Rangers, and practices it during rehearsals.

(l) The PL ensures that the assault and support elements deliver fire with the heaviest, most accurate volume possible on the enemy in the kill zone. In limited visibility, the PL may use infrared lasers to further define specific targets in the kill zone.

(m) Before assaulting the target, the PL gives the signal to lift or shift fires.

(n) The assault element--

- Assaults before the remaining enemy can react.
- Kills or captures enemy in the kill zone.
- Uses individual movement techniques or bounds by fire teams to move.
- Upon reaching the limit of advance, halts and establishes security. If needed, it reestablishes the chain of command and remains key weapon systems. All Soldiers will load a fresh magazine or drum of ammunition using the buddy system. ACE reports will be submitted through the chain of command. The PL will submit an initial contact report to higher.

(o) The PL directs special teams (EPW search, aid and litter, demo) to accomplish their assigned task once the assault element has established its LOA.

- Once the kill zone had been cleared, collect and secure all EPWs and move them out of the kill zone before searching bodies. Coordinate for an EPW exchange point to link up with higher to extract all EPWs and treat them IAW the five S's.
- Search from one side to the other and mark bodies that have been searched to ensure the area is thoroughly covered. Units should use the "*clear out, search in*" technique, clear from the center of the objective out ensuring the area is clear of all enemy combatants; then search all enemy personnel towards the center of the objective. Search all dead enemy personnel using two-man search techniques.
  - -- As the search team approaches a dead enemy Soldier, one-man guards while the other man searches. First, he kicks the enemy weapon away.
  - -- Second, he rolls the body over (if on the stomach) by lying on top and when given the go ahead by the guard (who is positioned at the enemy's head), the searcher rolls the body over on him. This is done for protection in case the enemy Soldier has a grenade with the pin pulled underneath him.
  - -- The searchers then conduct a systematic search of the dead Soldier from head to toe removing all papers and anything new (different type rank, shoulder boards, different unit patch, pistol, weapon, or NVD). They note if the enemy has a fresh or shabby haircut and the condition of his uniform and boots. They note the radio frequency, and then they secure the SOI, maps, documents, and overlays.
  - -- Once the body has been thoroughly searched, the search team will continue in this manner until all enemy personnel in and near the kill zone have been searched.
- Identify, collect, and prepare all equipment to be carried back or destroyed.
- Evacuate and treat friendly wounded first, then enemy wounded, time permitting.

- The demolition team prepares dual-primed explosives or incendiary grenades and awaits the signal to initiate. This is normally the last action performed before the unit departs the objective and may signal the security elements to return to the ORP.
- Actions on the objective with stationary assault line; all actions are the same with the exception of the search teams. To provide security within the teams to the far side of the kill zone during the search, they work in three-Ranger teams. Before the search begins, the Rangers move all KIAs to the near side of the kill zone.

(p) If enemy reinforcements try to penetrate the kill zone, the flank security will engage to prevent the assault element from being compromised.

(q) The platoon leader directs the unit's withdrawal from the ambush site:

- Elements normally withdraw in the reverse order that they established their positions.
- The elements may return to the RP or directly to the ORP, depending on the distance between elements.
- The security element of the ORP must be alert to assist the platoon's return to the ORP. It maintains security for the ORP while the rest of the platoon prepares to leave.
- If possible, all elements should return to the location at which they separated from the main body. This location should usually be the RP.

(r) The PL and PSG direct actions at the ORP, to include accountability of personnel and equipment and recovery of rucksacks and other equipment left at the ORP during the ambush.

(s) The platoon leader disseminates information, or moves the platoon to a safe location (no less than one kilometer or one terrain feature away from the objective) and disseminates information.

(t) As required, the PL and FO execute indirect fires to cover the platoon's withdrawal.

Figure 5-5. ACTIONS ON THE OBJECTIVE -- DELIBERATE AMBUSH

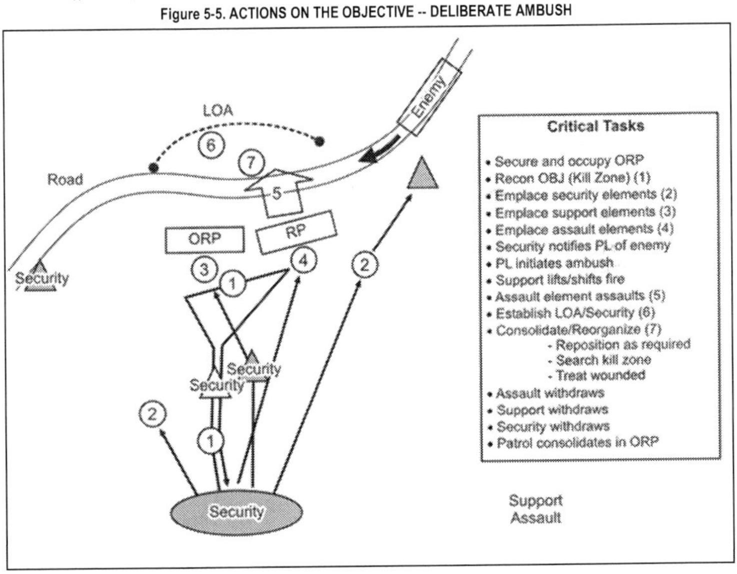

e. **Perform Raid**. The patrol initiates the raid NLT the time specified in the order, surprises the enemy, assaults the objective, and accomplishes its assigned mission within the commander's intent. The patrol does not become decisively engaged en route to the objective. The patrol obtains all available PIR from the raid objective and continues follow-on operations.

(1) ***Planning Considerations***. A Raid is a form of attack, usually small scale, involving a swift entry into hostile territory to secure information, confuse the enemy, or destroy installations followed by a planned withdrawal. Squads do not conduct raids. The sequence of platoon actions for a raid is similar to those for an ambush. Additionally, the assault element of the platoon may have to conduct a breach of an obstacle. It may have additional tasks to perform on the objective such as demolition of fixed facilities. Fundamentals of the raid include--

- Surprise and speed. Infiltrate and surprise the enemy without being detected.
- Coordinated fires. Seal off the objective with well-synchronized direct and indirect fires.
- Violence of action. Overwhelm the enemy with fire and maneuver.
- Planned withdrawal. Withdraw from the objective in an organized manner, maintaining security.

(2) ***Actions on the Objective (Raid)***. See Figure 5-6.

(a) The patrol moves to and occupies the ORP IAW the patrol SOP. The patrol prepares for the leader's recon.

(b) The PL, squad leaders, and selected personnel conduct a leader's recon.

- PL leaves a five point contingency plan with the PSG.
- PL establishes the RP, pinpoints the objective, contacts the PSG to prep men, weapons, and equipment, emplaces the surveillance team to observe the objective, and verifies and updates intelligence information.
- Leader's recon verifies location of and routes to security, support, and assault positions.
- Security teams are brought forward on the leader's reconnaissance and emplaced before the leader's recon leaves the RP
- Leaders conduct the recon without compromising the patrol.
- Leaders normally recon support by fire position first, then the assault position.

(c) The PL confirms, denies, or modifies his plan and issues instructions to his squad leaders.

- Assigns positions and withdrawal routes to all elements.
- Designates control measures on the objective (element objectives, lanes, limits of advance, target reference points, and assault line).
- Allows SLs time to disseminate information, and confirm that their elements are ready.

(d) Security elements occupy designated positions, moving undetected into positions that provide early warning and can seal off the objective from outside support or reinforcement.

(e) The support element leader moves the support element to designated positions. The support element leader ensures his element can place well-aimed fire on the objective.

(f) The PL moves with the assault element into the assault position. The assault position is normally the last covered and concealed position before reaching the objective. As it passes through the assault position the platoon deploys into its assault formation; that is, its squads and fire teams deploy to place the bulk of their firepower to the front as they assault the objective.

- Makes contact with the surveillance team to confirm any enemy activity on the objective.
- Ensures that the assault position is close enough for immediate assault if the assault element is detected early.
- Moves into position undetected, and establish local security and fire control measures.

(g) Element leaders inform the PL when their elements are in position and ready.

(h) The PL directs the support element to fire.

(i) Upon gaining fire superiority, the PL directs the assault element to move towards the objective.

- Assault element holds fire until engaged, or until ready to penetrate the objective.
- PL signals the support element to lift or shift fires. The support element lifts or shifts fires as directed, shifting fire to the flanks of targets or areas as directed in the FRAGO.

(j) The assault element attacks and secures the objective. The assault element may be required to breech a wire obstacle. As the platoon, or its assault element, moves onto the objective, it must increase the volume and accuracy of fires. Squad leaders assign specific targets or objectives for their fire teams. Only when these direct fires keep the enemy suppressed can the rest of the unit maneuver. As the assault element gets closer to the enemy, there is more emphasis on suppression and less on maneuver. Ultimately, all but one fire team may be suppressing to allow that one fire team to break into the enemy position. Throughout the assault, Soldiers use proper individual movement techniques, and fire teams retain their basic shallow wedge formation. The platoon does not get "*on-line*" to sweep across the objective.

- Assault element assaults through the objective to the designated LOA.
- Assault element leaders establish local security along the LOA, and consolidate and reorganize as necessary. They provide ACE reports to the PL and PSG. The platoon establishes security, operates key weapons, provides first aid, and prepares wounded Soldiers for MEDEVAC. They redistribute ammunition and supplies, and they relocate selected weapons to alternate positions if leaders believe that the enemy may have pinpointed them during the attack. They adjust other positions for mutual support. The squad and team leader provide ammunition, casualty, and equipment (ACE) reports to the platoon leader. The PL/PSG reorganizes the patrol based on the contact.
  -- On order, special teams accomplish all assigned tasks under the supervision of the PL, who positions himself where he can control the patrol.
  -- Special team leaders report to PL when assigned tasks are complete.

(k) On order or signal of the PL, the assault element withdraws from the objective. Using prearranged signals, the assault line begins an organized withdrawal from the objective site, maintaining control and security throughout the withdrawal. The assault element bounds back near the original assault line, and begins a single file withdrawal through the APL's choke point. All Rangers must move through the choke point for an accurate count. Once the assault element is a safe distance from the objective and the headcount is confirmed, the platoon can withdraw the support element. If the support elements were a part of the assault line, they withdraw together, and security is signaled to withdraw. Once the support is a safe distance off the objective, they notify the platoon leader, who contacts the security element and signals them to withdraw. All security teams link up at the release point and notify the platoon leader before moving to the ORP. Personnel returning to the ORP immediately secure their equipment and establish all-round security. Once the security element returns, the platoon moves out of the objective area as soon as possible, normally in two to three minutes.

- Before withdrawing, the demo team activates demo devices and charges.
- Support element or designated personnel in the assault element maintain local security during the withdrawal.
- Leaders report updated accountability and status (ACE report) to the PL and PSG.

(l) Squads withdraw from the objective in the order designated in the FRAGO to the ORP.

- Account for personnel and equipment.
- Disseminate information.
- Redistribute ammunition and equipment as required.

(m) The PL reports mission accomplishment to higher and continues the mission.

- Reports raid assessment to higher.
- Informs higher of any IR/PIR gathered.

NOTES

Figure 5-6. ACTIONS ON THE OBJECTIVE -- RAID

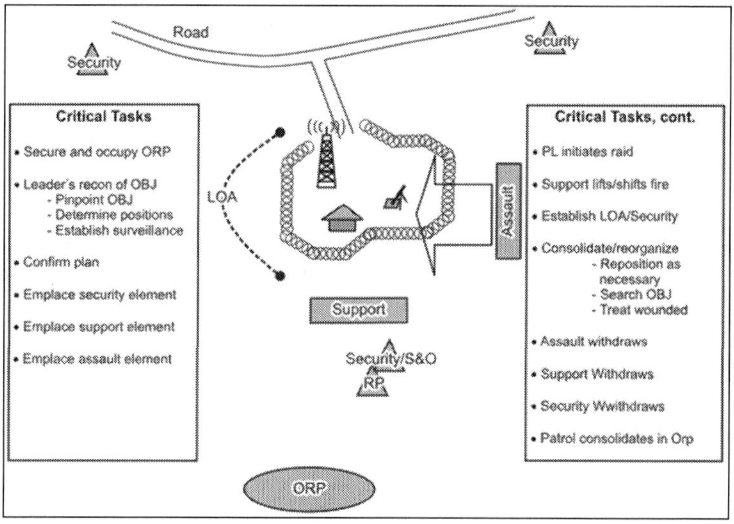

5-5. **SUPPORTING TASKS**. This section covers Linkup, Patrol Debriefing, and Occupation of an ORP.

a. **Linkup**. A linkup is a meeting of friendly ground forces. Linkups depend on control, detailed planning, communications, and stealth.

(1) ***Task Standard***. The units link up at the time and place specified in the order. The enemy does not surprise the main bodies. The linkup units establish a consolidated chain of command.

(2) ***Linkup Site Selection***. The leader identifies a tentative linkup site by map reconnaissance, other imagery, or higher headquarters designates a linkup site. The linkup site should have the following characteristics:

- Ease of recognition.
- Cover and concealment.
- No tactical value to the enemy.
- Location away from natural lines of drift.
- Defendable for a short period of time.
- N=Multiple access and escape routes.

(3) ***Execution***. Linkup procedure begins as the unit moves to the linkup point. The steps of this procedure are:

(a) The stationary unit performs linkup actions.

- Occupies the linkup rally point (LRP) NLT the time specified in the order.
- Establishes all-around security, establishes commo, and prepares to accept the moving unit.
- The security team clears the immediate area around the linkup point. It then marks the linkup point with the coordinated recognition signal. The security team moves to a covered and concealed position and observes the linkup point and immediate area around it.

(b) The moving unit--

- Performs linkup actions.
- The unit reports its location using phase lines, checkpoints, or other control measures.
- Halts at a safe distance from the linkup point in a covered and concealed position (the linkup rally point).

(c) The PL and a contact team--

- Prepare to make physical contact with the stationary unit.
- Issue a contingency plan to the PSG.
- Maintain commo with the platoon; verify near and far recognition signals for linkup (good visibility and limited visibility).
- Exchange far and near recognition signals with the linkup unit; conduct final coordination with the linkup unit.

(d) The stationary unit--

- Guides the patrol from its linkup rally point to the stationary unit linkup rally point.
- Linkup is complete by the time specified in the order.
- The main body of the stationary unit is alerted before the moving unit is brought forward.

(e) The patrol continues its mission IAW the order.

(4) ***Coordination Checklist***. The PL coordinates or obtains the following information from the unit that his patrol will link up with:

- Exchange frequencies, call signs, codes, and other communication information.
- Verify near and far recognition signals.
- Exchange fire coordination measures.
- Determine command relationship with the linkup unit; plan for consolidation of chain of command.
- Plan actions following linkup.
- Exchange control measures such as contact points, phase lines, contact points, as appropriate.

b. **Debrief**. Immediately after the platoon or squad returns, personnel from higher headquarters conduct a thorough debrief. This may include all members of the platoon or the leaders, RTOs, and any attached personnel. Normally the debriefing is oral. Sometimes a written report is required. Information on the written report should include--

- Size and composition of the unit conducting the patrol.
- Mission of the platoon such as type of patrol, location, and purpose.
- Departure and return times.
- Routes. Use checkpoints, grid coordinates for each leg or include an overlay.
- Detailed description of terrain and enemy positions that were identified.
- Results of any contact with the enemy.
- Unit status at the conclusion of the patrol mission, including the disposition of dead or wounded Soldiers.
- Conclusions or recommendations.

c. **Objective Rally Point**. The ORP is a point out of sight, sound, and small arms range of the objective area. It is normally located in the direction that the platoon plans to move after completion of actions on the objective. The ORP is tentative until the objective is pinpointed.

(1) ***Occupation of the ORP***. See Figure 5-7.

(a) The patrol halts beyond sight and sound of the tentative ORP (200-400m in good visibility, 100-200m in limited visibility).

(b) The patrol establishes a security halt IAW the unit SOP.

(c) After issuing a five-point contingency plan to the PSG, the PL moves forward with a recon element to conduct a leader's recon of the ORP.

(d) For a squad-sized patrol, the PL moves forward with a compass man and one member of each fire team to confirm the ORP.

- After physically clearing the ORP location, the PL leaves two Rangers at the 6 o'clock position facing in opposite directions.
- The PL issues a contingency plan and returns with the compass man to guide the patrol forward.
- The PL guides the patrol forward into the ORP, with one team occupying from 3 o'clock through 12 o'clock to 9 o'clock, and the other occupying from 9 o'clock through 6 o'clock to 3 o'clock.

(e) For a platoon-size patrol, the PL, RTO, WSL, three ammo bearers, a team leader, a SAW gunner, and riflemen go on the leaders recon for the ORP and position themselves at 10, 2, and 6 o'clock.

- The first squad in the order of march is the base squad, occupying from 10 to 2 o'clock.
- The trail squads occupy from 2 to 6 o'clock and 6 to 10 o'clock, respectively.
- The patrol headquarters element occupies the center of the triangle.

(2) ***Actions in the ORP***. The unit prepares for the mission in the ORP. Once the leader's recon pinpoints the objective, the PSG generally lines up rucks IAW unit SOP in the center of the ORP.

Figure 5-7. OCCUPATION OF THE ORP

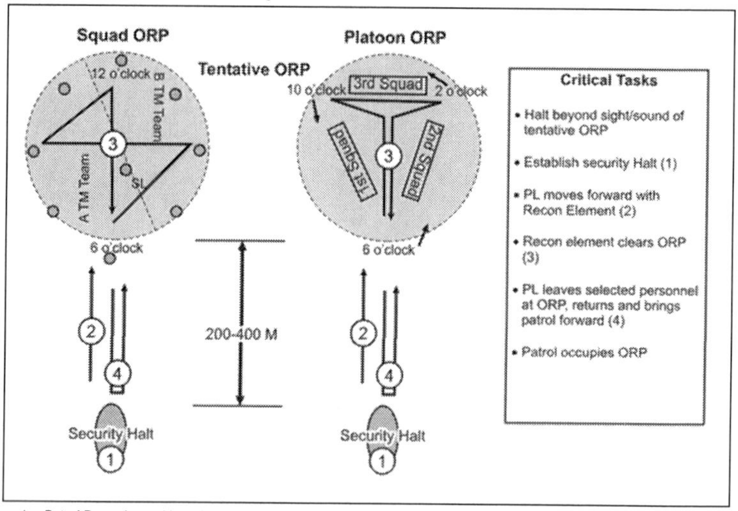

d. **Patrol Base**. A patrol base is a security perimeter that is set up when a squad or platoon conducting a patrol halts for an extended period. Patrol bases should not be occupied for more than a 24-hour period (except in emergency). A patrol never uses the same patrol base twice.

(1) ***Use***. Patrol bases are typically used--

- To avoid detection by eliminating movement.
- To hide a unit during a long detailed reconnaissance.
- To perform maintenance on weapons, equipment, eat and rest.
- To plan and issue orders.
- To reorganize after infiltrating on an enemy area.
- To establish a base from which to execute several consecutive or concurrent operations.

(2) ***Site Selection***. The leader selects the tentative site from a map or by aerial reconnaissance. The site's suitability must be confirmed and secured before the unit moves into it. Plans to establish a patrol base must include selecting an alternate patrol base site. The alternate site is used if the first site is unsuitable or if the patrol must unexpectedly evacuate the first patrol base.

(3) ***Planning Considerations***. Leaders planning for a patrol base must consider the mission and passive and active security measures. A patrol base (PB) must be located so it allows the unit to accomplish its mission.

- Observation posts and communication with observation posts.
- Patrol or platoon fire plan.
- Alert plan.
- Withdrawal plan from the patrol base to include withdrawal routes and a rally point, rendezvous point, or alternate patrol base.
- A security system to make sure that specific Soldiers are awake at all times.
- Enforcement of camouflage, noise, and light discipline.
- The conduct of required activities with minimum movement and noise.
- Priorities of Work.

(4) ***Security Measures***.

- Select terrain the enemy would probably consider of little tactical value.
- Select terrain that is off main lines of drift.
- Select difficult terrain that would impede foot movement, such as an area of dense vegetation, preferably bushes and trees that spread close to the ground.
- Select terrain near a source of water.
- Select terrain that can be defended for a short period and that offers good cover and concealment.
- Avoid known or suspected enemy positions.
- Avoid built-up areas.
- Avoid ridges and hilltops, except as needed for maintaining communications.
- Avoid small valleys.
- Avoid roads and trails.

(5) ***Occupation***. See Figure 5-8.

(a) A PB is reconned and occupied in the same manner as an ORP, with the exception that the platoon will typically plan to enter at a 90 degree turn. The PL leaves a two-man OP at the turn, and the patrol covers any tracks from the turn to the PB.

(b) The platoon moves into the PB. Squad-sized patrols will generally occupy a cigar-shaped perimeter; platoon-sized patrols will generally occupy a triangle-shaped perimeter.

(c) The PL and another designated leader inspect and adjust the entire perimeter as necessary.

(d) After the PL has checked each squad sector, each SL sends a two-man R&S team to the PL at the CP. The PL issues the three R&S teams a contingency plan, reconnaissance method, and detailed guidance on what to look for (enemy, water, built-up areas or human habitat, roads, trails, or possible rally points).

(e) Where each R&S team departs is based on the PLs guidance. The R&S team moves a prescribed distance and direction, and reenters where the PL dictates.

- Squad-sized patrols do not normally send out an R&S team at night.
- R&S teams will prepare a sketch of the area to the squad front if possible.
- The patrol remains at 100 % alert during this recon.
- If the PL feels the patrol was tracked or followed, he may elect to wait in silence at 100 % alert before sending out R&S teams.
- The R&S teams may use methods such as the "*I*," the "*Box*," or the "*T*." Regardless of the method chosen, the R&S team must be able to provide the PL with the same information.
- Upon completion of R&S, the PL confirms or denies the patrol base location, and either moves the patrol or begins priorities of work.

(6) ***Passive (Clandestine) Patrol Base (Squad)***.

- The purpose of a passive patrol base is for the rest of a squad or smaller size element.
- Unit moves as a whole and occupies in force.
- Squad leader ensures that the unit moves in at a 90-degree angle to the order of movement.
- A claymore mine is emplaced on route entering patrol base.
- Alpha and Bravo teams sit back-to-back facing outward, ensuring that at least one individual per team is alert and providing security.

(7) ***Priorities of Work (Platoon and Squad)***. Once the PL is briefed by the R&S teams and determines the area is suitable for a patrol base, the leader establishes or modifies defensive work priorities in order to establish the defense for the patrol base. Priorities of work are not a laundry list of tasks to be completed; to be effective, priorities of work must consist of a task, a given time, and a measurable performance standard. For each priority of work, a clear standard must be issued to guide the element in the successful accomplishment of each task. It must also be designated whether the work will be controlled in a centralized or decentralized manner. Priorities of work are determined IAW METT-TC. Priorities of Work may include, but are not limited to the following tasks:

(a) *Security (Continuous)*.

- Prepare to use all passive and active measures to cover all of the perimeter all of the time, regardless of the percentage of weapons used to cover that all of the terrain.
- Readjust after R&S teams return, or based on current priority of work (such as weapons maintenance).
- Employ all elements, weapons, and personnel to meet conditions of the terrain, enemy, or situation.
- Assign sectors of fire to all personnel and weapons. Develop squad sector sketches and platoon fire plan.
- Confirm location of fighting positions for cover, concealment, and observation and fields of fire. SLs supervise placement of aiming stakes and claymores.
- Only use one point of entry and exit, and count personnel in and out. Everyone is challenged IAW the unit SOP.
- Hasty fighting positions are prepared at least 18 inches deep (at the front), and sloping gently from front to rear, with a grenade sump if possible.

(b) *Withdrawal Plan*. The PL designates the signal for withdrawal, order of withdrawal, and the platoon rendezvous point and/or alternate patrol base.

(c) *Communication (Continuous)*. Commo must be maintained with higher headquarters, OP's, and within the unit. May be rotated between the patrol's RTOs to allow accomplishment of continuous radio monitoring, radio maintenance, act as runners for PL, or conduct other priorities of work.

(d) *Mission Preparation and Planning*. The PL uses the patrol base to plan, issue orders, rehearse, inspect, and prepare for future missions.

(e) *Weapons and Equipment Maintenance*. The PL ensures that machine guns, weapon systems, commo equipment, and night vision devices (as well as other equipment) are maintained. These items are not disassembled at the same time for maintenance (no more than 33 percent at a time), and weapons are not disassembled at night. If one machine gun is down, then security for all remaining systems is raised.

(f) *Water Resupply*. The PSG organizes watering parties as necessary. The watering party carries canteens in an empty rucksack or duffel bag, and must have commo and a contingency plan prior to departure.

(g) *Mess Plan*. At a minimum, security and weapons maintenance are performed prior to mess. Normally no more than half the platoon eats at one time. Rangers typically eat 1 to 3 meters behind their fighting positions.

- Rest/sleep plan management. The patrol conducts rest as necessary to prepare for future operations.
- Alert Plan and Stand-to. The PL states the alert posture and the stand-to time. He develops the plan to ensure all positions are checked periodically, OP's are relieved periodically, and at least one leader is always alert. The patrol typically conducts stand-to at a time specified by unit SOP (such as 30 minutes before and after BMNT or EENT).
- Resupply. Distribute or cross-load ammunition, meals, equipment, etc.
- Sanitation and personal hygiene. The PSG and medic ensure a slit trench is prepared and marked. All Soldiers will brush teeth, wash face, shave, wash hands, armpits, groin, feet, and darken (brush shine) boots daily. The patrol will not leave trash behind.

Figure 5-8. PATROL BASE

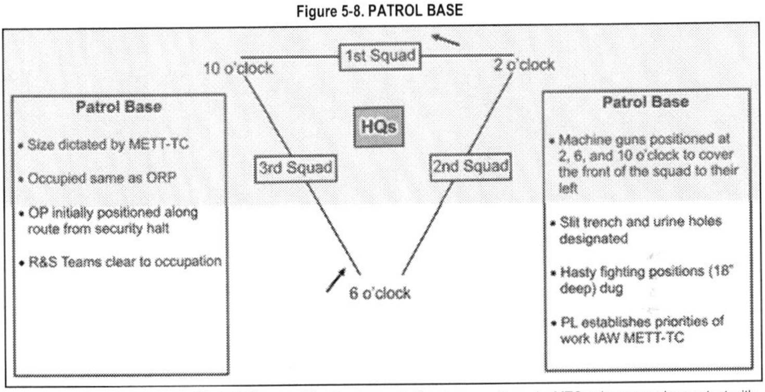

5-6. **MOVEMENT TO CONTACT**. The MTC is one of the five types of offensive operations. An MTC gains or regains contact with the enemy. Once contact is made, the unit develops the situation. Normally a platoon conducts an MTC as part of a larger force.

a. **Techniques**. The two techniques of conducting a movement to contact are search and attack and approach march.

(1) ***Search and Attack (S&A)***. This technique is used when the enemy is dispersed, is expected to avoid contact, disengage or withdraw, or you have to deny his movement in an area. The search and attack technique involves the use of multiple platoons, squads, and fire teams coordinating their actions to make contact with the enemy. Platoons typically try to find the enemy and then fix and finish him. They combine patrolling techniques with the requirement to conduct hasty or deliberate attacks once the enemy has been found.

(a) *Planning Considerations*.

- The factors of METT-TC.
- The requirement for decentralized execution.
- The requirement for mutual support.
- The length of operations.
- Minimize Soldier's load to improve stealth and speed.
- Resupply and MEDEVAC.
- Positioning key leaders and equipment.
- Employment of key weapons.
- Requirement for patrol bases.
- Concept for entering the zone of action.
- The concept for linkups while in contact.

(b) *Critical Performance Measures*.

- The platoon locates the enemy without being detected.
- Once engaged, fixes the enemy in position and maneuvers against the enemy.
- Maintains security throughout actions to avoid being flanked.

(2) ***Approach March***. The concept of the approach march is to make contact with the smallest element, allowing the commander the flexibility of destroying or bypassing the enemy. A platoon uses the approach march method as part of a larger unit. It can be tasked as the advance guard, move as part of the main body, or provide flank or rear security for the company or battalion. They may also receive on-order missions as part of the main body.

(a) *Fundamentals*. These basics are common to all movements to contact.

- Make enemy contact with smallest element possible.
- Rapidly develop combat power upon enemy contact.

- Provide all-round security for the unit.
- Support higher unit's concept.
- Reports all information rapidly and accurately and strives to gain and maintain contact with the enemy.
- Requires decentralized execution.

(b) *Planning Considerations*. The following issues should be considered heavily for MTC operations:

- Factors of METT-TC.
- Reduced Soldier's load.

(c) *Critical Performance Measures.*

- PL selects the appropriate movement formation based on likelihood of enemy contact.
- Maintains contact, once contact is made, until ordered to do otherwise.

b. **Task Standards**. The platoon moves NLT the time specified in the order, the platoon makes contact with the smallest element possible, and the main body is not surprised by the enemy. Once the platoon makes contact, it maintains contact. The platoon destroys squad and smaller-sized elements, and fixes elements larger than a squad. The platoon maintains sufficient fighting force capable of conducting further combat operations. Reports of enemy locations and contact are forwarded. If not detected by the enemy, the PL initiates a hasty attack. The platoon sustains no casualties from friendly fire. The platoon is prepared to initiate further movement within 25 minutes of contact, and all personnel and equipment are accounted for.

NOTES

Chapter 6
BATTLE DRILLS

Infantry battle drills describe how platoons and squads apply fire and maneuver to commonly encountered situations. They require leaders to make decisions rapidly and to issue brief oral orders quickly.

Section I. INTRODUCTION

This section defines and describes the format for battle drills.

6-1. **DEFINITION.** FM 7-1 defines a battle drill as "*a collective action rapidly executed without applying a deliberate decision-making process.*"

a. Characteristics of a battle drill are—

- They require minimal leader orders to accomplish and are standard throughout the Army.
- Sequential actions are vital to success in combat or critical to preserving life.
- They apply to platoon or smaller units.
- They are trained responses to enemy actions or leader's orders.
- They represent mental steps followed for offensive and defensive actions in training and combat.

b. A unit's ability to accomplish its mission often depends on Soldiers and leaders to execute key actions quickly. All Soldiers and their leaders must know their immediate reaction to enemy contact, as well as follow-up actions. Drills are limited to situations requiring instantaneous response; therefore, Soldiers must execute drills instinctively. This results from continual practice. Drills provide small units with standard procedures essential for building strength and aggressiveness.

(1) They identify key actions that leaders and Soldiers must perform quickly.

(2) They provide for a smooth transition from one activity to another; for example, from movement to offensive action to defensive action.

(3) They provide standardized actions that link Soldier and collective tasks at platoon level and below. (Soldiers perform individual tasks to CTT or SDT standard.)

(4) They require the full understanding of each individual and leader, and continual practice.

6-2. **FORMAT.** Drills include a ***title***, a ***situation*** to cue the unit or leader to start the drill, the ***required actions*** in sequence, and supporting illustrations. Where applicable, drills are cross-referenced with material in other chapters, or other drills, or both. Training standards for battle drills are in the mission training plan (MTP).

Section II. DRILLS

This section provides the battle drills for small units.

BATTLE DRILL 1
REACT TO CONTACT

SITUATION: A squad or platoon receives fires from enemy individual or crew-served weapons.

REQUIRED ACTIONS: See Figure 6-1.

1. Soldiers immediately assume the nearest covered positions.
2. Soldiers return fire immediately on reaching the covered positions.
3. Squad/team leaders locate and engage known or suspected enemy positions with well-aimed fire, and they pass information to the platoon/squad leader.
4. Fire team leaders control the fire of their Soldiers by using standard fire commands (initial and supplemental) containing the following elements:
   - Alert
   - Direction
   - Description of target
   - Range
   - Method of fire (manipulation and rate of fire)
   - Command to commence firing

5. Soldiers maintain contact (visual or oral) with the Soldiers on their left and right.

6. Soldiers maintain contact with their team leaders and indicate the location of enemy positions.

7. The leaders (visual or oral) check the status of their personnel.

8. The squad/team leaders maintain visual contact with the platoon/squad leader.

9. The platoon/squad leader moves up to the squad/fire team in contact and links up with its leader.
   a. The platoon leader brings his RTO, platoon FO, the nearest squad's squad leader, and one machine gun team.
   b. The squad leader of the trail squad moves to the front of his lead fire team.
   c. The PSG also moves forward with the second machine gun team and links up with the platoon leader, ready to assume control of the base-of-fire element.

10. The platoon/squad leader determines whether or not his unit must move out of an EA.

11. The platoon/squad leader determines whether his unit can gain and maintain suppressive fires with the element already in contact (based on the volume and accuracy of enemy fires against the element in contact). The platoon/squad leader assesses the situation. He identifies—

- The location of the enemy position and obstacles.
- The size of the enemy force engaging the unit in contact. The number of enemy automatic weapons, the presence of any vehicles, and the employment of indirect fires are indicators of the enemy strength.
- Vulnerable flanks.
- Covered and concealed flanking routes to the enemy position.

12. The platoon/squad leader determines the next COA, such as fire and movement, assault, breach, knock out bunker, enter and clear a building or trench.

13. The platoon/squad leader reports the situation to the company commander/platoon leader and begins to maneuver his unit.

14. Platoon leader directs platoon FO to call for and adjust indirect fires (mortars or artillery). Squad leaders relay requests through the platoon leader. The platoon/squad leader in conjunction with the platoon FO maintains accurate battle tracking of all friendly elements to facilitate quick clearance of fires.

15. Leaders relay all commands and signals from the platoon chain of command.

16. The PSG positions the base of fire element to observe and to provide supporting fires.

Note: Once the platoon has executed the React to Contact drill, the platoon leader quickly assesses the situation, for example, enemy size and location. He picks a COA. The platoon leader reports the situation to the company commander.

Figure 6-1. REACT TO CONTACT

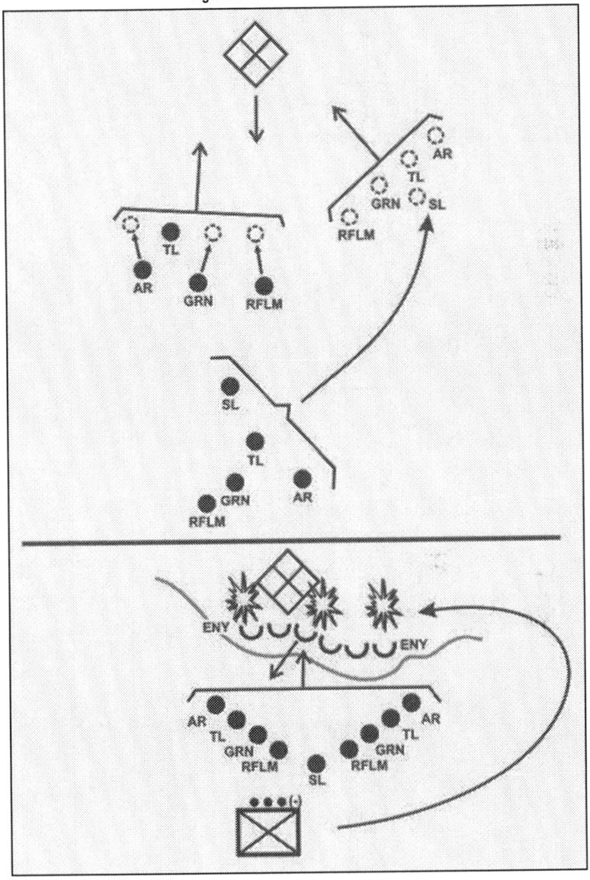

BATTLE DRILL 2
BREAK CONTACT

SITUATION: The squad or platoon is under enemy fire and must break contact.

REQUIRED ACTIONS: See Figure 6-2.

1. The platoon/squad leader directs fire support for the disengagement. He accomplishes this by—
   - Directing one squad/fire team in contact to support the disengagement of the remainder of the unit.
   - Considering the use of indirect fires for breaking contact.
   - Clearing the location, task, purpose, and method of conducting the fire mission with the platoon FO.

2. The platoon/squad leader orders a distance and direction, a terrain feature, or last ORP for the movement of the first squad/fire team. In conjunction with the platoon FO, the leader maintains accurate battle tracking of all friendly elements to facilitate quick clearance of fires.

3. The base of fire squad/team continues to suppress the enemy. The platoon/squad leader directs the platoon FO to execute the fire mission, if needed.

4. The moving squad/team assumes the overwatch position. The squad/team should use M203 grenade launchers, throw fragmentation and concussion grenades, and use smoke grenades to mask movement. The platoon/squad leader directs the platoon FO to execute smoke mission to screen friendly elements movement, if needed.

5. The moving squad/team takes up the designated position and engages the enemy position.

6. The platoon leader directs the base-of-fire element to move to its next location. Based on the terrain and the volume and accuracy of the enemy's fire, the moving fire team or squad may need to use fire and movement techniques.

7. The platoon/squad continues to bound away from the enemy until—
   - It breaks contact. It must continue to suppress the enemy as it does this.
   - It passes through a higher-level SBF position.
   - Its squad/fire teams are in the assigned position to conduct the next mission.

8. The leader should consider changing his unit's direction of movement once contact is broken. This will reduce the ability of the enemy to place effective indirect fires on the unit.

9. If the platoon/squad becomes disrupted, Soldiers stay together and move to the last designated rally point.

10. The platoon/squad leaders account for Soldiers, report, reorganize as necessary, and continue the mission. With the platoon FO, the platoon/squad develops a quick fire plan to support the hasty defense or new route of march.

Figure 6-2. BREAK CONTACT

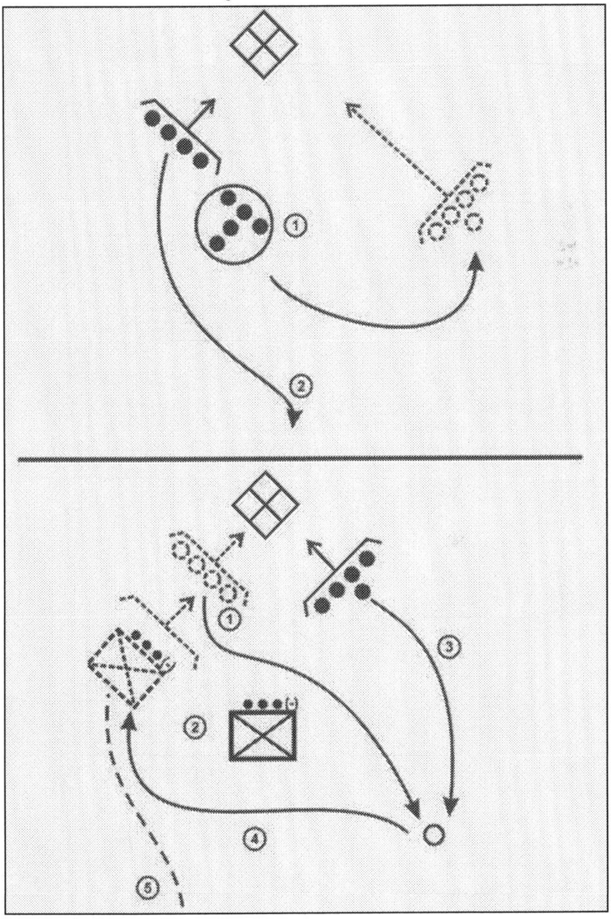

BATTLE DRILL 3
REACT TO AMBUSH

SITUATION: The squad or platoon enters a kill zone and the enemy initiates an ambush with a casualty-producing device and a high volume of fire.

REQUIRED ACTIONS: See Figure 6-3.

1. Near ambush (within hand-grenade range).
   a. Soldiers in the kill zone return fire immediately. How they go about this depends on the terrain. If—
      - Cover is not available, Soldiers immediately, without order or signal, assume prone position and throw concussion or fragmentation and smoke grenades.
      - Cover is available, without order or signal, Soldiers seek the nearest covered position, assume the prone position, and throw fragmentation or concussion and smoke grenades.

   b. Immediately after the explosion of the concussion or fragmentation grenades, Soldiers in the kill zone return fire and assault through the ambush position using fire and movement.

   c. Soldiers not in the kill zone identify the enemy's position. Fire is shifted as the personnel in the kill zone begin to assault.

   d. Soldiers in the kill zone continue the assault to eliminate the ambush or until contact is broken.

   e. The platoon/squad conducts consolidation and reorganization.

2. Far ambush (out of hand-grenade range). Soldiers receiving fire immediately return fire, take up covered positions, and suppress the enemy by—
   - Destroying or suppressing enemy crew-served weapons.
   - Sustaining suppressive fires.

3. Soldiers (squad/teams) not receiving fires move by a covered and concealed route to a vulnerable flank of the enemy position and assault using fire and movement techniques.

4. Soldiers in the kill zone continue suppressive fires and shift fires as the assaulting squad/team fights through the enemy position.

5. The platoon FO calls for and adjusts indirect fires as directed by the platoon leader. On order, he shifts or ceases fires to isolate the enemy position or to attack them with indirect fires as they retreat.

6. The platoon/squad leader reports, reorganizes as necessary, and continues the mission.

Figure 6-3. REACT TO AMBUSH

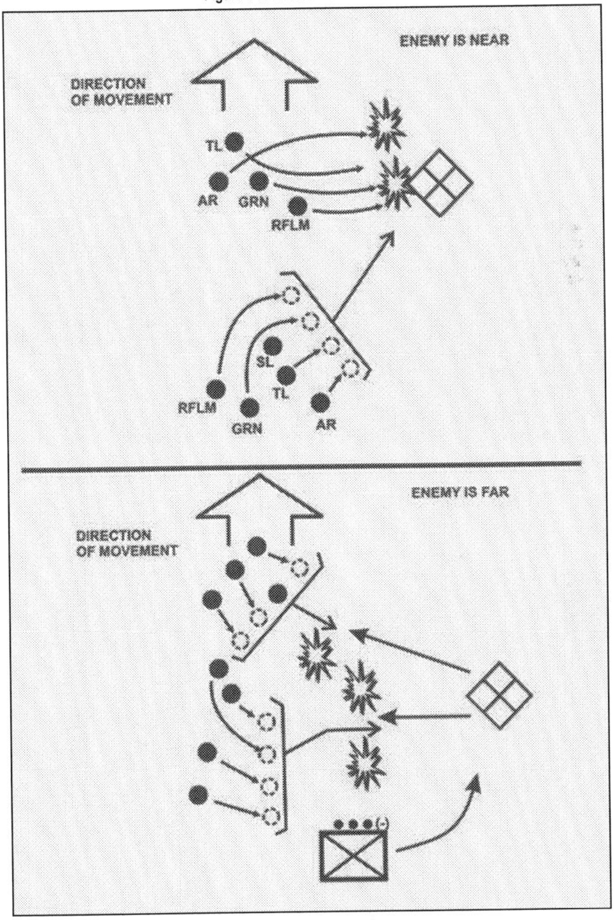

BATTLE DRILL 4
KNOCK OUT BUNKERS

SITUATION: While moving as part of a larger force, the platoon identifies the enemy in bunkers.

REQUIRED ACTIONS: See Figures 6-4 and 6-5.

1. The platoon initiates contact.
   a. The squad in contact establishes a base of fire.
   b. The platoon leader, his RTO, platoon FO, and one machine gun team move forward to link up with the squad leader of the squad in contact.
   c. The PSG moves forward with the second machine gun team and assumes control of the base-of-fire squad.
   d. The base-of-fire squad—
      - Destroys or suppresses enemy crew-served weapons first.
      - Obscures the enemy position with smoke (M203).
      - Sustains suppressive fires at the lowest possible level.
   e. The platoon leader directs platoon FO to call for and adjust indirect fires. The platoon leader in conjunction with the platoon FO maintains accurate battle tracking of all friendly elements to facilitate quick clearance of fires.
2. The platoon leader determines that he can maneuver by identifying—
   - Enemy bunkers, other supporting positions, and any obstacles.
   - Size of the enemy force engaging the platoon. The number of enemy automatic weapons, the presence of any vehicles, and the employment of indirect fires are indicators of enemy strength.
   - A vulnerable flank of at least one bunker.
   - A covered and concealed flanking route to the flank of the bunker.
3. The platoon leader determines which bunker is to be assaulted first and directs one squad (not in contact) to knock it out.
4. If necessary, the PSG repositions a squad, fire team, or machine gun team to isolate the bunker as well as to continue suppressive fires.
5. The assaulting squad, with the platoon leader and his RTO, move along the covered and concealed route and take action to knock out the bunker. The following occurs.
   a. On the platoon leader's signal, the support squad ceases or shifts fires to the opposite side of the bunker from which the squad is assaulting.
   b. At the same time, the platoon FO shifts indirect fires to isolate enemy positions.
6. The assaulting squad leader reports to the platoon leader and reorganizes his squad.
7. The platoon leader—
   - Directs the supporting squad to move up and knock out the next bunker.
   - Directs the assaulting squad to continue and knock out the next bunker.
   - Rotates squads as necessary.
8. The platoon leader reports, reorganizes as necessary, and continues the mission. The company follows up the success of the platoon attack and continues to assault enemy positions.

Figure 6-4. KNOCK OUT A BUNKER (SQUAD)

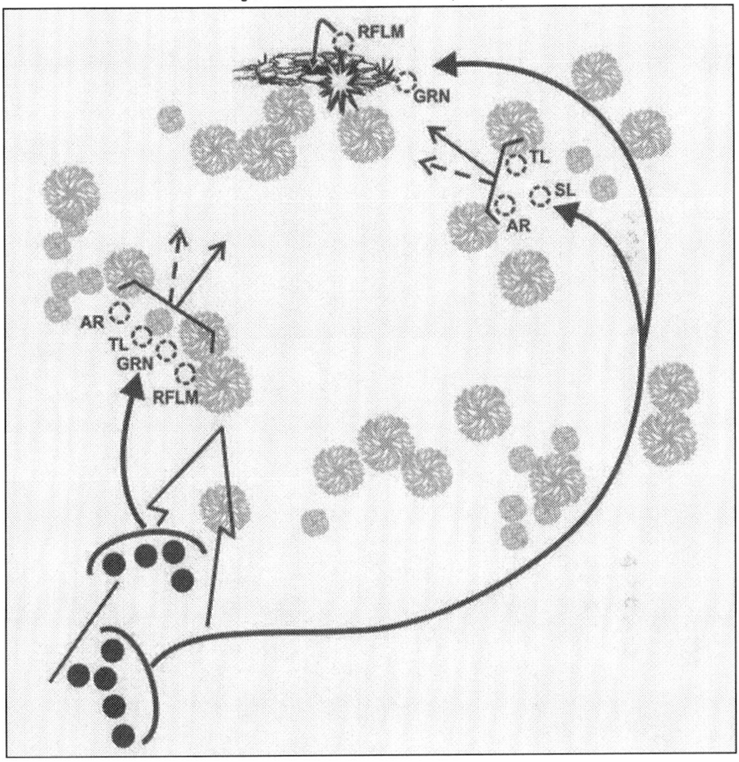

Figure 6-5. KNOCK OUT BUNKERS (PLATOON)

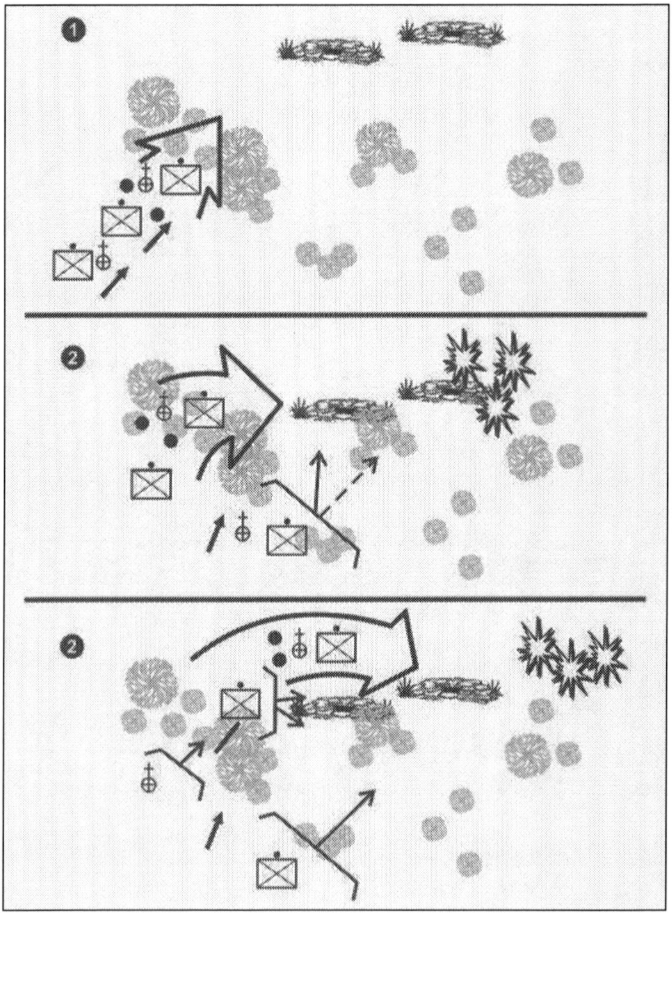

BATTLE DRILL 5
ENTER BUILDING/CLEAR ROOM

SITUATION: Operating as part of a larger force (in daylight or darkness), the squad is tasked to participate in clearing a building. The platoon leader directs the squad to enter the building or to clear a room. An entry point or breach has already been identified or will be created before initiating the entry. For a detailed discussion of urban entry breaching techniques, see FM 3.06-11. Enemy forces and noncombatants may or may not be present in the room and or building to be cleared.

REQUIRED ACTIONS: See Figures 6-6 through 6-14.

1. Platoon and squad leaders must consider the task and purpose they have been given and the method they are to use to achieve the desired results. They must operate IAW the ROE and must be aware of the affects that platoon weapons will have on the type and composition of the buildings.

2. To seize or gain control of a building may not always require committing troops into the structure or closing with the enemy. Before initiating this action and exposing members of the clearing squad to direct enemy contact and risking casualties, the platoon leader should consider/direct employment of all organic, crew-served, and supporting weapon systems onto the objective area to suppress and neutralize the threat, providing the mission, purpose, building composition, and ROE permit.

3. When conducting urban operations, Soldiers must be equipped at all times with a properly-mounted and immediately-useable NVD or light source to illuminate the immediate area.

Note: The following discussion assumes that only the platoon's organic weapons are supporting the infantry squad. Urban situations may require precise application of firepower. This situation is especially true of an urban environment in which the enemy is mixed with noncombatants. Noncombatants may be found in the room, which can restrict the use of fires and reduce the combat power available to a squad leader. His squad may have to operate in no-fire areas. ROE can prohibit the use of certain weapons until a specific hostile action takes place. All Soldiers must be aware of the ROE. Leaders must include the specific use of weapons in their planning for precision operations in urban terrain. Leaders should always consider the use of snipers or designated marksmen to apply precise fires to the objective.

4. Clearing team members must approach the entry point quickly, quietly, and in standard order. The squad leader must ensure he is in a position to control the actions of both fire teams. This approach preserves the element of surprise and allows for quick entry and domination of the room. If a breach is required, the order may be slightly modified based on the breach technique (FM 3-06.11). The members of the fire team are assigned numbers one through four. The rifleman is # 1 and the grenadier is # 3. If one member of the clearing team is armed with the SAW rather than an M16 rifle or carbine, he should be designated # 4. The team leader is normally the # 2 man, because he will have the most immediate decision to make as he enters the room.

5. The entire team enters the room as quickly and smoothly as possible and clears the doorway immediately. If possible, the team moves from a covered or concealed position already in their entry order. Ideally, the team arrives and passes through the entry point without having to stop. If the team must stop to effectively "*stack*" outside the entry point, it must do so only momentarily, and it must provide cover.

6. The door is the focal point of anyone in the room. It is known as the "*fatal funnel*," because it focuses attention at the precise point where the individual team members are the most vulnerable. Moving into the room quickly reduces the chance of anyone being hit by enemy fire directed at the doorway.

7. For this battle to be effectively employed, each member of the team must know his sector of fire and how his sector overlaps and links with the sectors of the other team members. Team members do not move to the points of domination and then engage their targets. Rather, they engage targets as they calmly and quickly move to their designated points. Engagements must not slow movement to their points of domination. Team members may shoot from as short a range as 1 to 2 inches. They engage the most immediate threat first and then the less immediate threats in sector. Immediate threats are personnel who—

- Are armed and prepared to return fire immediately.
- Block movement to the position of domination.
- Are within arm's reach of a clearing team member.
- Are within 3 to 5 feet of the breach point.

8. The squad leader designates the assault team and identifies the location of the entry point for them.

9. The squad leader positions the follow-on assault team to provide overwatch and supporting fires for the initial assault team.

10. Assault team members use available cover and concealment, and move as close to the entry point as possible.

a. If an explosive breach or a ballistic breach is to be performed by a supporting element, the assault team remains in a covered position until the breach is made. They may provide overwatch and fire support for the breaching element if necessary.

b. All team members must signal one another that they are ready before the team moves to the entry point.

c If stealth is a consideration, team members avoid the use of verbal signals, which may alert the enemy and remove the element of surprise.

d. Assault team members must move quickly from the covered position to the entry point, minimizing the time they are exposed to enemy fire.

11. The assault team enters through the entry point or breach. Unless a grenade will be thrown prior to entry, the team should avoid stopping outside the point of entry.

a. If required, the # 2 man throws a grenade of some type (fragmentary, concussion, stun) into the room before entry.

b. The use of grenades should be consistent with the ROE and building structure. The grenade should be cooked off before being thrown, if applicable to the type of grenade used.

c. If stealth is not a factor, the thrower should sound off with a verbal indication (*"Frag out," "Concussion out," "Stun out"*) that a grenade of some type is being thrown. If stealth is a factor, only visual signals are given as the grenade is thrown.

CAUTION

If walls and floors are thin, fragments from fragmentation grenades and debris created by concussion grenades can injure Soldiers outside the room. If the structure has been stressed by previous explosive engagements, the use of these grenades could cause it to collapse. Leaders must determine the effectiveness of these types of grenades compared to possibilities of harm to friendly troops.

12. On the signal to go, or after the grenade detonates, the assault team moves through the entry point (Figure 6-6 through Figure 6-9) and quickly takes up positions inside the room that allow it to completely dominate the room and eliminate the threat. Unless restricted or impeded, team members stop movement only after they have cleared the door and reached their designated point of domination. In addition to dominating the room, all team members are responsible for identifying possible loopholes and mouseholes in the ceiling, walls, and floor.

Note: Where enemy forces may be concentrated and the presence of noncombatants is highly unlikely, the assault team can precede their entry by throwing a fragmentation or concussion grenade (structure dependent) into the room, followed by bursts of automatic small-arms fire by the # 1 man as he enters. Carefully consider the ROE and building composition before employing this method.

13. The # 1 and # 2 men are initially concerned with the area directly to their front, then along the wall on either side of the door or entry point. This area is in their path of movement, and it is their primary sector of fire. Their alternate sector of fire is from the wall they are moving toward, back to the opposite far corner.

14. The # 3 and # 4 men start at the center of the wall opposite their point of entry and clear to the left if moving toward the left, or to the right if moving toward the right. They stop short of their respective team member (either the # 1 man or the # 2 man).

15. The team members move toward their points of domination, engaging all threat or hostile targets in sequence in their sector. Team members must exercise fire control and discriminate between hostile and noncombatant room occupants. (The most practical way to do this is to identify whether or not the target has a weapon in his or her hands.) Shooting is done without stopping, using reflexive shooting techniques. Because the Soldiers are moving and shooting at the same time, they must move using the careful hurry. Figure 6-10, shows all four team members at their points of domination for a room with a center door and their overlapping sectors of fire.

16. The first man (rifleman) enters the room and eliminates the immediate threat. He has the option of going left or right, normally moving along the path of least resistance to one of two corners. When using a doorway as the point of entry, the team uses the path of least resistance, which they determine initially based on the way the door opens. If the door opens inward, the first man plans to move away from the hinges. If the door opens outward, he plans to move toward the hinged side. Upon entering, his direction is influenced by the size of the room, the enemy situation, and furniture or other obstacles that hinder or channel his movement.

17. The direction each man moves in should not be preplanned unless the exact room layout is known. Each man should go in a direction opposite the man in front of him (Figure 6-6). Every team member must know the sectors and duties of each position.

18. As the first man goes through the entry point, he can usually see into the far corner of the room. He eliminates any immediate threat and continues to move along the wall if possible and to the first corner, where he assumes a position of domination facing into the room.

Figure 6-6. FIRST MAN ENTERS A ROOM, FOLLOWED BY TEAM LEADER

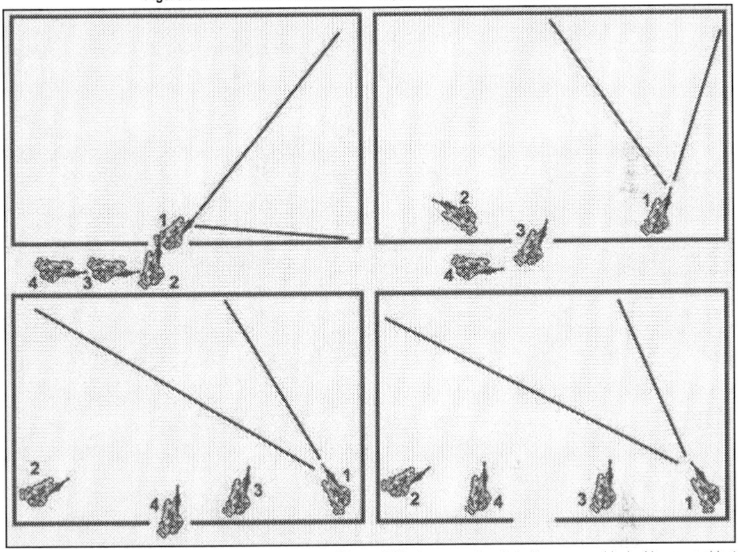

Note: Team members must always stay within 1 meter of the wall. If a team member finds his progress blocked by some object that will force him more than 1 meter from the wall, he must either step over it (if able) or stop where he is and clear the rest of his sector from where he is. If this action creates dead space in the room, the team leader directs which clearing actions to take once other members of the team have reached their points of domination.

19. The second man (normally the team leader), entering almost simultaneously with the first, moves in the opposite direction, following the wall (Figure 6-7). The second man must clear the entry point, clear the immediate threat area, clear his corner, and move to a dominating position on his side of the room. The second man must also immediately determine if he is entering a center door or corner door and act accordingly (Figure 6-7 and Figure 6-10).

Figure 6-7. SECOND MAN (TEAM LEADER) ENTERS A ROOM

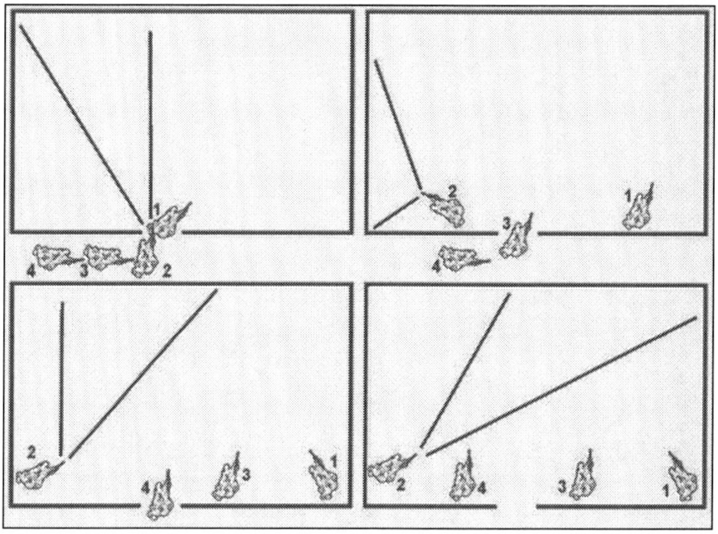

20. The third man (normally the grenadier) simply goes opposite of the second man inside the room, moves at least 1 meter from the entry point, and takes a position that dominates his sector (Figure 6-8).

Figure 6-8. THIRD MAN ENTERS A ROOM

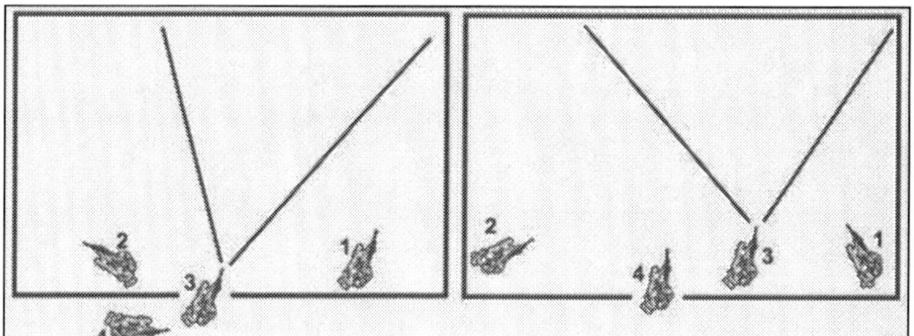

21. The fourth man (normally the SAW gunner) moves opposite of the third man, clears the doorway by at least 1 meter, and moves to a position that dominates his sector (Figure 6-9).

Figure 6-9. FOURTH MAN IN A ROOM

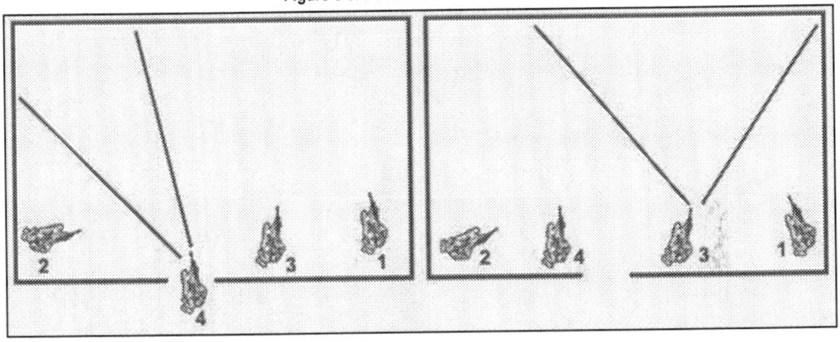

Note: If the path of least resistance takes the first man to the left, then all points of domination are the mirror image of those shown in the diagrams.

22. Points of domination should not be in front of doors or windows so team members are not silhouetted to the outside of the room (Figure 6-10). No movement should mask the fire of any of the other team members.

Figure 6-10. POINTS OF DOMINATION AND SECTORS OF FIRE, CENTER DOOR VERSUS CORNER DOOR

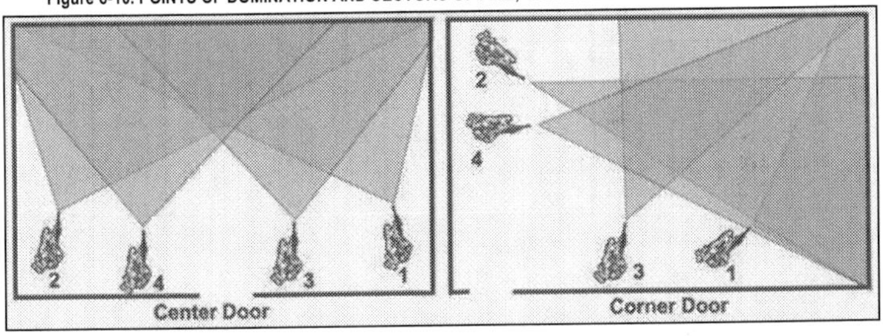

23. On order, any member of the assault team may move deeper into the room overwatched by the other team members. The team leader must control this action.

24. Once the room is cleared, the team leader signals to the squad leader that the room has been cleared. The squad leader marks the room IAW unit SOP (see FM 3.06-11). The squad leader determines whether or not his squad can continue to clear through the building.

25. The squad reorganizes as necessary. Leaders redistribute the ammunition. The squad leader reports to the platoon leader when the room is clear.

Note: If at any point a team member experiences a weapon malfunction in the presence of enemy combatants, he has to make an immediate decision. If the enemy is outside immediate danger distance from the Soldier, then the Soldier should drop to one knee, which indicates that he has experienced a weapon malfunction and prevents fratricide by ensuring that the next Soldier's fields of fire are clear. Once on a knee, the Soldier must remain there until the team leader directs him to stand up. If the kneeling Soldier corrects his weapon malfunction, he can continue to engage targets from his kneeling position. If, however, the enemy Soldier is so close that he presents an immediate threat, and if dropping to a knee would expose the US Soldier to immediate harm, then the US Soldier must try to subdue or disable the enemy Soldier. He can strike the enemy with the muzzle of his weapon or a bayonet to the face, throat, or chest. Or, he could grapple with the enemy soldier and take him to the ground immediately to clear the other team members' fields of fire. Once they have cleared their sectors and eliminated any other threats in the room, the other members of the clearing team assist.

26. Although this battle drill is effective, leaders might have to modify it for the situation. Some reasons and methods for modifying the technique are shown in Table 6-1.

Table 6-1. REASONS FOR AND METHODS OF MODIFYING ENTRY TECHNIQUES

REASON	METHOD
Objective rooms are consistently small	Clear in teams of 2 or 3
Shortage of personnel	Clear in teams of 2 or 3
Enemy poses no immediate threat	2 or 3 men search rooms to ensure no enemy or noncombatants are present
No immediate threat, speed is critical	1 to 3 men visually search each room

27. When full four-man teams are not available for room clearing, two- and three-man teams can be used. If the # 1 or # 2 man discovers that the room is very small, he can yell, "*Short room*" or "*Short,*" which tells the # 3 or # 4 man (whoever following the # 1 or # 2 man) to stay outside the room. Figures 6-11 and 6-12 show the points of domination and sectors of fire for a two-man clearing team. Figures 6-13 and 6-14 show the actions for a three-man team.

DANGER

RICOCHETS POSE A HAZARD. ALL SOLDIERS MUST BE AWARE OF THE TYPE OF WALL CONSTRUCTION OF THE ROOM BEING CLEARED. THE WALLS OF AN ENCLOSED ROOM PRESENT MANY RIGHT ANGLES. COMBINED WITH HARD SURFACES SUCH AS CONCRETE, A BULLET MAY CONTINUE TO RICOCHET AROUND A ROOM UNTIL ITS ENERGY IS SPENT. AFTER HITTING THREAT PERSONNEL, BALL AMMUNITION MAY PASS THROUGH THE BODY AND RICOCHET. BODY ARMOR AND THE KEVLAR HELMET PROVIDE SOME PROTECTION FROM THIS HAZARD.

Figure 6-11. POINTS OF DOMINATION AND SECTORS OF FIRE (TWO-MAN TEAM, CENTER DOOR)

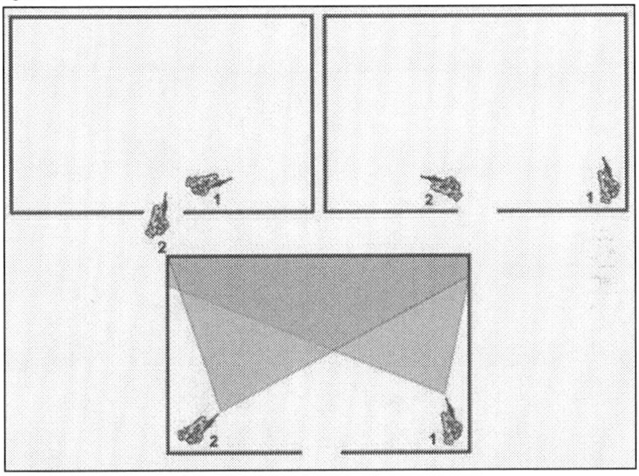

Figure 6-12. POINTS OF DOMINATION AND SECTORS OF FIRE (TWO-MAN TEAM, CORNER DOOR)

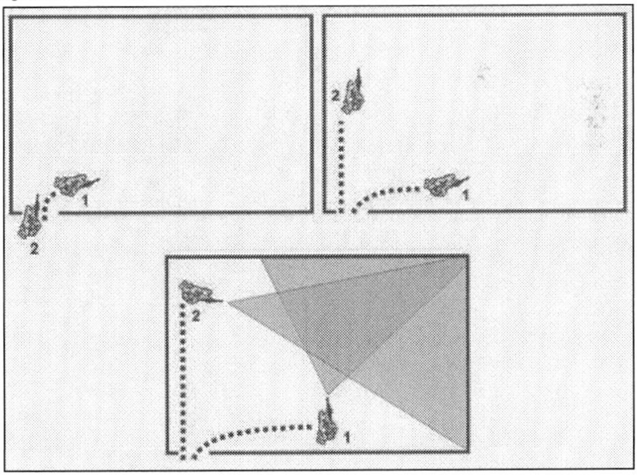

Figure 6-13. POINTS OF DOMINATION AND SECTORS OF FIRE (THREE-MAN TEAM, CENTER DOOR)

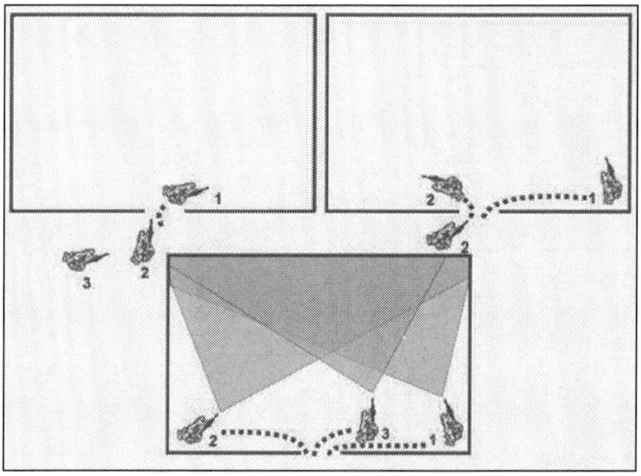

Figure 6-14. POINTS OF DOMINATION AND SECTORS OF FIRE (THREE-MAN TEAM, CORNER DOOR)

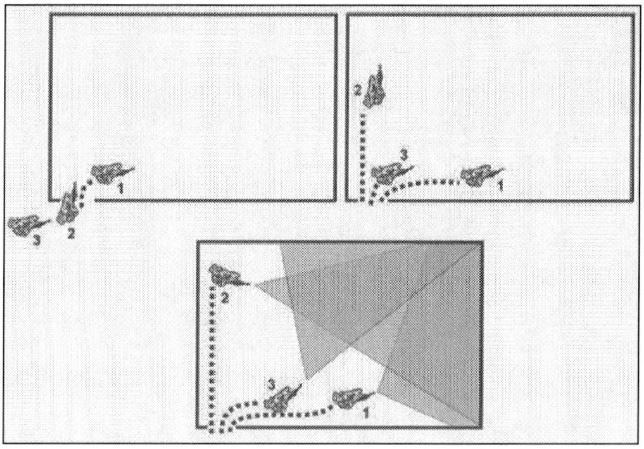

28. While moving through a building Soldiers may encounter the following architectural building features.

a. **Multiple Team/Multiple Rooms**. Figure 6-15 shows this.

STEP 1: First team enters and clears Room 1.

STEP 2: Squad leader determines the direction from which the second clearing team must enter Room 1, based on the location of the Room 2 entry point.

STEP 3: First team collapses inward to allow the second team to move into the room.

STEP 4: Second team 'stacks left' and prepares to enter Room 2.

Figure 6-15. MULTIPLE TEAMS / MULTIPLE ROOMS

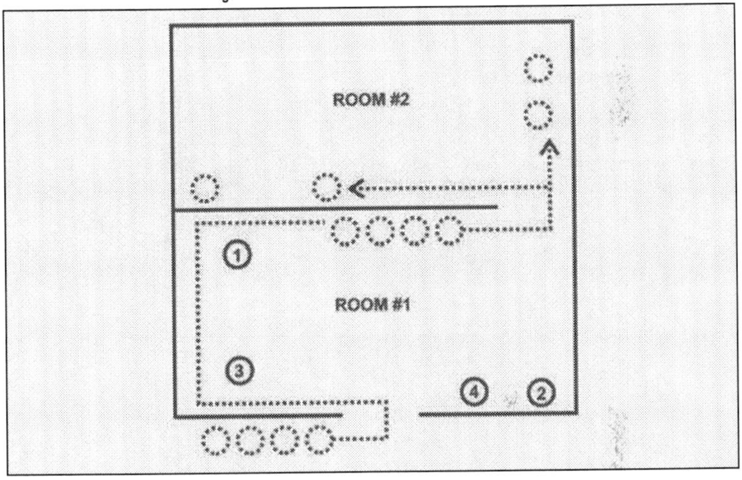

b. **Open Stairwell**. This is a gap between flights of stairs that allows a person to look up and down between the flights. Figure 6-16 shows an open stairwell.

STEP 1: The # 1 man pulls security on the highest point he can see / engage.

STEP 2: The # 2 man moves up the stairs on the inside with the # 3 man to a point that he can see / engage the next landing, where he turns around and continues to move up to the next landing.

STEP 3: The # 3 man moves up the stairs with the # 2 man on the outside and engages the threat on the immediate landing.

STEP 4: The # 4 man moves up the stairs with the # 1 man, on the squeeze, the # 2 man turns around to engage the next landing.

STEP 5: The flow continues with the # 2 man picking up the sector of the # 1 man had. The # 3 man picks up where the # 2 man was. The # 4 man picks up where the # 3 man was. The # 1 man picks up where the # 4 man was.

Note: Most stairwells require a second team.

Figure 6-16. OPEN STAIRWELL

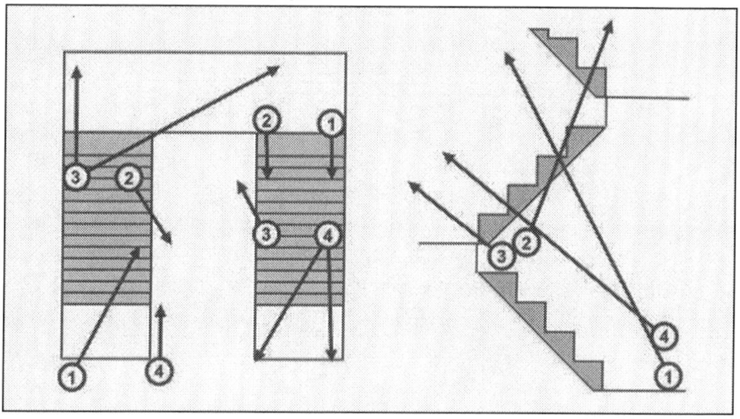

c. **Closed Stairwell**. This is any stairwell with walls separating the flights of stairs. Figure 6-17 shows a closed stairwell.

STEP 1: The # 1 man checks high to ensure there is no opening on the landing or between the stairs.

STEP 2: The # 2 man pulls long security to the next bend or landing.

STEP 3: The # 1 or # 3 man moves up the steps with the # 2 man. As they approach the corner, the # 2 man signals his presence by tapping the # 1 man on the shoulder.

STEP 4: Keying off the # 1 man's movement, the other men simultaneously break around the corner.

STEP 5: If no fire is received, the # 2 man moves to the opposite wall. Both men continue to move up until they reach their objective.

STEP 6: The # 3 and # 4 men will continue to move 3 to 4 steps behind.

Note: Avoid getting locked into a security position such as inside a stairwell. Also, avoid spreading yourselves thin or getting separated by more than one floor of stairs.

Figure 6-17. CLOSED STAIRWELL

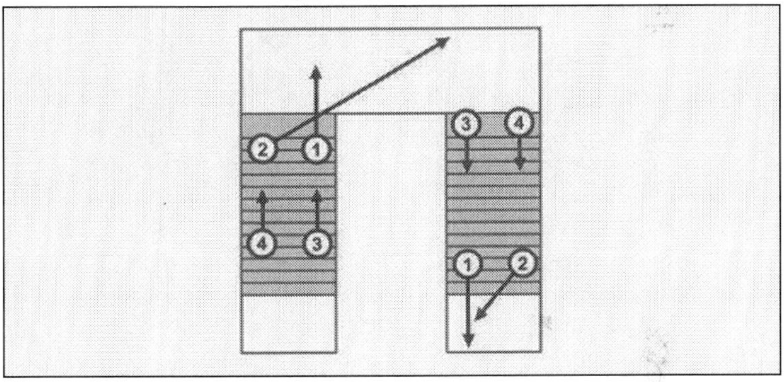

d. **Hallway Movement**. Clearing team(s) move down the hallway using the frontal security (cross cover) technique (Figure 6-18).

Figure 6-18. HALLWAY MOVEMENT

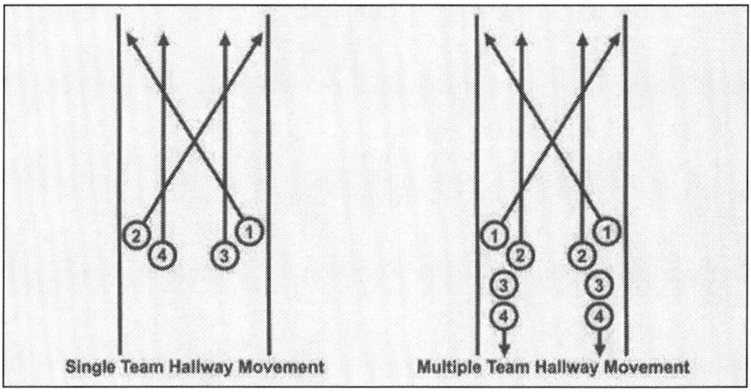

e. **T-Shaped Intersection**. See Figure 6-19. This technique can incorporate the dynamic corner clear (Figure 6-20).

STEP 1: Each # 1 man goes to a knee covering his sector.

STEP 2: On a predetermined signal, each two-man team will break the corner, picking up their sectors of fire.

Figure 6-19. T-SHAPED INTERSECTION

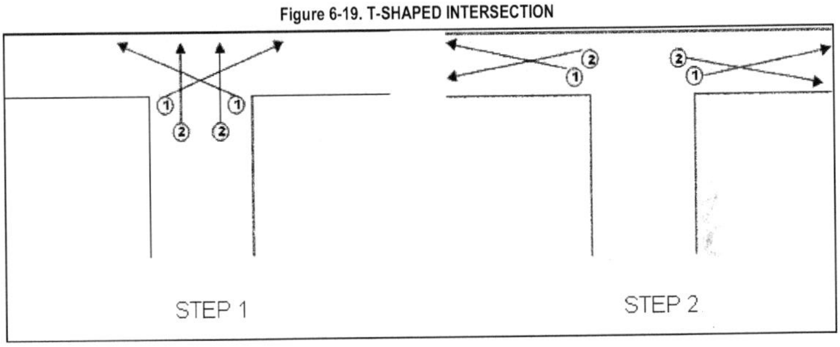

f. **Dynamic Corner**. (See Figure 6-20.)

STEP 1: The #1 and #2 man, as they approach the corner, have to clear. Do not slow down.

STEP 2: The #2 man will tap the #1 man on the shoulder about 2-3 meters away from the corner letting the #1 man know the #2 man is with him.

STEP 3: Keying off of the #1 man's movement, they both break the corner simultaneously.

STEP 4: The #1 man goes low to a knee, the #2 man stays high.

STEP 5: If the Rangers are not receiving fire the #2 man rabbits, or moves, to the far side.

STEP 6: The #1 and #2 man take up sectors of fire.

STEP 7: The #3 and #4 man take long security in the direction of movement.

Figure 6-20. DYNAMIC CORNER

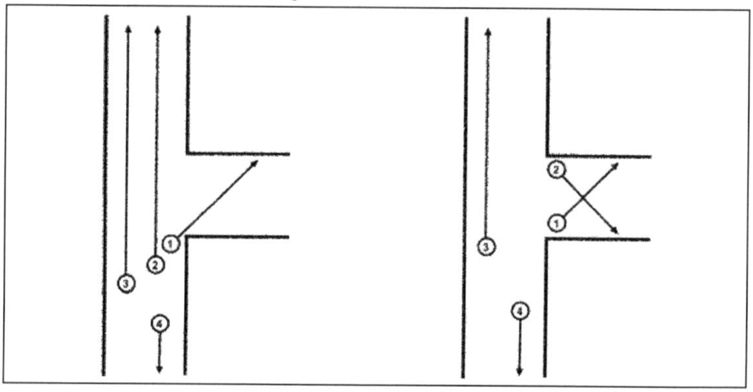

g. **Three-Way Intersection**. (See Figure 6-21.)

STEP 1: The #1 through #4 man will use one of the corner clearing techniques to clear the intersection.

STEP 2: When the intersection is secure, the #5 and #6 man move through the intersection and enter either more hallway or a room. If they are entering more hallway, the #1 through #4 man provide cross coverage as the #5 and #6 man move through the intersection into more hallway. If they are moving into a room, the #5 and #6 man move to one side of the door and signal the #1 and #3 or #2 and #4 man to enter the stack depending on which side of the door they stack on. The remaining members of the team, not in the stack, will continue to provide security down the hallway.

STEP 3: If the #5 and #6 man move into more hallway the #1 through #4 man will enter the stack and proceed down the hallway with the #3 and #4 man providing rear security.

Figure 6-21. THREE WAY INTERSECTION

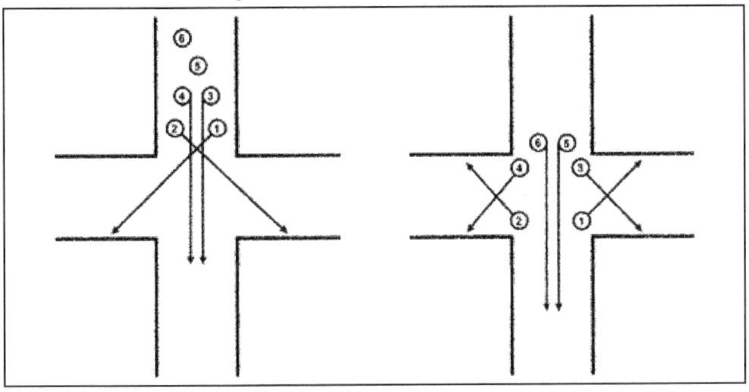

BATTLE DRILL 6
ENTER/CLEAR A TRENCH

SITUATION: The platoon is attacking as part of a larger force and identifies the enemy in a trench line. The platoon deploys and establishes a base of fire. The platoon leader determines that he has sufficient combat power to maneuver and assault the trench line.

REQUIRED ACTIONS: See Figure 6-22 and Figure 6-23.

1. The platoon leader directs one squad to enter the trench and secure a foothold.

2. The platoon leader designates the entry point of the trench line and the direction of movement once the platoon begins clearing.

3. The PSG positions Soldiers and machine guns to suppress the trench and isolate the entry point.

4. Platoon leader directs platoon FO to initiate fire mission, if necessary, in support of assault. They maintain accurate battle tracking of all friendly elements to facilitate quick clearance of fires. The platoon FO ceases or shifts fires to isolate the OBJ as assault team advances.

5. The assaulting squad executes actions to enter the trench and establish a foothold. The squad leader signals to the platoon leader that the foothold is secure, and the follow on elements can move into the trench. The squad leader remains at the entry point and marks it. The platoon follows the success of the seizure of the foothold with the remainder of the platoon as part of the platoon actions to clear a trench line.

6. The platoon leader moves into the trench with the assaulting squad.

7 The platoon leader directs one of the base-of-fire squads to move into the trench and begin clearing it in the direction of movement from the foothold.

8. The base-of-fire element repositions as necessary to continue suppressive fires.

9. The assaulting squad passes the squad that has secured the foothold and executes actions to take the lead and clear the trench.

a. The squad leader designates a lead fire team and a trail fire team.

b. The lead fire team and the squad leader move to the forward most secure corner or intersection. The squad leader tells the team securing that corner or intersection that his squad is ready to continue clearing the trench. The trail fire team follows, maintaining visual contact with the last Soldier of the lead team.

Notes: 1. The fire support element must be able to identify the location of the lead fire team in the trench at all times.

2. Throughout this battle drill, the team leader positions himself at the rear of the fire team to have direct control (physically, if necessary) of his Soldiers. Other Soldiers in the fire team rotate the lead to change magazines and prepare grenades. Rotating the lead provides constant suppressive fires down the trench and maintains the momentum of the attack as the squad clears the trench.

c. The lead fire team passes the element securing the foothold. The following then occurs.

(1) The lead Soldier of the fire team moves abreast of the Soldier securing the corner or intersection, taps him, and announces, "*Taking the lead.*"

(2) The Soldier securing the corner or intersection acknowledges that he is handing over the lead by shouting, "*Okay!*" He allows the fire team to pass him.

d. The lead fire team starts clearing in the direction of movement. They arrive at a corner or intersection. The following then occurs.

(1) Allowing for cook-off (2 seconds maximum) and shouting, "*Frag out,*" the second Soldier prepares and throws a grenade around the corner.

(2) Upon detonation of the grenade, the lead Soldier moves around the corner firing three-round bursts and advancing as he fires. The entire fire team follows him to the next corner or intersection.

e. The squad leader—

- Follows behind the team.
- Ensures that the trailing fire team moves up and is ready to pass the lead at his direction.
- Rotates fire teams as necessary to keep his Soldiers fresh and to maintain the momentum of the attack.
- Requests indirect fires, if required, through the platoon leader. The squad leader also directs the employment of the M203 to provide immediate suppression against positions along the trench line.
- Ensures fire teams maintain sufficient interval to prevent them from being engaged by the same enemy fires.

f. At each corner or intersection, the lead fire team performs the same actions previously described.

g. If the lead Soldier finds that he is nearly out of ammunition before reaching a corner or intersection, he announces, "*Ammo,*" and then--

(1) The lead Soldier stops and moves against one side of the trench, ready to let the rest of the team pass. He continues to aim his weapon down the trench in the direction of movement.

(2) The next Soldier ensures that he has a full magazine, moves abreast of the lead Soldier, taps him, and announces, "*Taking the lead.*"

(3) The lead Soldier acknowledges that he is handing over the lead by shouting, "*Okay.*" Positions rotate and the squad continues forward.

h. The trailing fire team secures intersections and marks the route within the trench as the squad moves forward. The trailing fire team leader ensures that follow-on squads relieve his buddy teams to maintain security.

i. The squad leader reports the progress of the clearing operation. The base-of-fire element must be able to identify the location of the lead fire team in the trench at all times.

10. The platoon leader rotates squads to keep the Soldiers fresh and to maintain the momentum of the assault.

11. The PSG calls forward ammunition resupply and organizes teams to move it forward into the trench.

12. The base-of-fire element ensures that all friendly forces move into the trench only through the designated entry point to avoid fratricide.

13. The platoon leader reports to the company commander that the trench line is secured, or he is no longer able to continue clearing. If trench line is secured, then platoon leader directs platoon FO to develop a fire plan to support the defense of the platoon position.

DANGER

FIRE TEAMS: MAINTAIN SUFFICIENT INTERVALS TO AVOID ENGAGEMENT BY THE SAME ENEMY FIRES.

Figure 6-22. ENTER A TRENCH (SQUAD)

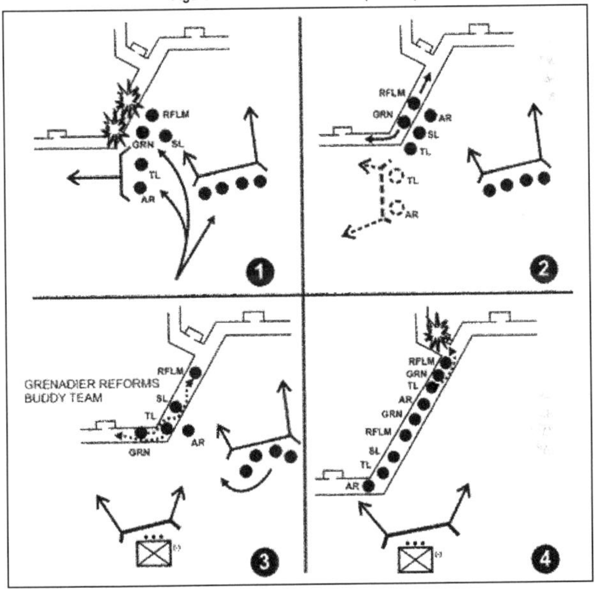

Figure 6-23. CLEAR A TRENCH LINE (PLATOON)

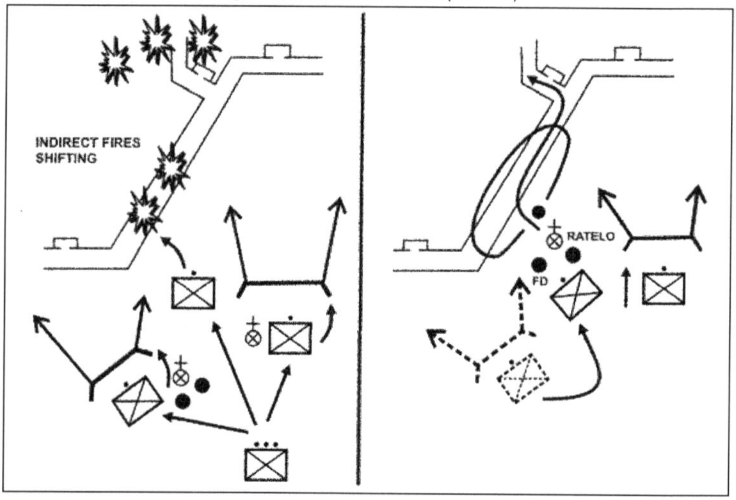

BATTLE DRILL 7

CONDUCT INITIAL BREACH OF A MINED WIRE OBSTACLE (PLATOON)

SITUATION: The platoon is operating as part of a larger force. The lead squad identifies a wire obstacle reinforced with mines that cannot be bypassed and enemy positions on the far side of the obstacle.

REQUIRED ACTIONS: See Figure 6-24 and Figure 6-25.

1. The squad in contact reacts to contact.
2. The platoon gains suppressive fires. The following then occurs.
   a. The squad in contact establishes a base-of-fire position.
   b. The platoon leader, his RTO, platoon FO, and the squad leader of the next squad with one machine gun team move forward to link up with the squad leader of the squad in contact.
3. The platoon leader determines that he can maneuver by identifying—
   - The obstacle and enemy positions.
   - The size of the enemy force engaging the squad, for example, the number of enemy automatic weapons, the presence of any vehicles, and the employment of indirect fires are indicators of enemy strength.
   - A breach point.
   - A covered and concealed route to the breach point.
4. The platoon leader directs the squad in contact to support the movement of another squad to the breach point. He indicates the base-of-fire position and the route to it, the enemy position to be suppressed, and the breach point and route the rest of the platoon will take to it. The platoon leader also clears the location, task, purpose, and method of conducting the fire mission with the platoon FO.

5. On the platoon leader's signal, the base-of-fire squad—
   - Destroys or suppresses enemy weapons that are firing effectively against the platoon.
   - Obscures the enemy position with smoke.
   - Continues suppressive fires at the lowest possible level.

6. The platoon leader designates one squad as the breach squad and the remaining squad as the assault squad once the breach has been made. The assault squad may add its fires to the base-of-fire squad. Normally, it follows the covered and concealed route of the breach squad and assaults through immediately after the breach is made.

7. The base-of-fire squad moves to the breach point and establishes a base of fire.

8. The PSG moves forward to the base-of-fire squad with the second machine gun and assumes control of the squad.

9. The platoon leader leads the breach and assault squads along the covered and concealed route.

10. The platoon FO calls for and adjusts indirect fires as directed by the platoon leader. The platoon leader in conjunction with the platoon FO maintains accurate battle tracking of all friendly elements to facilitate quick clearance of fires.

11. The breach squad executes actions to breach the obstacle.
   a. The squad leader directs one fire team to support the movement of the other fire team to the breach point.
   b. The squad leader identifies the breach point.
   c. The base-of-fire element continues to provide suppressive fires and isolates the breach point.
   d. The breaching fire team, with the squad leader, moves to the breach point using the covered and concealed route. From there, the following takes place:

(1) The squad leader and breaching fire team leader employ smoke grenades to obscure the breach point. The platoon base-of-fire element shifts direct fires away from the breach point and continues to suppress key enemy positions. The platoon FO ceases indirect fires or shifts them beyond the obstacle.

(2) The breaching fire team leader positions himself and the automatic rifleman on one flank of the breach point to provide close-in security.

(3) The grenadier and rifleman of the breaching fire team probe for mines and cut the wire obstacle, marking their path as they proceed. If available, a bangalore is preferred.

(4) Once the obstacle has been breached, the breaching fire team leader and the automatic rifleman move to the far side of the obstacle and take up covered and concealed positions with the rifleman and grenadier. The team leader signals to the squad leader when they are in position and ready to support.

(5) The squad leader signals the base-of-fire team leader to move his fire team up and through the breach. He then moves through the obstacle and joins the breaching fire team, leaving the grenadier (or anti-armor specialist) and rifleman of the supporting fire team on the near side of the breach to guide the rest of the platoon through.

(6) Using the same covered and concealed route as the breaching fire team, the base-of-fire team moves through the breach and takes up covered and concealed positions on the far side.

12. The squad leader reports the situation to the platoon leader and posts guides at the breach point.

13. The platoon leader leads the assault squad through the breach in the obstacle and positions them beyond the breach to support the movement of the remainder of the platoon or assaults the enemy position covering the obstacle. Platoon leader directs the platoon FO to shift indirect fires off the enemy position.

14. The breaching squad continues to widen the breach to allow vehicles to pass through.

15. The platoon leader reports the situation to the company commander, and then he directs his base-of-fire squad to move up and through the obstacle. The platoon leader appoints Soldiers to guide the company through the breach point.

16. The PSG brings the remaining elements forward and through the breach on the platoon leader's command.

17. The company follows up the success of the platoon as it conducts the breach and continues the assault against the enemy positions.

Figure 6-24. CONDUCT INITIAL BREACH OF A MINED WIRE OBSTACLE (SQUAD)

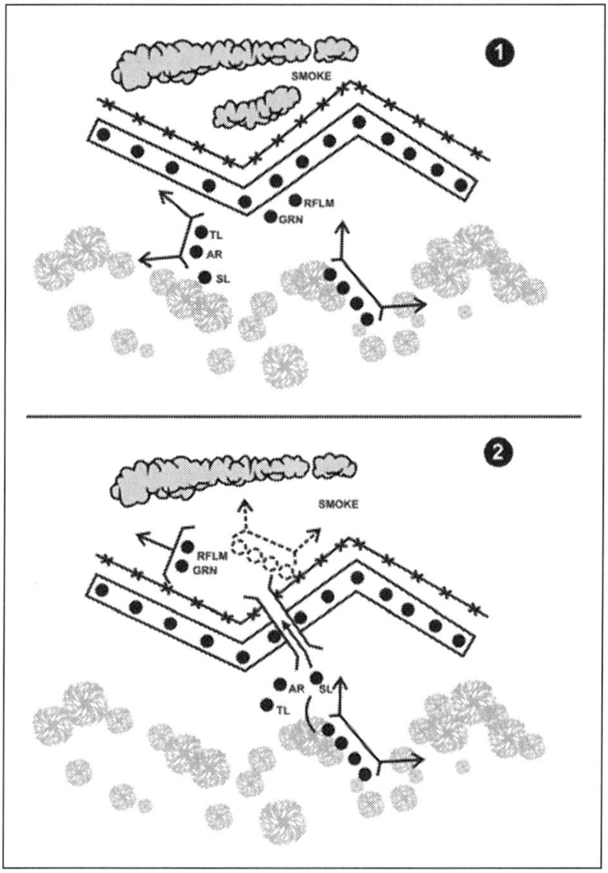

Figure 6-25. CONDUCT INITIAL BREACH OF A MINED WIRE OBSTACLE (PLATOON)

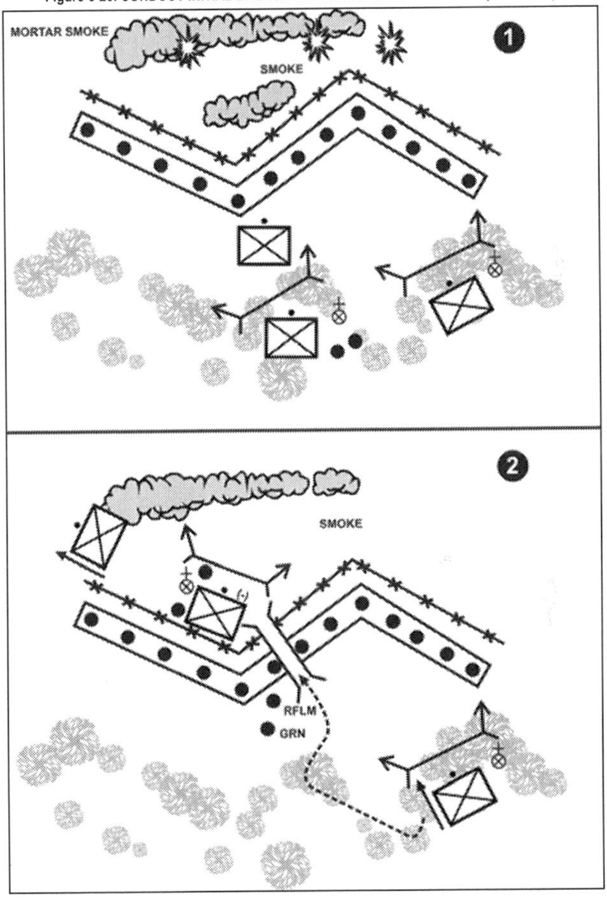

BATTLE DRILL 8
REACT TO INDIRECT FIRE

SITUATION: A platoon or squad while moving or at the halt (not dug in) receives indirect fire.

REQUIRED ACTIONS: NA

Any Soldier announces, "*Incoming.*" Then, if the platoon/squad is moving, the following takes place:

1. The platoon/squad leader gives direction and distance for the platoon/squad to move to a rally point by ordering direction and distance, for example, "*Four o'clock, three hundred meters.*"

2. Platoon/squad members move rapidly along the direction and distance to the rally point.

3. At the rally point, the leader immediately accounts for personnel and equipment, and forms the platoon/squad for a move to an alternate position.

4. The senior leader submits a SITREP to higher headquarters.

5. If the platoon/squad is halted (not dug in) or is preparing to move because they hear incoming artillery, the following takes place:

   a. The platoon/squad leader gives direction and distance for the platoon/squad to move to a rally point by ordering direction and distance, for example, "*Four o'clock, three hundred meters.*"

   b. Platoon/squad members secure all mission-essential equipment and ammunition and move rapidly along the direction and distance to the rally point.

   c. At the rally point, the leader immediately accounts for personnel and equipment, and forms the platoon/squad for a move to an alternate position.

   d. The senior leader submits a SITREP to higher headquarters.

Note: If platoon/squad members are in defensive (dug in) positions, then members will remain in those positions if appropriate. Senior leader submits SITREP.

NOTES

Chapter 7
COMMUNICATIONS

The basic requirement of combat communications is to provide rapid, reliable, and secure interchange of information. Communications are a vital aspect in successful mission accomplishment. The information in this chapter helps the Ranger squad/platoon maintain effective communications and correct any radio antenna problems.

Section I. INTRODUCTION

This section discusses military frequency modulated radios and automated net control devices (ANCDs).

7-1. **MILITARY FREQUENCY MODULATED RADIOS**. Military frequency modulated (FM) radios are composed of a receiver and a transmitter. Together, the receiver-transmitter has many capabilities and features that enable you to perform your mission more effectively. The radio can operate in Single Channel mode or frequency hopping mode. The radio has about 2,320 SC Channels and includes voice and digital communication. The operating voltage for the manpack radio is 13.5 volts from the primary battery. The range of the manpack radio is 5-10 KM on Hi power. This range is based upon line of sight and is derived from averages achieved under normal atmospheric and weather conditions. Ranges depend upon location, sighting, weather, and surrounding noise level, among other factors. the automated net control device (ANCD) is an essential tool. It replaces the signal operating instructions booklet. The ANCD has many capabilities. The main purpose of the ANCD for your mission is to obtain signal operating instructions electronically. The following discusses assembly, operation, and troubleshooting for the AN/PRC-119; provides step-by-step instructions to obtain SOI from an ANCD; and explains how to repair broken antennas:

a. **Manpack Radio Assembly**. To assemble a manpack radio you must first check and install a battery.
   (1) Inspect the battery box for dirt or damage.
   (2) Stand RT on front panel guards.
   (3) Check battery life condition (you will be using the rechargeable BB-390 batteries).
   (4) Place battery in box.
   (5) Close battery cover, and secure using latches.
   (6) Return radio to upright position.
   (7) If used battery was installed, enter the battery life condition into radio by performing the following:
      (a) Set **FCTN** to **LD**.
      (b) Press **BAT**, then **CLR**.
      (c) Enter number recorded on side of battery.
      (d) Press **STO**.
      (e) Set **FCTN** to **SQ ON**.

b. **Antenna**.
   (1) Inspect whip antenna connector on antenna and on radio for damage.
   (2) Screw whip antenna into base.
   (3) Hand tighten.
   (4) Carefully mate antenna base with RT ANT connector.
   (5) Hand tighten.
   (6) Position antenna as needed by bending goose neck.

Note: Keep antenna straight, if possible. If the antenna is bent to a horizontal position, it may be necessary to turn the radio in order to receive and transmit messages.

c. **Handset**.
   (1) Inspect the handset for damage.
   (2) Push handset on **AUD/DATA** and twist clockwise to lock in place.

d. **Field Pack**.
   (1) Place radio transmitter (RT) in field pack with antenna on the left shoulder.
   (2) Fold top flap of field over RT and secure flap to field pack using straps and buckles.

e. **Presets**.
   (1) Set **CHAN** to **1**.
   (2) Set **MODE** to **SC**.
   (3) Set **RF PWR** to **HI**.

(4) Set **VOL** to mid range.
(5) Set **DIM** full clockwise.
(6) Set **FCTN** to **LD**.
(7) Set **DATA RATE** to **OFF**.

f. **Single Channel Loading Frequencies**.
(1) Obtain Ranger SOI.
(2) Set **FCTN** to **LD**.
(3) Set **MODE** to **SC**.
(4) Set **CHAN** to **MAN**, **Cue**, or desired channel (**1** to **6**) where frequency is to be stored.
(5) Press **FREQ** (display will show "**00000**," or the frequency the RT is currently set to).
(6) Press **CLR** (display will show five lines).
(7) Enter the number of the new frequency.
(8) If you make a mistake with a number, press **CLR**.
(9) Press **STO** (display will blink).
(10) Set **FCTN** to **SQ ON**.

g. **Clearing of Frequencies**.
(1) Set **MODE** to **SC**.
(2) Set **CHAN** to **MAN**, **Cue** or the channel whose frequency is to be cleared.
(3) Press **FREQ**.
(4) Press **CLR**.
(5) Press **Load**; then press **STO**.
(6) Set **FCTN** to **SQ ON**.

h. **Scanning of Multiple Frequencies**.
(1) Load all desired frequencies using "Single Channel Loading Frequencies" instructions (see 7-2h).
(2) Set **CHAN** to **CUE**.
(3) Set **SC** to **FH**.
(4) Set **FCTN** to **SQ ON**.
(5) Press **STO** (display will say **SCAN**).
(6) Press **8**. You will now be able to scan more than one frequency.

7-2. **AUTOMATED NET CONTROL DEVICE**. The RTO retrieves all necessary COMSEC and information from the ANCD. To retrieve the SOI from an ANCD:

a. Press the **ON** button on the ANCD keypad.
b. Press the letter lock button to unlock the keys on the ANCD.
c. Press **Main Menu** key.
d. Once you press the main menu key, this is what should show on the screen.

Appl Date Time Setup
Util Bit.

e. Using the arrow keys, scroll over to **Appl** (as shown above) and press the enter button on the keypad.

f. Once you have entered into the **Appl** (applications) menu using the arrow keys scroll over to **SOI** and press the enter button on the key pad. For example--

RDS SOI Radio.
(Appl) <--*(Indicates that you are in the applications menu).*

SOI menu will look like this:
Qref Group Net Sufx Pyro
Tmpd Set C/S Find Memo

g. Using the arrow keys scroll over to **Set** and press the enter button on the keypad.
h. Press enter on choose once in the **Set** menu.

> *EXAMPLE:*
> SELECT:
> **Choose Send Receive**.

i. Once in the **Choose** menu press the enter key on set **1-5** or **6-10**.

j. After you choose your selection out of the **Set** menu, the ANCD will automatically return back to the **SOI** menu.

k. Using the keypad scroll over to **Tmpd** and enter the number out of the set you need for that day. After this operation, the ANCD will automatically return back to the **SOI** menu.

l. Using the arrow keys on the keypad scroll over to **Net** and press the enter button on the keypad. The Net menu should look like the following example:

> *EXAMPLE:*
> T05 <--(time period) 1 <--(platoon) / A <--(company) /A3Q <--(prefix).
> C 60.000.

Note: When in **Net** menu, this is where you will obtain your frequency and the first part of your call sign.

m. After you obtain the information you need out of **Net**, press the **Abort** button on the keypad. This will return you to the **SOI** menu.

n. Using the keypad, scroll over to **Sufx** and press the **Enter** button.

o. Once you are in the **Sufx** menu, use the up and down arrow keys on the ANCD to locate your two-digit designator. These two digits will go at the end of your prefix; your prefix plus the two digits equal your call sign.

p. Once you are done, press the **OFF** button on the ANCD to end operation.

Section II. ANTENNAS

This section discusses repair techniques, construction and adjustment, field-expedient antennas, antenna length and orientation, and improvement of marginal communications.

7-3. **REPAIR**. Antennas are sometimes broken or damaged, causing either a communications failure or poor communications. If a spare antenna is available, the damaged antenna is replaced. When there is no spare, the Ranger squad/platoon may have to construct an emergency antenna. The following paragraphs contain suggestions for repairing antennas and antenna supports and the construction and adjustment of emergency antennas.

> **DANGER**
>
> **SERIOUS INJURY OR DEATH CAN RESULT FROM CONTACT WITH THE RADIATING ANTENNA OF A MEDIUM-POWER OR HIGH-POWER TRANSMITTER. TURN THE TRANSMITTER OFF WHILE ADJUSTING THE ANTENNA.**

a. **Whip Antennas**. When a whip antenna is broken into two sections, the part of the antenna that is broken off can be connected to the part attached to the base by joining the sections.

(1) Use the method shown in A, Figure 7-1 when both parts of the broken whip are available and usable.

(2) Use the method in B, Figure 7-1, when the part of the whip that was broken off is lost, or when the whip is so badly damaged that it cannot be used.

(3) To restore the antenna to its original length, a piece of wire is added that is nearly the same length as the missing part of the whip. The pole support is then lashed securely to both sections of the antenna. The two antenna sections are cleaned thoroughly to ensure good contact before connecting them to the pole support. If possible, the connections are soldered.

b. **Wire Antennas**. Emergency repair of a wire antenna may involve the repair or replacement of the wire used as the antenna or transmission line; or the repair or replacement of the assembly used to support the antenna.

(1) When one or more wires of an antenna are broken, the antenna can be repaired by reconnecting the broken wires. To do this, lower the antenna to the ground, clean the ends of the wires, and twist the wires together. Whenever possible, solder the connection.

(2) If the antenna is damaged beyond repair, construct a new one. Make sure that the lengths of the wires on the substitute antenna match the length of the original.

(3) Antenna supports may also require repair or replacement. A substitute item may be used in place of a damaged support and, if properly insulated, can be of any material of adequate strength. If the radiating element is not properly insulated, field antennas may be shorted to ground and be ineffective. Many commonly found items can be used as field-expedient insulators.

(4) The best of these items are plastic or glass to include plastic spoons, buttons, bottle necks, and plastic bags. Though less effective than plastic or glass but still better than no insulator at all are wood and rope. The radiating element--the actual antenna wire-should touch only the antenna terminal and should be physically separated from all other objects, other than the supporting insulator. (Figure 7-2 shows how to make emergency insulators.)

Figure 7-1. USE OF BROKEN ANTENNA

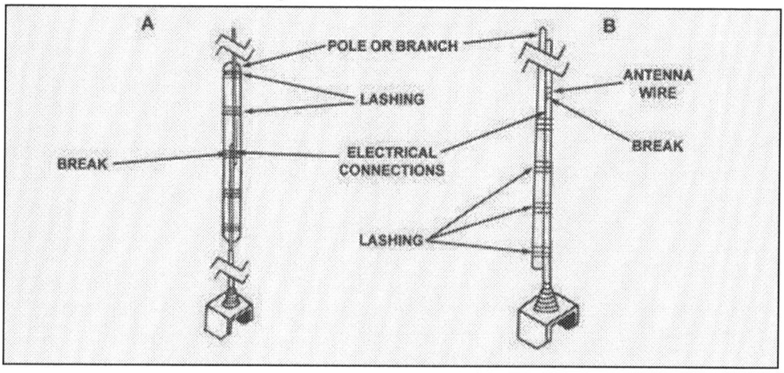

Figure 7-2. METHODS OF CONSTRUCTING EMERGENCY INSULATORS

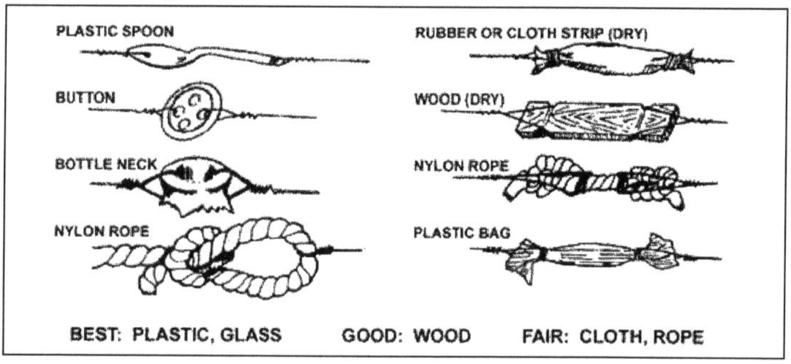

7-4. **CONSTRUCTION AND ADJUSTMENT**. Ranger squad/platoons may use the following methods to construct and adjust antennas:

a. **Construction**. The best kinds of wire for antennas are copper and aluminum. In an emergency, however, Rangers use any type of wire that is available.

(1) The exact length of most antennas is critical. The emergency antenna should be the same length as the antenna it replaces.

(2) Antennas supported by trees can usually survive heavy wind storms if the trunk of a tree or a strong branch is used as a support. To keep the antenna taut and to prevent it from breaking or stretching as the trees sway, the Ranger attaches a spring or old inner tube to one end of the antenna. Another technique is to pass a rope through a pulley or eyehook. The rope is attached to the end of the antenna and loaded with a heavyweight to keep the antenna tightly drawn.

(3) Guidelines used to hold antenna supports are made of rope or wire. To ensure the guidelines will not affect the operation of the antenna, the Ranger cuts the wire into several short lengths and connects the pieces with insulators.

b. **Adjustment**. An improvised antenna may change the performance of a radio set. The following methods can be used to determine if the antenna is operating properly:

(1) A distant station may be used to test the antenna. If the signal received from this station is strong, the antenna is operating satisfactorily. If the signal is weak, the Ranger adjusts the height and length of the antenna and the transmission line to receive the strongest signal at a given setting on the volume control of the receiver. This is the best method of tuning an antenna when transmission is dangerous or forbidden.

(2) In some radio sets, the Ranger uses the transmitter to adjust the antenna. First, he sets the controls of the transmitter to normal; then, he tunes the system by adjusting the antenna height, the antenna length, and the transmission line length to obtain the best transmission output.

7-5. **FIELD-EXPEDIENT OMNIDIRECTIONAL ANTENNAS**. Vertical antennas are omnidirectional. The omnidirectional antenna transmits and receives equally well in all directions.

a. **Vertical Antennas**. Most tactical antennas are vertical. For example, the man-pack portable radio uses a vertical whip and so do the vehicular radios in tactical vehicles. A vertical antenna can be made by using a metal pipe or rod of the correct length, held erect by means of guidelines. The lower end of the antenna should be insulated from the ground by placing it on a large block of wood or other insulating material. A vertical antenna may also be a wire supported by a tree or a wooden pole (Figure 7-3).

Figure 7-3. VERTICAL ANTENNA MADE FROM A WIRE SUPPORTED BY A TREE OR WOODEN POLE

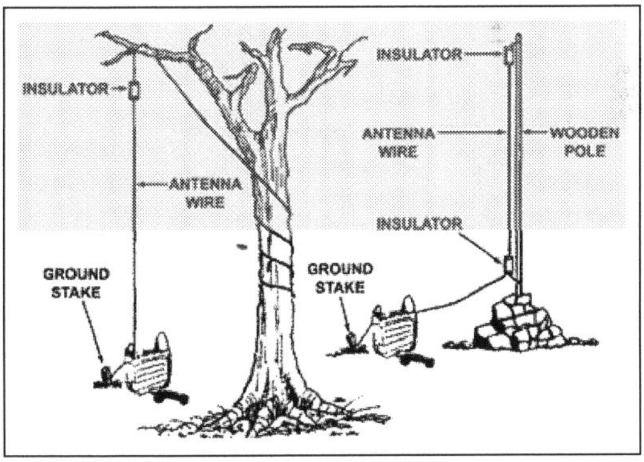

b. **Modification of Connection on Shorter Vertical Antennas**. For short vertical antennas, the pole may be used without guidelines (if properly supported at the base). If the length of the vertical mast is not long enough to support the wire upright, it may be necessary to modify the connection at the top of the antenna (Figure 7-4; see FM 24-18).

Figure 7-4. MODIFIED CONNECTION AT TOP OF ANTENNA

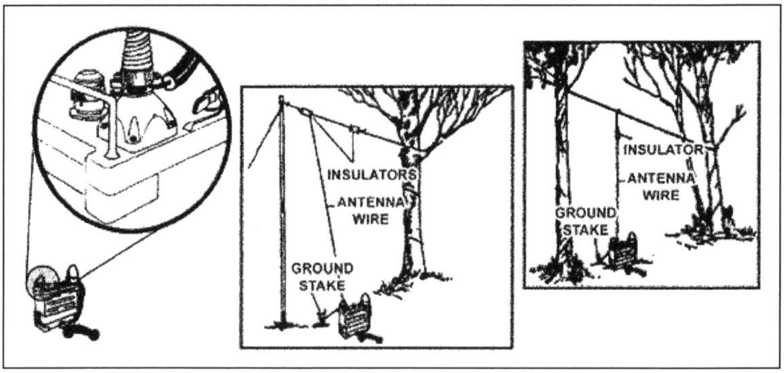

c. **End-Fed, Half-Wave Antenna**. An emergency, end-fed half-wave antenna (Figure 7-5) can be constructed from available materials such as field wire, rope, and wooden insulators. The electrical length of this antenna is measured from the antenna terminal on the radio set to the far end of the antenna. The best performance can be obtained by constructing the antenna longer than necessary and then shortening it, as required, until the best results are obtained. The ground terminal of the radio set should be connected to a good earth ground for this antenna to function efficiently.

Figure 7-5. END-FED, HALF-WAVE ANTENNA

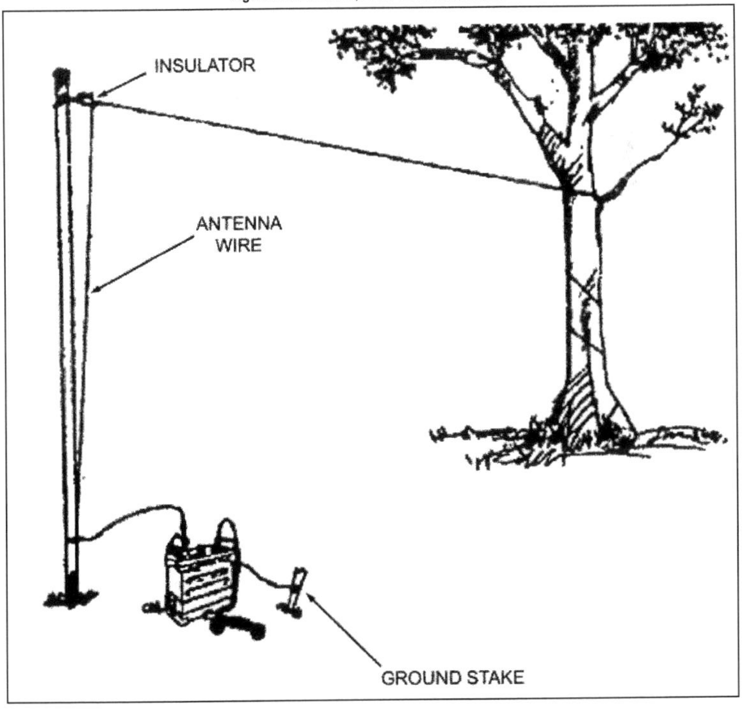

d. **Center-Fed, Doublet Antenna**. The center-fed doublet is a half-wave antenna consisting of two quarter wavelength sections on each side of the center (Figure 7-6). Doublet antennas are directional broadside to their length, which makes the vertical doublet antenna omnidirectional. This is because the radiation pattern is doughnut-shaped and bidirectional.

Figure 7-6. CENTER-FED, DOUBLET ANTENNA

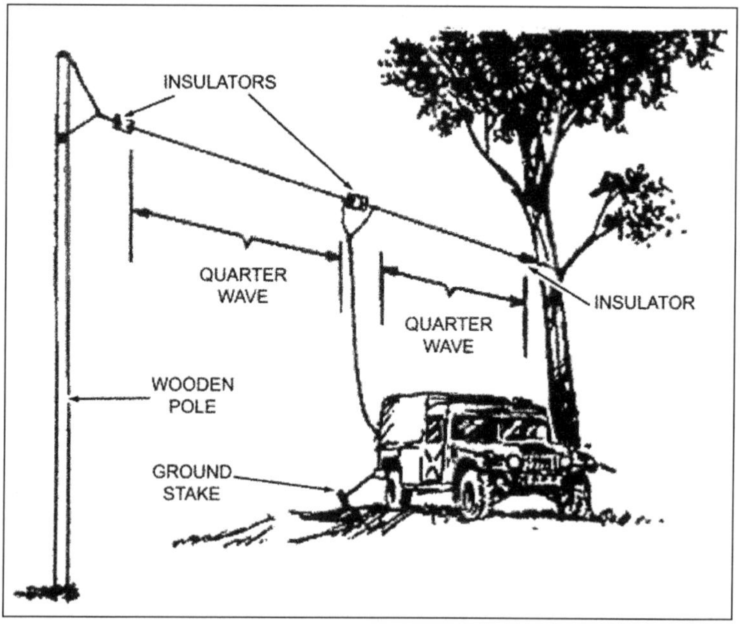

(1) Compute the length of a half-wave antenna by using the formula previously discussed. Cut the wires as close as possible to the correct length; this is very important.

(2) Uses transmission line for conducting electrical energy from one point to another and for transferring the output of a transmitter to an antenna. Although it is possible to connect an antenna directly to a transmitter, the antenna is usually located some distance away.

(3) Support center-fed half-wave FM antennas entirely with pieces of wood. (A horizontal antenna of this type is shown in A, Figure 7-7 and a vertical antenna in B, Figure 7-7.) Rotate these antennas to any position to obtain the best performance.

(a) If the antenna is erected vertically, bring out the transmission line horizontally from the antenna for a distance equal to at least one-half of the antenna's length before it is dropped down to the radio set.

Figure 7-7. CENTER-FED ANTENNAS SUPPORTED WITH WOOD

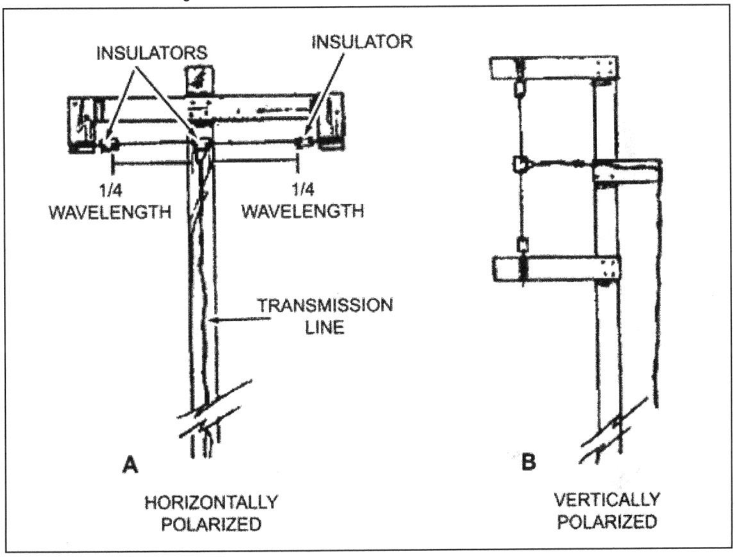

(b) The half-wave antenna is used with FM radios (Figure 7-8). It is effective in heavily wooded areas to increase the range of portable radios. Connect the top guidelines to a limb or pass it over the limb and connect it to the tree trunk or a stake.

Figure 7-8. HALF-WAVE ANTENNA USED WITH FM RADIO

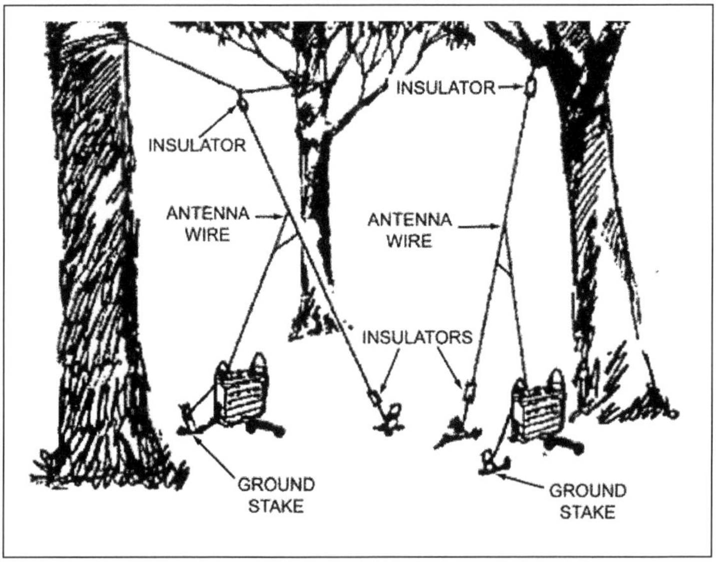

c. **Expedient 292-Type Antenna**. The expedient 292-type antenna was developed for use in the jungle and, if used properly, can increase the team's ability to communicate. In its entirety, the antenna is bulky, heavy, and not acceptable for sniper team operations. The team can, however, carry only the mast head and antenna sections, mounting these on wood poles or hanging them from trees; or, the team can make a complete expedient 292-type antenna (Figure 7-9), using WD-1, wire, and other readily available material. The team can also use almost any plastic, glass, or rubber objects for insulators. Dry wood is acceptable when nothing else is available. (See Figure 7-2 for types of insulators that may be used.) The following describes how to make this antenna:

(1) Use the quick-reference table (Table 7-2) to determine the length of the elements (one radiating and three ground planes) for the frequency that will be used. Cut these elements (A, Figure 7-9) from WD-1 field wire (or similar wire). Cut spacing sticks (B, Figure 7-9) the same length. Place the ends of the sticks together to form a triangle and tie the ends with wire, tape, or rope. Attach an insulator to each corner. Attach a ground-plane wire to each insulator. Bring the other ends of the ground-plane wires together, attach them to an insulator (C, Figure 7-9), and tie securely. Strip about 3 inches of insulation from each wire and twist them together (D, Figure 7-9).

(2) Tie one end of the radiating element wire to the other side of insulator C and the other end to another insulator (D, Figure 7-9). Strip about 3 inches of insulation from the radiating element at insulator C.

(3) Cut enough WD-1 field wire to reach from the proposed location of the antenna to the radio set. Keep this line as short as possible, because excess length reduces the efficiency of the system. Tie a knot at each end to identify it as the *hot* lead. Remove insulation from the *hot* wire and tie it to the radiating element wire at insulator C. Remove insulation from the other wire and attach it to the bare ground-plane element wires at insulator C. Tape all connections and do not allow the radiating element wire to touch the ground-plane wires.

(4) Attach a rope to the insulator on the free end of the radiating element and toss the rope over the branches of a tree. Pull the antenna as high as possible, keeping the lead-in routed down through the triangle. Secure the rope to hold the antenna in place.

(5) At the radio set, remove about 1 inch of insulation from the *hot* lead and about 3 inches of insulation from the other wire. Attach the *hot* line to the antenna terminal (doublet connector, if so labeled). Attach the other wire to the metal case-the handle, for example. Be sure both connections are tight or secure.

(6) Set up correct frequency, turn on the set, and proceed with communications.

Figure 7-9. COMPLETE EXPEDIENT 292-TYPE ANTENNA

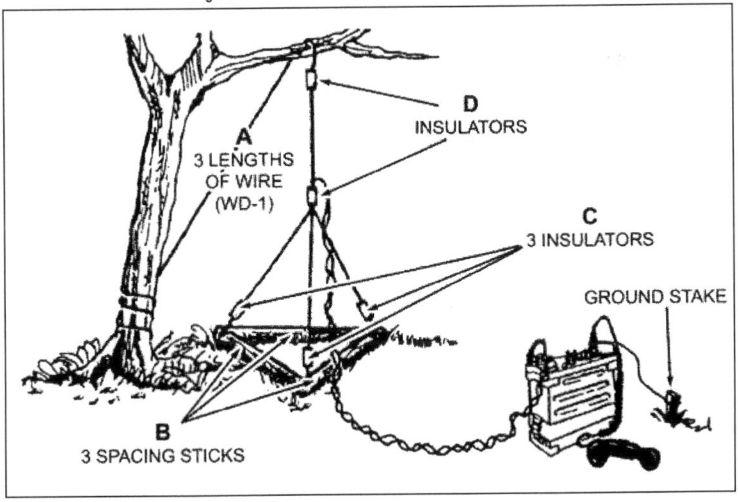

Table 7-2. QUICK REFERENCE TABLE

OPERATING FREQUENCY IN MHz	ELEMENT LENGTH (RADIATING ELEMENTS AND GROUND-PLANE ELEMENTS)	
30	2.38 m	(7 ft 10 in)
32	2.23 m	(7 ft 4 in)
34	2.1 m	(6 ft 11 in)
36	1.98 m	(6 ft 6 in)
38	1.87 m	(6 ft 2 in)
40	1.78 m	(5 ft 10 in)
43	1.66 m	(5 ft 5 in)
46	1.55 m	(5 ft 1 in)
49	1.46 m	(4 ft 9 in)
52	1.37 m	(4 ft 6 in)
55	1.3 m	(4 ft 3 in)
58	1.23 m	(4 ft 0 in)
61	1.17 m	(3 ft 10 in)
64	1.12 m	(3 ft 8 in)
68	1.05 m	(3 ft 5 in)
72	0.99 m	(3 ft 3 in)
76	0.94 m	(3 ft 1 in)

7-6. **FIELD-EXPEDIENT DIRECTIONAL ANTENNAS**. The vertical half-rhombic antenna (Figure 7-10) and the long-wire antenna (Figure 7-11) are two field-expedient directional antennas. These antennas consist of a single wire, preferably two or more wavelengths long, supported on poles at a height of 3 to 7 meters (10 to 20 feet) above the ground. The antennas will, however, operate satisfactorily as low as 1 meter (about 3 feet) above the ground--the radiation pattern is directional. The antennas are used mainly for either transmitting or receiving high-frequency signals.

Figure 7-10. VERTICAL HALF-RHOMBIC ANTENNA

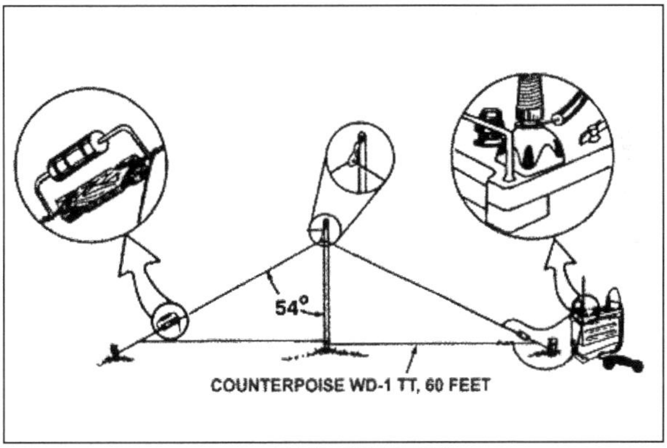

Figure 7-11. LONG-WIRE ANTENNA

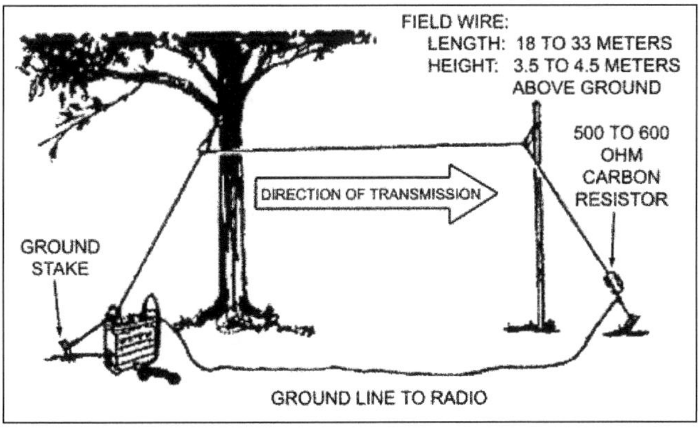

a. The vee antenna (Figure 7-12) is another field-expedient directional antenna. It consists of two wires forming a V with the open area of the "V" pointing toward the desired direction of transmission or reception. To make construction easier, the legs should slope downward from the apex of the "V." This is called a sloping vee antenna (Figure 7-13). The angle between the legs varies with the length of the legs to achieve maximum performance. (Use Table 7-1 to determine the angle and the length of the legs.)

Figure 7-12. VEE ANTENNA

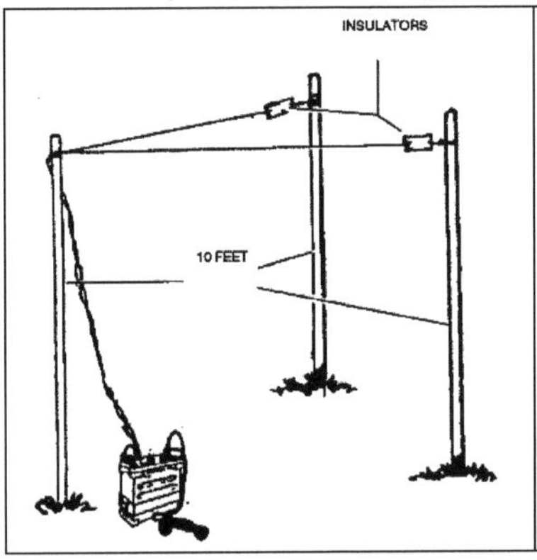

Figure 7-13. SLOPING VEE ANTENNA

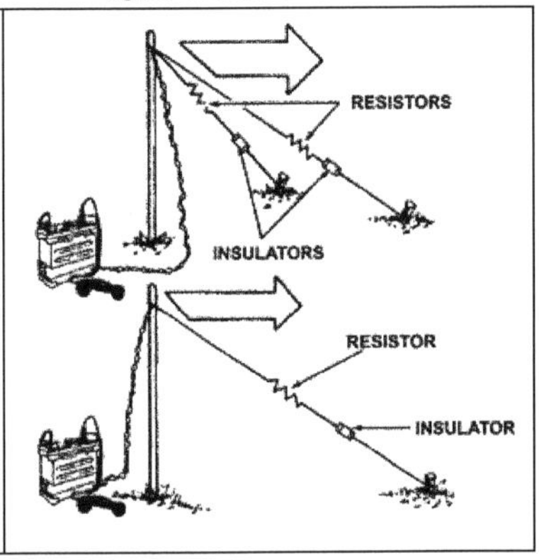

b. When the antenna is used with more than one frequency or wavelength, use an apex angle that is midway between the extreme angles determined by the chart. To make the antenna radiate in only one direction, add noninductive terminating resistors from the end of each leg (not at the apex) to ground. (See TM 11-666.)

7-7. **ANTENNA LENGTH**. The length of an antenna must be considered in two ways: both a physical and an electrical length. These two lengths are never the same. The reduced velocity of the wave on the antenna and a capacitive effect (known as end *effect*) make the antenna seem longer electrically than it is physically. The contributing factors are the ratio of the diameter of the antenna to its length and the capacitive effect of terminal equipment, such as insulators and clamps, used to support the antenna.

a. To calculate the physical length of an antenna, use a correction of 0.95 for frequencies between 3.0 and 50.0 MHz The figures given below are for a half-wave antenna.

b. Use the following formula to calculate the length of a long-wire antenna (one wavelength or longer) for harmonic operation:

EXAMPLE:

If N = the number of half-wavelengths in the total length of the antenna = 3,
and if the frequency in MHz = 7, then--

$$\text{Length (meters)} = \frac{150(N-0.05)}{\text{Frequency in MHz}} = \frac{150(3-.05)}{7} =$$

$$\frac{150 \times 2.95}{7} = \frac{442.50 = 63.2 \text{ meters}}{7}$$

7-8. **ANTENNA ORIENTATION**. If the azimuth of the radio path is not provided, the azimuth should be determined by the best available means. The accuracy required in determining the azimuth of the path depends on the radiation pattern of the directional antenna. In transportable operation, the rhombic and V antennas may have such a narrow beam as to require great accuracy in azimuth determination. The antenna should be erected for the correct azimuth. Great accuracy is not required in erecting broad-beam antennas. Unless a line of known azimuth is available at the site, the direction of the path is best determined by a magnetic compass.

7-9. **IMPROVEMENT OF MARGINAL COMMUNICATIONS**. Under certain situations, it may not be feasible to orient directional antennas to the correct azimuth of the desired radio path. As a result, marginal communications may suffer. To improve marginal communications, the following procedure can be used:

a. Check, tighten, and tape cable couplings and connections.
b. Return all transmitters and receivers in the circuit.
c. Ensure antennas are adjusted for the proper operating frequency.
d. Change the heights of antennas.
e. Move the antenna a short distance away and in different locations from its original location.

NOTES

Chapter 8
ARMY AVIATION

Army aviation and infantry units can be fully integrated with other members of the combined arms team to form powerful and flexible air assault task forces that can project combat power throughout the entire depth, width, and breadth of the modern battlefield with little regard for terrain barriers. These combat operations are deliberate, precisely planned, and vigorously executed. They are assigned to strike the enemy when and where he is most vulnerable (See page 2-28, *Air Movement Annex* and page 2-37, *Army Aviation Coordination Checklist*).

8-1. **REVERSE PLANNING SEQUENCE**. Successful air assault execution is based on a careful analysis of METT-TC and detailed, precise reverse planning. Five basic plans that comprise the reverse planning sequence are developed for each air assault operation. The battalion is the lowest level that has sufficient personnel to plan, coordinate, and control air assault operations. When company size or lower operations are conducted, the bulk of the planning takes place at battalion or higher headquarters. They are:

a. **Ground Tactical Plan**. The foundation of a successful air assault operation is the commander's ground tactical plan. All additional plans must support this plan. The plan specifies actions in the objective area to ultimately accomplish the mission and address subsequent operations.

b. **Landing Plan**. The landing plan must support the ground tactical plan. This plan outlines a sequence of events that allows elements to move into the area of operations, and ensures that units arrive at designated locations at prescribed times prepared to execute the ground tactical plan.

c. **Air Movement Plan**. The air movement plan is based on the ground tactical and landing plans. It specifies the schedule and provides instructions for air movement of troops, equipment, and supplies from PZs to LZs.

d. **Loading Plan**. The loading plan is based on the air movement plan. It ensures that troops, equipment, and supplies are loaded on the correct aircraft. Unit integrity is maintained when aircraft loads are planned. Cross-loading may be necessary to ensure survivability of command and control assets, and that the mix of weapons arriving at the LZ is ready to fight.

e. **Staging Plan**. The staging plan is based on the loading plan and prescribes the arrival time of ground units (troops, equipment and supplies) at the PZ in the order of movement

8-2. **SELECTION AND MARKING OF PICKUP AND LANDING ZONES**

a. Small unit leaders should consider the following when selecting a PZ/LZ:

(1) ***Size***. Minimal circular landing point separation from other aircraft and obstacles is needed:

- OH-58D – 25 meters.
- UH-1, AH-1 – 35 meters.
- UH-60, AH-64 – 50 meters.
- Cargo helicopters – 80 meters.

(2) ***Surface Conditions***. Avoid potential hazards such as sand, blowing dust, snow, tree stumps, or large rocks.

(3) ***Ground Slope***.

- 0% - 6 % -- land upslope.
- 7% - 15% -- land sideslope.
- over 15% -- no touchdown (aircraft may hover).

(4) ***Obstacles***. An obstacle clearance ratio of 10 to 1 is used in planning approach and departure of the PZ and LZ, for example, a 10-foot tall tree requires 100 feet of horizontal distance for approach or departure. Obstacles will be marked with a red chemlight at night, or red panels in daytime. Avoid using markings if the enemy would see them.

(5) ***Approach/Departure***. Approach and departure are made into the wind and along the long axis of the PZ/LZ.

(6) ***Loads***. The greater the load, the larger the PZ/LZ must be to accommodate the insertion or extraction.

b. **Marking of PZs and LZs**.

(1) ***Day***. A ground guide will mark the PZ or LZ for the lead aircraft by holding an M16/M4 rifle over his head, by displaying a folded VS-17 panel chest high, or by other coordinated and identifiable means.

(2) ***Night***. The code letter "Y" (inverted "Y") is used to mark the landing point of the lead aircraft at night (Figure 8-1). Chemical lights or "beanbag" lights are used to maintain light discipline. A swinging chemlight may also be used to mark the landing point.

Figure 8-1. INVERTED "Y"

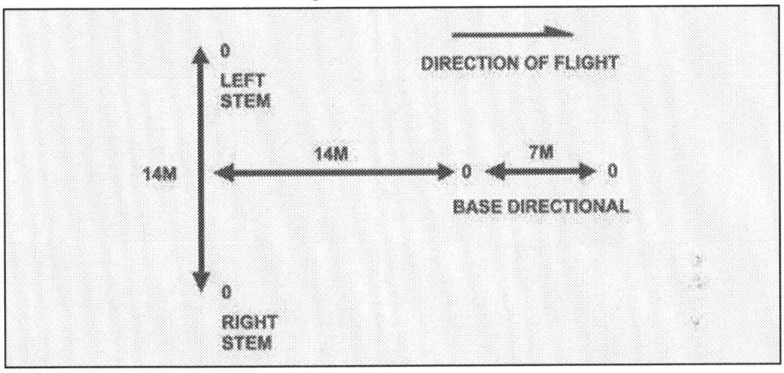

8-3. **AIR ASSAULT FORMATIONS**. Aircraft supporting an operation may use any of the following PZ/LZ configurations which are prescribed by the air assault task force (AATF) commander working with the air mission commander (AMC):

a. ***Heavy Left or Right***. Requires a relatively long, wide landing area; presents difficulty in pre-positioning loads; restricts suppressive fire by inboard gunners; provides firepower to front and flank (Figure 8-2).

Figure 8-2. HEAVY LEFT/HEAVY RIGHT

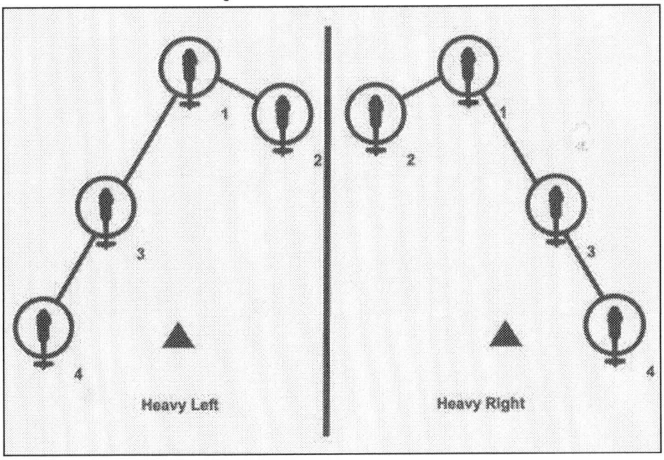

b. **Diamond**. Allows rapid deployment for all-round security; requires small landing area; presents some difficulty in pre-positioning loads; restricts suppressive fire of inboard gunners (Figure 8-3).

Figure 8-3. DIAMOND

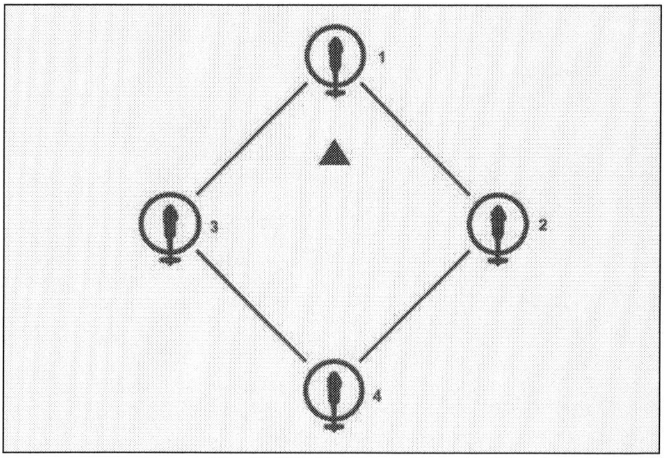

c. **Vee**. Requires a relatively small landing area; allows rapid deployment of forces to the front; restricts suppressive fire of inboard gunners; presents some difficulty in pre-positioning loads (Figure 8-4).

Figure 8-4. VEE

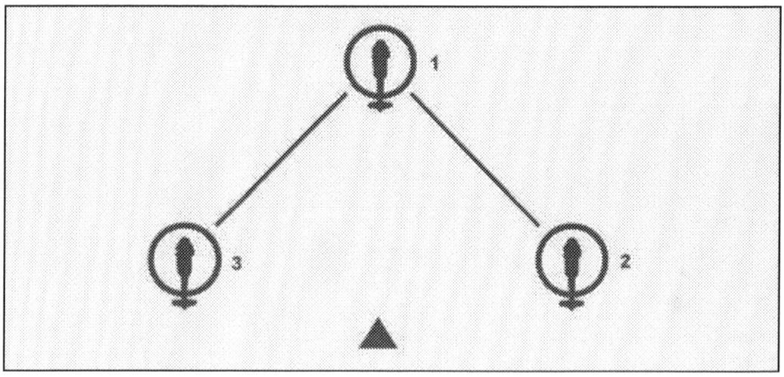

d. **Echelon Left or Right**. Requires a relatively long, wide landing area; presents some difficulty in pre-positioning loads; allows rapid deployment of forces to the flank; allows unrestricted suppressive fire by gunners (Figure 8-5).

Figure 8-5. ECHELON LEFT/ECHELON RIGHT

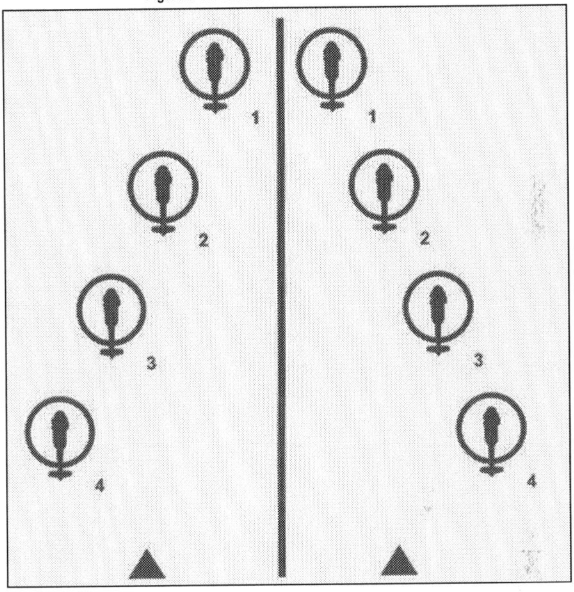

e. **Trail**. Requires a relatively small landing area; allows rapid deployment of forces to the flank; simplifies pre-positioning loads; allows unrestricted suppressive fire by gunners (Figure 8-6).

Figure 8-6. TRAIL

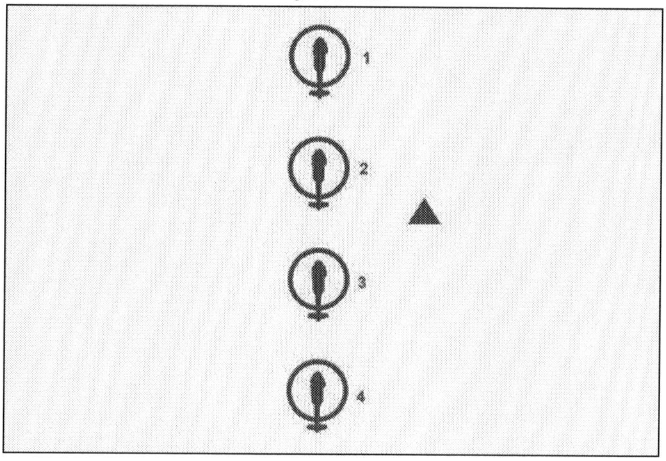

f. **Staggered Trail Left or Right**. Requires a relatively long, wide landing area; simplifies pre-positioning loads; allows rapid deployment for all round security; gunners' suppressive fire restricted somewhat (Figure 8-7).

Figure 8-7. STAGGERED TRAIL LEFT/STAGGERED TRAIL RIGHT

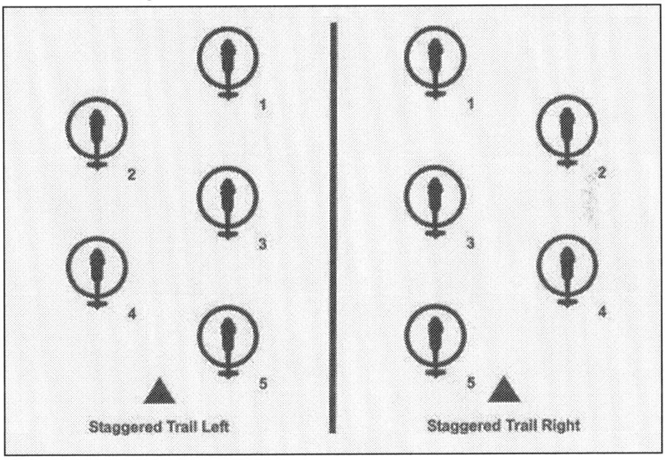

8-4. **PICKUP ZONE OPERATIONS**. Prior to arrival of aircraft, the PZ is secured, PZ control party is positioned, and the troops and equipment are positioned in platoon/squad assembly areas. In occupying a patrol/squad assembly area, the patrol/squad leader does the following. Figure 8-8 shows an example of a large, one-sided PZ. Figures 8-8 through 8-12 show loading/unloading procedures and techniques:

- Maintain all-around security of the assembly area.
- Maintain communications.
- Organize personnel and equipment into chalks and loads.
- Conduct safety briefing and equipment check of troops.

Figure 8-8. LARGE, ONE SIDED PZ

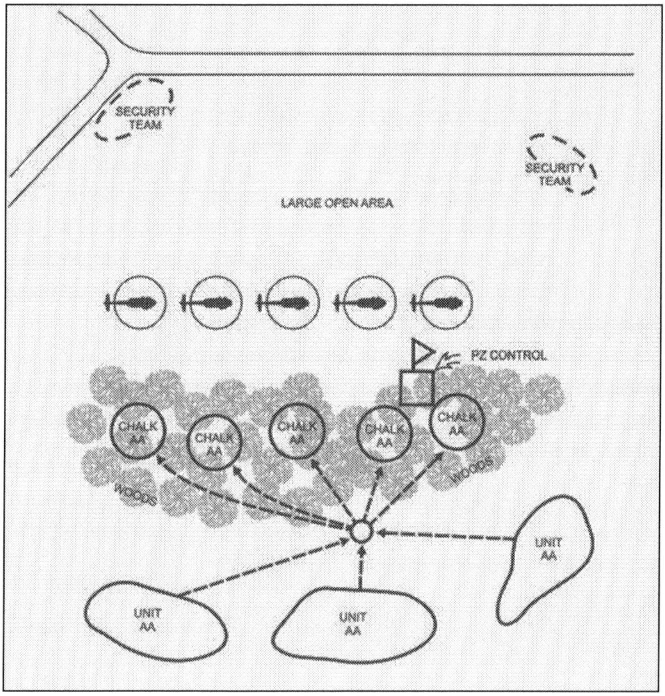

Figure 8-9. UH-60 LOADING SEQUENCE

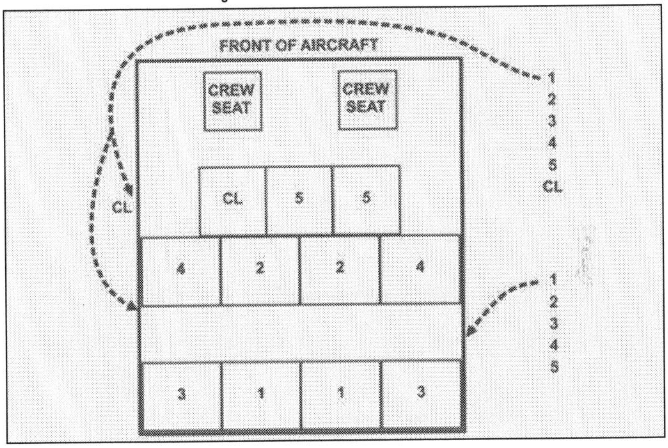

Figure 8-10. UNLOADING SEQUENCE

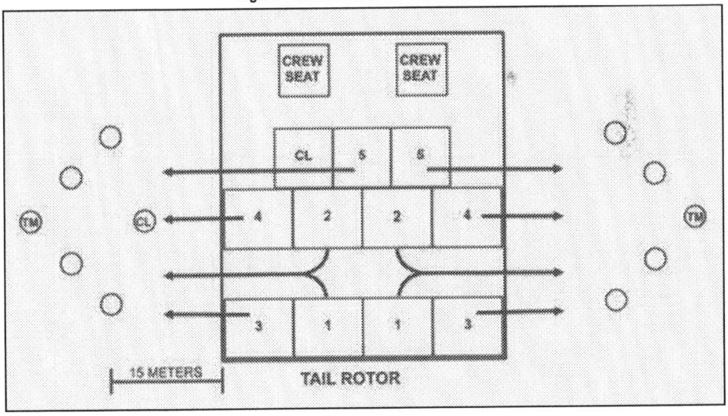

Figure 8-11. TACTICAL LOADING SEQUENCE

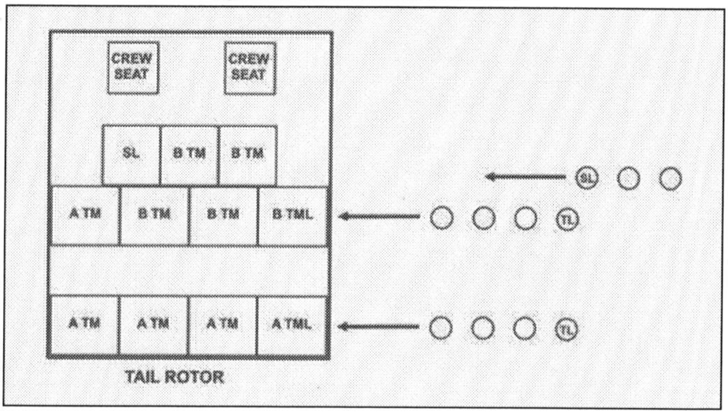

Figure 8-12. TACTICAL UNLOADING USING DOOR NEAREST COVER, CONCEALMENT

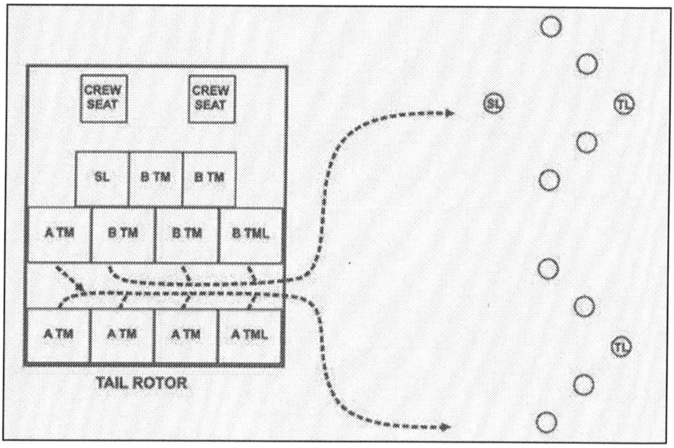

8-5. **SAFETY**. Safety is the primary concern of all leaders when operating in/around aircraft. The inclusion of aircraft into Ranger operations carries an inherent high-risk factor. The following guidelines are to be considered.

a. Approach the aircraft from 90 degrees to 45 degrees off the nose.

b. Weapons with blank firing adapters are pointed muzzles up.

c. Weapons loaded with live ammunition are muzzles down.

d. Wear the ballistic helmet.

e. When possible, conduct an air crew safety brief with all personnel. At a minimum, cover loading/off-loading, emergencies, and egress procedures.

f. Leaders need to carry a manifest and turn a copy into higher.

8-6. **CLOSE COMBAT ATTACK AVIATION**. Close combat attack is defined as a hasty or deliberate attack in support of units engaged in close combat. During CCA, armed helicopters engage enemy units with direct fire that impacts near friendly forces. Targets may range from a few hundred meters to a few thousand meters. Close combat attack is coordinated and directed by a team, platoon or company level ground unit soldiers using standardized CCA procedures in unit SOP's.

a. **Aircraft Capabilities and Limitations**.

(1) AH-64D (Apache):
- Air speed (knots): 164 max, 120 Cruise combat radius: 200 KM
- 30 mm chain gun: 1200 rds, 3,500 m range
- 2.75" rockets: 19 per pod (76), 3-5 km range
- Hellfire Missile: 8 per side (16), 5-8 km range

(2) Capabilities.
- MOBILITY - AO can cover corps or division area
- SPEED - 100-120 knots day/80-100 Knots night
- VERSATILITY - scout-weapons teams vs. pure attack
- LETHALITY - Attack bn can engage 288 targets
- VIDEO RECONNAISSANCE - Provide near real time intelligence

(3) ***Limitations***.
- Threat ID with FLIR
- Low ceilings (Clouds) <500' AGL degrades hellfire capability
- Combat service support consume large amounts of class III, V, and IX

(4) **OH-58D (Kiowa)**.
- Air speed (knots): 125 max, 100 cruise combat radius: 120 KM
- .50 cal MG: 500 rds, 2,000 m range
- 2.75" rockets: 7 per pod (14), 3-5 km range
- Hellfire: 2 per side (4), 5-8 km range
- Stinger: 2 per side (4), 4 km range

(5) ***Capabilities***.
- Mast-mounted sight
- Ability to designate targets while remaining masked
- Thermal imaging system (day/night)
- Laser designator / Aim1 laser
- Video image crosslink (VIXL)
- Moving map display
- Video

(6) ***Limitations***.
- Power limited
- Infrared crossover
- Battlefield obscurants
- Low ceilings - Hellfire
- Remote designation constraints
- Instrument meteorological conditions

c. **CCA Call-for-Fire Format**.

(1) IP/BP/ABF or friendly location________
- Grid/Lat Long

(2) Known point
- Terrain feature

(3) HDG to TGT ________ (magnetic)
- Specify from IP/BP/ABF or friendly location

(4) DST to TGT _________ (m) (Specify from IP/BP/ABF or friendly location.)

(5) TGT elevation________ (feet, mean sea level)

(6) TGT Description ________________

(7) TGT location ____________
- Grid
- Known point
- Terrain Feature

(8) Type TGT Mark _________ (day/night)

(9) location of friendly _________
- Omit if previously given
- Grid/Lat Long
- Known point
- Terrain feature

(10) Egress direction__________
- Cardinal direction not to over fly threats

d. **Marking Method Chart**. Table 8-1 shows this chart.

Table 8-1. MARKING METHOD CHART

Method	D	N	NVG	NVS	Friendly	Target
Smoke	X				X	X
Smoke IR	X		X		X	X
Signal Mirror	X				X	
IR Laser			X		X	X
Tracer	X	X	X			X
Glint Tape				X	X	
CID	X				X	
Strobe		X	X		X	
IR Strobe			X		X	
IR Panel				X	X	
Chem Light		X	X		X	
AN/PAQ-4			X			X
VS-17 Panel	X				X	
Spot Light		X	X		X	
MRE Heater				X	X	
AN/PEQ-2			X			X
Hydra 70 Illum	X	X	X	X		X
Ref FM 3-04.111 Appendix Q for more detail						

e. **Risk-Estimate Distances**. REDs for CCA munitions are shown in Table 8-2.

Table 8-2. RISK ESTIMATE DISTANCES FOR CCA MUNITIONS

WEAPON	*DANGER CLOSE*	*MAX EFF RANGE)*
50 cal	150 meters	2,000 meters
30 mm	150 meters	3,500 meters
2.75 rocket	175 meters	8,000 meters
Hellfire Missile	500 meters	8,000 meters
Danger Close = Risk-Estimate Distances for 0.1% (1/10th of a percent) probability of incapacitation (PI) per FM 3-09.32.		

Chapter 9
WATERBORNE OPERATIONS

This chapter discusses rope bridges, poncho rafts, and watercraft.

9-1. **ROPE BRIDGE TEAM.** The Ranger patrol seldom has ready-made bridges, so they must know how to employ field-expedient stream crossing techniques.

a. **Personnel**.

(1) # 1 man................Lead safety swimmer and far side lifeguard.

(2) # 2 man................Rope puller, swims water obstacle pulling 150-foot rope, ties off rope on far side anchor point.

(3) # 3 man................Near side lifeguard is the last man to cross water obstacle.

(4) # 4 man................Bridge team commander (BTC; most knowledgeable person on team).

(5) # 5/6 men............Rope tighteners.

b. **One-Rope Bridge (Wet Crossing).** Special equipment follows:

- Two snaplinks per piece of heavy equipment.
- Two snaplinks for every 120 feet of rope.
- One 14-foot utility rope per person (Swimmer safety line).
- Two snaplinks per person.
- One waterproof bag per RTO.
- Two B-7 life preservers.
- Three flotation work vests.
- Two 150-foot nylon ropes.

c. **Planning**. A stream crossing annex is prepared in conjunction with the unit's operation order. Special organization is accomplished at this time. For a platoon size patrol, a squad is normally given the task of providing the bridge team. The squad leader designates the most technically proficient soldier in the squad as the bridge team commander (BTC).

d. **Rehearsals and Inspections**.

(1) Rehearse the entire stream crossing, emphasizing--

- Security and actions on enemy contact.
- Actual construction of the rope bridge within the 8-minute time standard on dry land.
- Individual preparation.
- Order of crossing.
- All signals and control measures.
- Reorganization.

(2) Conduct rehearsals as realistically as possible.

(3) Ensure personnel are proficient in the mechanics of a stream crossing operation.

(4) Inspect for equipment completeness, correct rigging and preparation, personnel knowledge and understanding of the operation.

(5) Actions of the # 4 man (BTC) during the preparation phase.

(a) Rehearse the bridge team.

(b) Account for all equipment in the bridge kit.

(c) Ensure 150-foot rope is coiled.

e. **Execution Phase**.

(1) Steps for the establishment and conduct of bridge stream crossing:

(a) Leader halts short of the river, local security is established, and a recon is conducted of the area for the presence of the enemy and for crossing site suitability/ necessity. He directs the BTC to construct the bridge.

(b) Security is established up and downstream while unit leader briefs BTC on anchor points. The unit leader counts individuals across.

(c) The bridge team begins to establish the rope bridge while unit members begin individual preparation.

(d) The BTC is responsible for construction of one-rope bridge and selection of the near side anchor point as well as the far side anchor point if visibility permits. He will tie a swimmer's safety line to anchor himself into the bridge. Tying an end of the line bowline around the waist secured with an overhand knot, and on the free running end, an end of line bowline with an overhand knot. A snaplink will be placed on the loop portion of the end of line bowline, which will extend no further than arms length away from the soldier (standard waterborne uniform). This will ensure that the soldier is never more than one arms length away from the rope bridge should he lose his grip.

(e) Noise and light discipline are enforced and security is maintained.

(f) Individual soldiers put a snaplink in their end of the line bowline and the front sight assembly of every M-4/M16 or M203. M240 gunners put a snaplink through the front sight assembly and rear swivel of their M240 MG. RTOs (and others with heavy rucksacks) place an additional snaplink on their rucksack frame, top center.

(2) The responsibilities of the bridge team while constructing the rope bridge are as follows:

(a) # 1 man (Lead safety swimmer and far side lifeguard). He grounds his rucksack (with snaplink through top of frame) to the rear of the near side anchor point. He wears equipment in the following order (body out). He carries a safety line to assume duties of far side lifeguard:

-- Waterborne uniform (top button buttoned, pants unbloused)
-- B7 life preserver or engineer work vest
-- LBV
-- Weapon (across the back)

(b) # 1 man enters the water upstream from the # 2 man and stays an arms length away from the # 2 man on the upstream side. He identifies the far side anchor point upon exiting the water and once the # 2 man has exited the water moves to his far side lifeguard position downstream of the rope bridge with knotted safety line on wrist, LBV LBV/weapon grounded, and work vest held in throwing hand. He continues to wear the B7.

(c) # 2 man (rope puller) in waterborne uniform (same as # 1 man) wears his equipment in the following order. He grounds his rucksack (with snaplink through top of frame) to the rear of the near side anchor point. His duties are to swim across the water obstacle pulling the rope. He ties off the rope on the anchor point identified by the # 1 man with a round turn and two half hitches with a quick release. The direction of the round turn is the same direction as the flow of water (current) to facilitate exit off the rope bridge:

- Work vest.
- LBV.
- Weapon (across the back).
- Swimmer safety line.

(d) # 3 man (near side lifeguard) in the same waterborne uniform as the far side lifeguard. The # 3 man positions himself on the downstream side of the bridge before the # 1/2 men enter the water. He (# 1 man) grounds his rucksack (with snaplink through top of frame) on rear of near-side anchor point. After the PSG crosses and verifies the headcount, the # 1 man unties the quick release at the near side anchor point. The # 3 man reties his safety line into an Australian rappel seat, hooks the end of line bowline into his snaplink, and connects his snaplink to the one on the end-of-line bowline . # 3 man is the last pulled across the water obstacle. Before crossing the water obstacle, he dons his equipment in the following order:

- B7.
- Work vest.
- LBV.
- Weapon.

(e) # 4 man (BTC). He is in the standard waterborne uniform with LBV and sling rope tied in safety line (round the waist bowline with end of line bowline no more than one arms length). He is responsible for construction of rope bridge and organization of bridge team. He is also responsible for back feeding the rope and tying end of line bowlines. He designates the near side anchor point, ties the Figure Eight slip of the transport tightening system, and hooking all personnel to the rope bridge. He ensures that the transport tightening knot is on the upstream side of the rope bridge. He ensures that all individuals are in the waterborne uniform, hooked into the rope facing the current with the safety line routed through the trailing shoulder of the individual's LBV and rucksack. He ensures that the weapon is hooked onto the rope. He controls the flow of traffic on the bridge. He is responsible for crossing with the # 1 man's rucksack. He is generally the next to the last man to cross (follows PSG who is keeping head count).

(f) # 5 and # 6 men (rope tighteners) wear the waterborne uniform with LBV and safety line. They tighten the transport tightening knot. They also take the rucksacks of the # 2 and # 3 men across. Once they reach the far side, # 5 and # 6 men pull the last man (# 3 man) across.

(g) The rucksacks of # 1/2/3 men are transported across by # 4/5/6 men. The rucksacks of the # 1/2/3 men are hooked into the rope by the snaplink through the top of the frames and the # 4/5/6 men pull them across. The weapons of # 4/5/6 men are attached between the # 4/5/6 men and the rucksack that they are pulling across the bridge.

(h) BTC rehearses the bridge team during the planning sessions and directs construction and emplacement. The unit leader selects the crossing site which complements the tactical plan.

- # 3 man positions himself downstream of crossing site.
- # 1 man enters water upstream of # 2. He stays one arm's length from the # 2 man and is prepared to render any assistance to the # 2 man. Both swim in conjunction upstream to compensate for the current. BTC feeds rope out of rucksack positioned on the downstream side of near side anchor point.

(i) The # 1 man exits and identifies the far side anchor point (if BTC cannot identify it for the # 2 man). # 2 man exits on the upstream side of the far side anchor point. The rope is now routed to facilitate movement onto and off the bridge.

(j) Radios and heavy equipment are waterproofed and rigged. All individuals don waterborne uniform and tie safety lines. PSG moves to anchor point and maintains accountability through headcount.

(k) The # 2 man signals the BTC that the rope is temporarily attached to the far side anchor point, and the BTC pulls out excess slack and ties the transport tightening system using a Figure Eight slip. The BTC signals the # 2 man to pull the knot twelve to fifteen feet from the near side anchor point. After this is done the # 2 two man ties round turns 18 to 24 inches off the water with the remaining rope and secures the rope, to itself, with a snaplink. The # 2 man signals the BTC and the pulling team (# 4/5/6) tightens the bridge, pulling the transport tightening system as close as possible to the near side anchor point.

(l) The # 1 man moves downstream and assumes his duties as the far side lifeguard. The bridge team commander will tie off the rope with a round turn and two half hitches around the near side anchor point. The BTC will place himself on the upstream side of the bridge (facing downstream). He begins to hook individuals into the rope inspecting them for safety.

Note: Any Ranger identified as a weak swimmer will cross with no other personnel on the rope bridge. The weak swimmer crosses individually to allow the near and far side lifeguards to focus their attention exclusively on the weak swimmer and not be distracted by other personnel crossing the bridge.

(m) The # 2 man moves to the upstream side of the rope bridge, assists personnel off the rope on the far side, and maintains head count. The # 5 and # 6 men cross with the # 1 and # 2 men's rucksacks..

(n) The BTC maintains the flow of traffic ensuring that no more than three men are on the bridge at any one time (one hooking up, one near the center, and one being unhooked). Once the PSG has accounted for everyone on the near side, he withdraws left and right (L/R) security and sends them across. PSG follows security across. The # 3 man hooks the BTC (with # 3's ruck) onto the rope. Once the BTC has crossed, the # 3 man unhooks near side anchor point and the BTC unties far side anchor point. The # 3 man ties an Australian rappel seat with snaplink to the front, hooks onto the snaplink that is in the end of the line bowline on the 120-foot rope, and then signals the # 4/5/6 men to take in slack. The # 3 man extends arms in front of his head, slightly upstream to fend off debris and is pulled across by the # 4/5/6. All individuals (except # 1/2/3 and RTOs) wear rucksacks across. The # 4/5/6 men hook the rucksacks of the # 1/2/3 men onto the bridge by the snaplink. All the men cross facing upstream.

(o) Once the far side headcount, weapons and equipment are verified (between PSG and # 5 man), personnel reorganize, and then they continue the mission.

(p) Personnel with heavy equipment:

M240 All major groups are tied together with 1/4-inch cord. An anchor line bowline runs through the rear swivel, down the left side of gun. Tie a round turn through the trigger guard. Route the cord down the right side and tie off two half hitches around the forearm assembly with a round turn and two half hitches through the front sight posts. The remainder of the working end is tied off with an end of the rope bowline approximately one foot from the front sight post large enough to place leading hand through. The M240 is secured to the bridge by snaplinks on the front sight post and rear swivel. The M240 is pulled across by the trailing arm of the M240 gunner.

AN-PRC/II9s These are waterproofed before crossing a one-rope bridge. Once far side FM commo is established, the near side RTO breaks down and waterproofs his radio, and prepares to cross the bridge. He puts a snaplink in the top center of the rucksack frame (same as for # 1/2/3 men). The BTC will hook the rucksack to the rope.

Note: Using two snaplinks binds the load on the rope. Adjust arm straps all the way out. RTO pulls radio across the rope bridge.

9-2. **PONCHO RAFT.** Normally a poncho raft is constructed to cross rivers and streams when the current is not swift. A poncho raft is especially useful when the unit is still dry and the platoon leader desires to keep the individuals equipment dry.

a. **Equipment Requirements**.

(1) Two serviceable ponchos.

(2) Two weapons (poles can be used in lieu of weapons).

(3) Two rucksacks per team.

(4) 10 feet of utility cord per team.

(5) One sling rope per team.

b. **Conditions**. Poncho rafts are used to cross water obstacles when any or all of the following conditions are found:

(1) The water obstacle is too wide for 150-foot rope.

(2) No sufficient near or far shore anchor points are available to allow rope bridge construction.

(3) Under no circumstances will poncho rafts be used as a means to cross a water obstacle if an unusually swift current is present.

c. **Choosing a Crossing Site**. Before a crossing site is used, a thorough reconnaissance of the immediate area is made. Analyzing the situation using METT-TC, the patrol leader chooses a crossing site that offers as much cover and concealment as possible and has entrance and exit points that are as shallow as possible. For speed of movement, it is best to choose a crossing site that has near and far shore banks that are easily traversed by an individual Ranger.

d. **Execution Phase**. Steps for the construction of a poncho raft:

(1) Pair off the unit/patrol in order to have the necessary equipment.

(2) Tie off the hood of one poncho and lay out on the ground with the hood up.

(3) Weapons are then placed in the center of the poncho, approximately 18 inches apart, muzzle to butt.

(4) Next, rucksacks and LBV are placed between the weapons with the two individuals placing their rucksacks as far apart as possible.

(5) The two will then start to undress (bottom to top), first with their boots, taking the laces completely out for subsequent use as tie downs if necessary).

(6) The boots are then placed over muzzle/butt of weapon toe in.

(7) Members continue to undress, folding each item neatly and placing it on top of their boots.

(8) Once all of the equipment is placed between the two weapons or poles, the poncho is snapped together. The snapped portion of the poncho is then lifted into the air and tightly rolled down to the equipment. Start at the center and work out to the end of the raft creating pigtails at the end. This is accomplished much easier if done by both soldiers together. The pigtailed ends are then folded in toward the center top of the raft and tied off with a single boot lace.

(9) The other poncho is then laid out on the ground with the hood up and the first poncho with equipment is placed in its center. The second poncho is then snapped, rolled and tied in the same manner as the first poncho. The third and fourth boot laces (or utility cord) are then tied around the raft approximately one foot from each end for added security. The poncho raft is now complete.

Note: The patrol leader must analyze the situation using METT-TC and make a decision on the uniform for crossing the water obstacle such as weapons inside the poncho raft or slung across the back, remaining dressed or stripped down with clothes inside raft.

9-3. **WATERCRAFT.** Use of inland and coastal waterways may add flexibility, surprise, and speed to tactical operations. Use of these waterways will also increase the load carrying capacity of normal dismounted units. Watercraft are employed in reconnaissance and assault operations.

a. **Inflation Method**. Inflatable with foot pumps using four separate valves located on the inside of the buoyancy tubes. Each of the four valves are used to section off the assault boat into eight separate airtight compartments. To pump air into the boat, turn all valves into the "orange" or "inflate" section of the valve. Once the assault boat is filled with air, turn all valves onto the "green" or "navigation" section. This will section the assault boat into eight separate compartments.

b. **Characteristics**.

- Overall length: 15 feet, 5 inches.
- Overall width: 6 feet, 3 inches.
- Weight: 265 pounds.
- Maximum payload: 2,756 pounds.
- Crew : 1 coxswain + either 10 paddlers or a 65 HP short-shaft outboard motor.

c. **Preparation**.

(1) ***Rubber Boat***.

- All rubber boats will have a twelve-foot bowline secured to the front starboard D-Ring. This rope will be tied with an anchor-line bowline, and the knot will be covered with 100-MPH tape.
- All rubber boats will have a fifteen foot center line tied to the rear floor D-Ring. The same procedure for securing the bowline will be used for the centerline.
- All rubber boats will be filled to 240 millibars of air, and checked to ensure that all valve caps are tight, and set in the NAVIGATE position.
- All rubber boats will be equipped with one foot pumps. These will be placed in the front pouches of the rubber boat. If no pouches are present, the foot pumps will be placed on the floor.
- All rubber boats will be inspected with the maintenance chart.

(2) ***Personnel and Equipment***.

- All personnel will wear work vest or kapoks (or another suitable positive flotation device).
- LBV is worn over the work vest, unbuckled at the waist.
- Individual weapon is slung across the back, muzzle pointed down and facing toward the inside of the boat.
- Crew served weapons, radios, ammunition and other bulky equipment is lashed securely to the boat to prevent loss if the boat should overturn. Machine guns with hot barrels are cooled prior to being lashed inside the boats.
- Radios and batteries are waterproofed.
- Pointed objects are padded to prevent puncturing the boat.

NOTES

d. **Positions.** Assign each individual a specific boat position (Figure 9-1).

e. **Duties**.

(1) Designate a commander for each boat, (normally coxswain)

(2) Designate a navigator (normally a leader within the platoon) - observer team as necessary.

(3) Crew is positioned as shown in Figure 9-2.

(4) Duties of the Coxswain.

(a) Responsible for control of the boat and actions of the crew.

(b) Supervises the loading, lashing and distribution of equipment.

(c) Maintains the course and speed of the boat.

(d) Gives all commands.

(5) # 2 paddler (long count) is responsible for setting the pace.

(6) # 1 paddler is the observer, stowing and using the bowline unless another observer is assigned.

f. **Embarkation and Debarkation Procedures**.

(1) When launching, the crew will maintain a firm grip on the boat until they are inside it: similarly, when beaching or debarking, they hold on to the boat until it is completely out of the water. Loading and unloading is done using the bow as the entrance and exit point.

(2) Keep a low center of mass when entering and existing the boat to avoid capsizing. Maintain 3 points of contact at all times.

(3) The long count is a method of loading and unloading by which the boat crew embarks or debarks individually over the bow of the boat. It is used at river banks, on loading ramps, and when deep water prohibits the use of the short count method.

(4) The short count is a method of loading or unloading by which the boat crew embarks or debarks in pairs over the sides of boat while the boat is in the water. It is used in shallow water allowing the boat to be quickly carried out of the water. The short count method of organization is primarily used during surf operations.

(5) Beaching the boat is a method of debarking the entire crew at once into shallow water and quickly carrying the boat out of the water.

g. **Commands**. Commands are issued by the coxswain to ensure the boat is transported over land and controlled in the water. All crew members learn and react immediately to all commands issued by the coxswain. T he various commands are as follows:

(1) "Short Count...count off," Crew counts off their position by pairs, such as 1,2,3,4,5 (passenger # 1, # 2, if applicable) coxswain.

(2) "Long Count...count off," Crew counts off the position by individual, such as 1,2,3,4,5,6,7,8,9,10, (Passenger # 1, # 2, if applicable), coxswain.

(3) "Boat Stations," Crew takes positions along side the boat.

(4) "High Carry...Move," (used for long distance move overland).

(a) On the preparatory command of "High Carry," the crew faces the rear of the boat and squats down grasping carrying handles with the inboard hand.

(b) On the command "Move," the crew swivels around, lifting the boat to the shoulders so that the crew is standing and facing to the front with the boat on their inboard shoulders.

(c) Coxswain guides the crew during movement.

(5) "Low Carry...Move," (used for short distance moves overland).

(a) On preparatory command of "Low Carry," the crew faces the front of the boat, bent at the waist, and grasps the carrying handles with their inboard hands.

(b) On the command of "Move," the crew stands up straight raising the boat approximately six to eight inches off the ground.

(c) Coxswain guides the crew during movement.

(6) "Lower the Boat----Move," Crew lowers the boat gently to the ground using the carrying handles.

(7) "Give Way Together," Crew paddles to front with # 2 setting the pace.

(8) "Hold," Entire crew keeps paddles straight downward motionless in the water stopping the boat.

(9) "Left side hold (Right)," Left crew holds, right continues with previous command.

(10) "Back paddle," Entire crew paddles backward propelling the boat to the rear.

Figure 9-1. BOAT POSITIONS

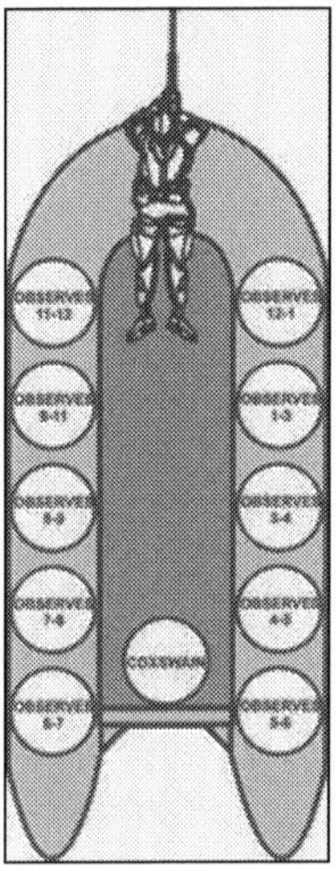

Figure 9-2. CREW POSITIONS, LONG COUNT AND SHORT COUNT

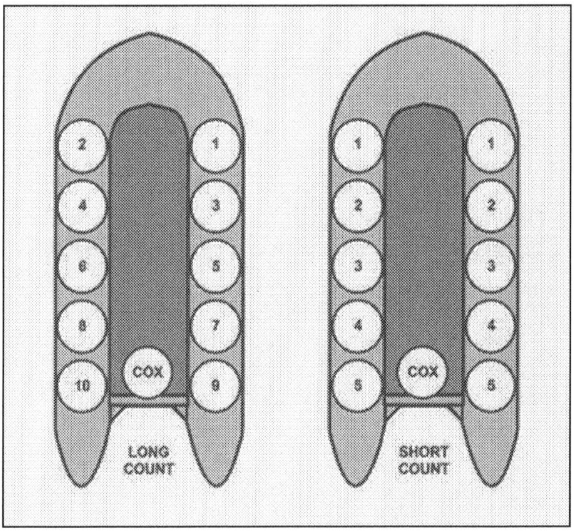

(11) "Back Paddle Left" (Right), Left crew back paddles causing the boat to turn left, right crew continues with previous command.

(12) "Rest Paddles," Crew members place paddles on their laps with blades outboard. This command may be given in pairs such as "# 1s, rest paddles").

h. **Securing of Landing Site.**

(1) If the landing site cannot be secured prior to the waterborne force landing, some form of early warning, such as scout swimmers, is considered. These personnel swim to shore from the assault boats and signal the boats to land. All signals and actions are rehearsed prior to the actual operation.

(2) If the patrol is going into an unsecured landing site, it can provide security by having a security boat land, reconnoiter the landing site and then signal to the remaining boats to land. This is the preferred technique.

(3) The landing site can be secured by force with all the assault boats landing simultaneously in a line formation. While this is the least desirable method of securing a landing site, it is rehearsed in the event the tactical situation requires its use.

(4) Arrival at the debarkation point.

(a) Unit members disembark according to leaders order (Figure 9-3).

(b) Local security is established.

(c) Leaders account for personnel and equipment.

(d) Unit continues movement.

- Soldiers pull security initially with work vest on.
- Coxswains and two men unlash and de-rig rucksacks.
- Soldiers return in buddy teams to secure rucksack and drop off work vest.
- Boats are camouflaged/cached if necessary prior to movement.

Figure 9-3. DEBARKATION

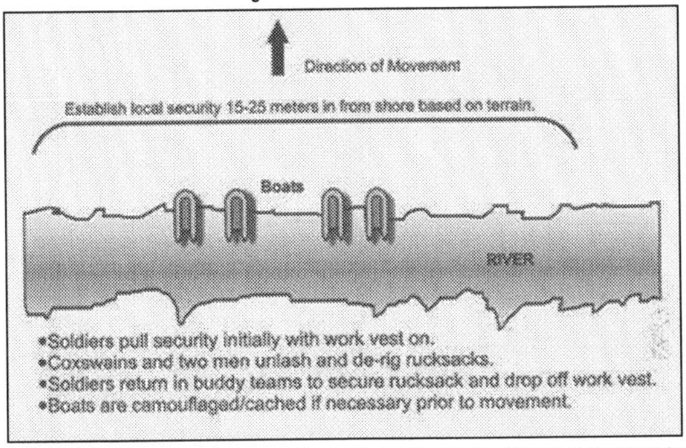

i. **Capsize Drill.** The following commands and procedures are used for capsize drills or to right an over-turned boat,

(1) ***PREPARE TO CAPSIZE***. This command alerts the crew and they raise paddles above their heads, with the blade pointed outward. Before capsizing, the coxswain will conduct a long count.

(2) ***PASS PADDLES***. All paddles are passed back and collected by the # 9/10 men.

(3) ***CAPSIZE THE BOAT***. All personnel slide into the water except the # 3/5/7 men. The # 1 man secures the bowline. The three men in the water grasp the capsize lines (ensuring the lines are routed under the safety lines) and stand on the buoyancy tubes opposite the capsize lines anchor points. The boat is then turned over by the # 3/5/7 men, who lean back and straighten their legs while pulling back on the capsize lines. As the boat lifts off the water, the # 4 man grasps the center carrying handle and rides the boat over. Once the boat is over, the # 4 man helps the # 3/7 men back onto the boat. The # 5 man holds onto the center carrying handle and turns the boat over the same way. The # 5 man rides the boat back over and helps the rest of the crew into the boat. As soon as the boat is capsized, the coxswain commands a long count to ensure that no one sank or was trapped under the boat. Every time the boat is turned over, he conducts another long count.

j. **River Movement.**

(1) ***Characteristics of River***.

(a) Know local conditions prior to embarking on river movement.

(b) A bend is a turn in the river course.

(c) A reach is a straight portion of river between two curves.

(d) A slough is a dead end branch from a river. They are normally quite deep and can be distinguished from the true river by their lack of current

(e) Dead water is a part of the river, due to erosion and changes in the river course that has no current. Dead water is characterized by excessive snags and debris.

(f) An island is usually a pear-shaped mass of land in the main current of the river. Upstream portions of islands usually catch debris and are avoided.

(g) The current in a narrow part of a reach is normally greater than in the wide portion.

(h) The current is greatest on the outside of a curve; sandbars and shallow water are found on the inside of the curve.

(i) Sandbars are located at those points where a tributary feeds into the main body of a river or stream.

(j) Because the # 1 and # 2 men are sitting on the front left and right sides of the boat, they observe for obstacles as the boat moves downriver. If either notices an obstacle on either side of the boat, he notifies the coxswain. The coxswain then adjusts steering to avoid the obstacle.

(2) ***Navigation***. The patrol leader is responsible for navigation. The three acceptable methods of river navigation are--

(a) Checkpoint and general route. These methods are used when the drop site is marked by a well-defined checkpoint and the waterway is not confused by a lot of branches and tributaries. They are best used during daylight hours and for short distances.

(b) Navigator-observer method. This method is the most accurate means of river navigation and is used effectively in all light conditions.

(c) *Equipment Needed.*

- Compass
- GPS
- Photo map (1st choice)
- Topographical map (2nd choice)
- Poncho (for night use)
- Pencil/Grease pencil
- Flashlight (for night use)

(d) *Procedure.* Navigator is positioned in center of boat and does not paddle. During hours of darkness, he uses his flashlight under the poncho to check his map. The observer (or # 1 man) is at the front of the boat.

(a) The navigator keeps his map and compass oriented at all time.

(b) The navigator keeps the observer informed of the configuration of the river by announcing bends, sloughs, reaches, and stream junctions as shown on his map.

(c) The observer compares this information with the bends, sloughs, reaches, and stream junctions he actually sees. When these are confirmed, the navigator confirms the boat's location on his map.

(d) The navigator also keeps the observer informed of the general azimuths of reaches as shown on his map and the observer confirms these with actual compass readings of the river.

(e) The navigator announces only one configuration at a time to the observer and does not announce another until it is confirmed and completed.

(f) A strip-map drawn on clear acetate backed by luminous tape may be used. The drawing is to scale or a schematic. It should show all curves and the azimuth and distance of all reaches. It may also show terrain features, stream junctions, and sloughs.

k. **Formations.** Various boat formations are used (day and night) for control, speed, and security (Figure 9-4). The choice of which is used depends on the tactical situation and the discretion of the patrol leader. He should use hand and arm signals to control his assault boats. The formations are:

- Wedge
- Line
- File
- Echelon
- Vee

Figure 9-4. FORMATIONS

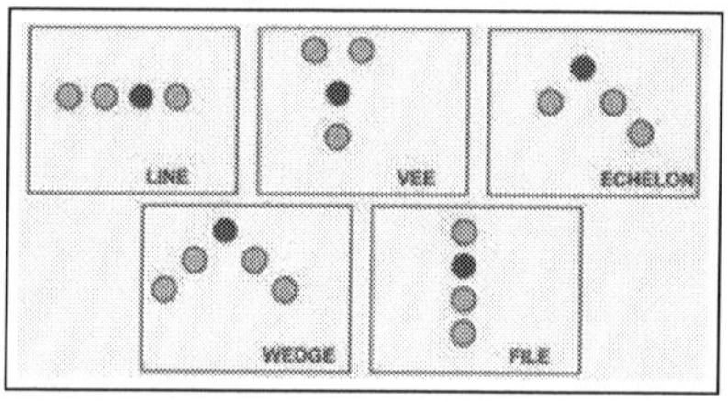

Chapter 10
MILITARY MOUNTAINEERING

In the mountains, commanders face the challenge of maintaining their units' combat effectiveness and efficiency. To meet this challenge, commanders conduct training that provides Rangers with the mountaineering skills necessary to apply combat power in a rugged mountain environment, and they develop leaders capable of applying doctrine to the distinct characteristics of mountain warfare.

10-1. **TRAINING**. Military mountaineering training provides tactical and fundamental mobility skills needed to move units safely and efficiently in mountainous terrain. Rangers should receive extensive training prior to executing combat operations in mountainous environments. Some of the areas to train are as follows:

- Characteristics of the mountain environment.
- Care and use of basic mountaineering equipment.
- Mountain bivouac techniques.
- Mountain communications.
- Mountain travel and walking techniques.
- Mountain navigation, hazard recognition and route selection.
- Rope management and knots.
- Natural and artificial anchors.
- Belay and rappel techniques.
- Installation construction and use such as rope bridges.
- Rock climbing fundamentals.
- Rope bridges and lowering systems.
- Individual movement on snow and ice.
- Mountain stream crossings (to include water survival techniques).

10-2. **DISMOUNTED MOBILITY**. Movement in more technical terrain demands specialized skills and equipment. Before Rangers can move in such terrain, a technical mountaineering team might have to secure the high ground.

10-3. **MOUNTAINEERING EQUIPMENT**. Mountaineering equipment is everything that allows the trained Ranger to accomplish many tasks in the mountains. The importance of this gear to the mountaineer is no less than that of the rifle to the Infantry Soldier.

a. **Ropes and Cords**. Ropes and cords are the most important pieces of mountaineering equipment. They secure climbers and equipment during steep ascents and descents. They are also used to install other ropes and to haul equipment. From WWII until the 1980's, US armed forces mainly used 7/16 *nylon laid rope*, often referred to as *green line* for all mountaineering operations. Since kernmantle ropes were introduced, other, more specialized ropes have in many cases replaced the all-purpose green line. Kernmantle ropes are made much like parachute cord, with a smooth sheath around a braided or woven core. Laid ropes are still used, but should be avoided whenever rope failure could cause injury or loss of equipment.

(1) ***Types***. Kernmantle rope is classified as *dynamic* or *static*:

(a) *Dynamic Ropes*. Ropes used for climbing are classified as *dynamic ropes*. These rope stretch or elongate 8 to 12 percent once subjected to weight or impact. This stretching is critical in reducing the impact force on the climber, anchors, and belayer during a fall by softening the catch. An 11-mm by 150-meter rope is standard for military use. Other sizes are available.

(b) *Static Ropes*. Static ropes are used in situations where rope stretch is undesired, and when the rope is subjected to heavy static weight. Static ropes should never be used while climbing, since even a fall of a few feet could generate enough impact force to injure climber and belayer, and/or cause anchor failure. Static Ropes are usually used when constructing rope bridges, fixed rope installations, vertical haul lines, and so on.

(2) ***Sling Ropes and Cordelettes***. A short section of static rope or static cord is referred to as a sling rope or cordelette. These are a critical piece of personal equipment during mountaineering operations. They are usually 7 to 8mm diameter and up to 21 feet long. The minimum Ranger standard is 8mm by 15 feet.

(3) ***Rope Care***. Ropes used daily should be used no longer that one year. An occasionally used rope can be used generally up to 4 to 5 years if properly cared for.

- Inspect ropes thoroughly before, during, and after use for cuts, frays, abrasion, mildew, and soft or worn spots.
- Never step on a rope or drag it on the ground unnecessarily.
- Avoid running rope over sharp or rough edges (pad if necessary).

- Keep ropes away from oil, acids, and other corrosive substances.
- Avoid running ropes across one another under tension (nylon to nylon contact will damage ropes).
- Do not leave ropes knotted or under tension longer than necessary.
- Clean in cool water, loosely coil, and hang to dry out of direct sunlight. Ultraviolet light rays harm synthetic fibers. When wet, hang rope to drip-dry on a rounded wooden peg, at room temperature (do not apply heat).

(4) ***Webbing and Slings***. Loops of tubular webbing or cord, called slings or runners, are the simplest pieces of equipment and some of the most useful. The uses for these simple pieces are endless, and they are a critical link between the climber, the rope, carabiners, and anchors. Runners are usually made from either 9/16- or 1-inch tubular webbing and are either tied or sewn by a manufacturer.

b. **Carabiners**. The carabiner is one of the most versatile pieces of equipment available in the mountains. This simple piece of gear is the critical connection between the climber, his rope, and the protection attaching him to the mountain. New metal alloys are stronger to hold hard falls; and, they are lighter than steel, which allows the Ranger to carry many at once. Consequently, steel carabiners are used less and less often. Basic carabiners come in several different shapes.

c. **Protection**. This is the generic term used to describe a piece of equipment, natural or artificial, that is used to construct an anchor. Protection, along with a climber, belayer, and climbing rope, forms the lifeline of the climbing team. The rope connects two climbers, and the protection connects them to the rock or ice. The two types of artificial protection are traditional (removable) and fixed (usually permanent) (Figures 10-1 and 10-2).

Figure 10-1. EXAMPLES OF TRADITIONAL (REMOVABLE) PROTECTION USED ON ROCK

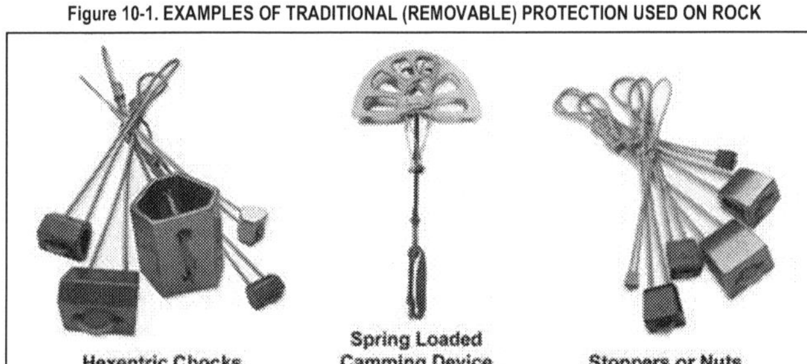

Figure 10-2. EXAMPLES OF FIXED (PERMANENT OR SEMI) PROTECTION USED ON ROCK

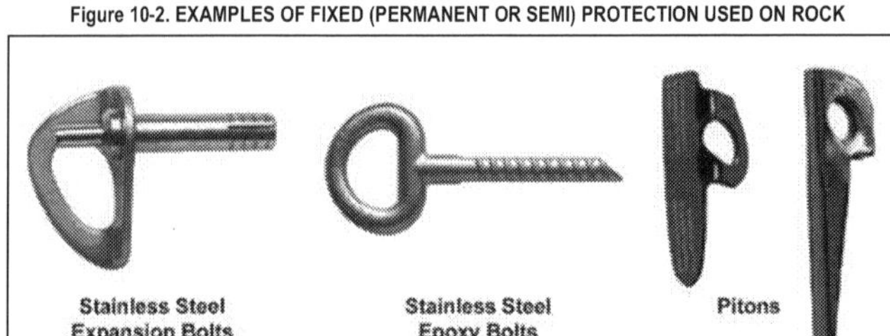

10-4. **ANCHORS**. Anchors form the base for all installations and roped mountaineering techniques. Anchors must be strong enough to support the entire weight of the load or of the impact placed upon them. Several pieces of artificial or natural protection may be incorporated together to make one multipoint anchor. Anchors are classified as artificial or natural:

a. **Artificial Anchors**. Artificial anchors are made from man-made materials. The most common ones use traditional or fixed protection (Figure 10-3).

Figure 10-3. CONSTRUCTION OF THREE-POINT, PRE-EQUALIZED ANCHOR WITH FIXED ARTIFICIAL PROTECTION

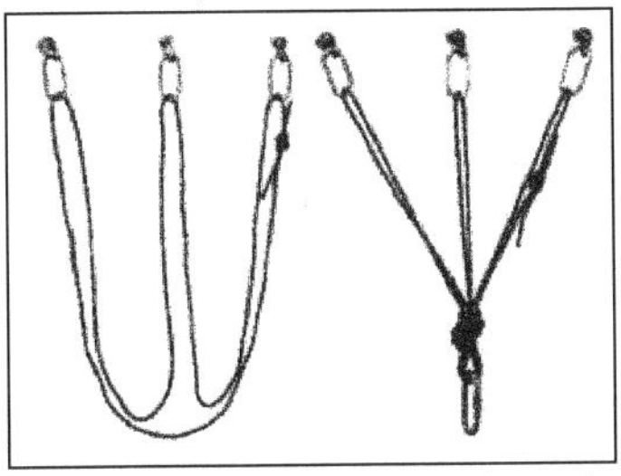

b. **Natural Anchors**. Natural anchors are usually very strong and often simple to construct with little equipment. Trees, shrubs, and boulders are the most common anchors. All natural anchors require is a method of attaching a rope. Regardless of the type of natural anchor used, it must support the weight of the load.

(1) ***Trees***. These are probably the most widely used of all anchors. In rocky terrain, trees usually have a very shallow root system. Check this by pushing or tugging on the tree to see how well it is rooted. Anchor as low as possible to keep the pull on the base of the tree, where it is nearer its own anchor and therefore stronger. Use padding on soft, sap-producing trees to keep sap off ropes and slings.

(2) ***Rock Projections and Boulders***. You can use these, but they must be heavy enough, and they must have a stable enough base to support the load.

(3) ***Bushes and Shrubs***. If no other suitable anchor is available, route a rope around the bases of several bushes. As with trees, place the anchoring rope as low as possible. Make sure all vegetation is healthy and well rooted to the ground.

(4) ***Tensionless Anchor***. Use this to anchor rope on high-load installations such as bridging. The wraps of the rope around the anchor (Figure 10-4) absorb the tension, keeping it off the knot and carabiner. Tie this anchor with at least four wraps around the anchor. A smooth anchor, such as a pipe or rail, might require several more wraps. Wrap the rope from top to bottom. Place a fixed loop into the end of the rope, and attach it loosely back onto the rope with a carabiner.

Figure 10-4. TENSIONLESS NATURAL ANCHOR

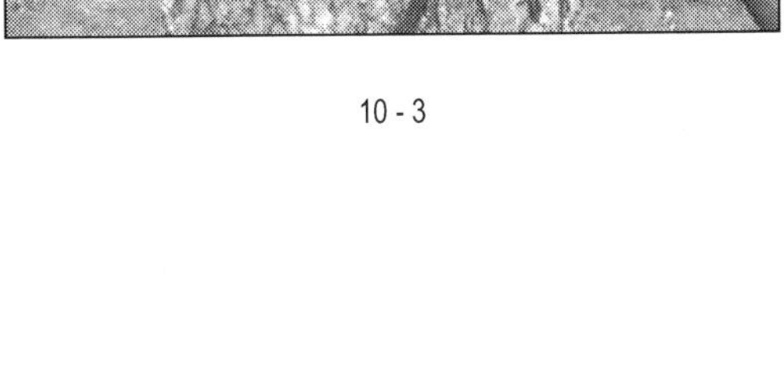

10-5. **KNOTS**.

a. **Square Knot**. This joins two ropes of equal diameter (Figure 10-5). It has two interlocking bights; the running ends exit on the same side of the standing portion of the rope. Each tail is secured with an overhand knot on the standing end. When you dress the knot, leave at least a 4-inch tail on the working end.

Figure 10-5. SQUARE KNOT

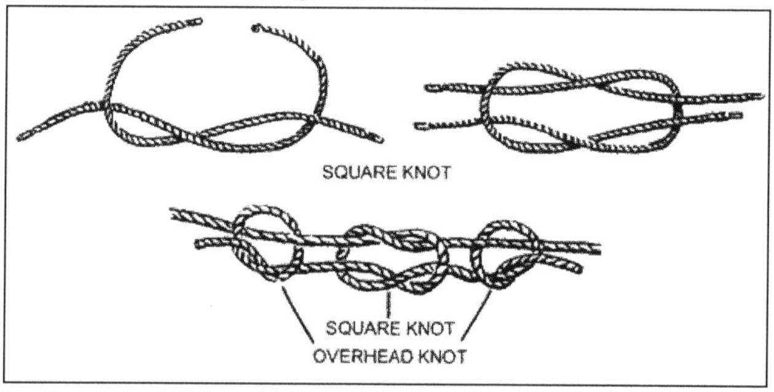

b. **Round-Turn with Two Half Hitches**. This is a constant tension anchor knot (Figure 10-6). The rope forms a complete turn around the anchor point (thus the name round turn), with both ropes parallel and touching, but not crossing. Both half hitches are tightly dressed against the round turn, with the locking bar on top. When you dress the knot, leave at least a 4-inch tail on the working end.

Figure 10-6. ROUND-TURN WITH TWO HALF-HITCHES

c. **End-of-the-Rope Clove Hitch**. The end-of-the-rope clove hitch (Figure 10-7) is an intermediate anchor knot that requires constant tension. Make two turns around the anchor. A locking bar runs diagonally from one side to the other. Leave no more than one rope width between turns of rope. Locking bar is opposite direction of pull. When you dress the knot, leave at least a 4-inch tail on the working end.

Figure 10-7. END-OF-THE-ROPE CLOVE HITCH

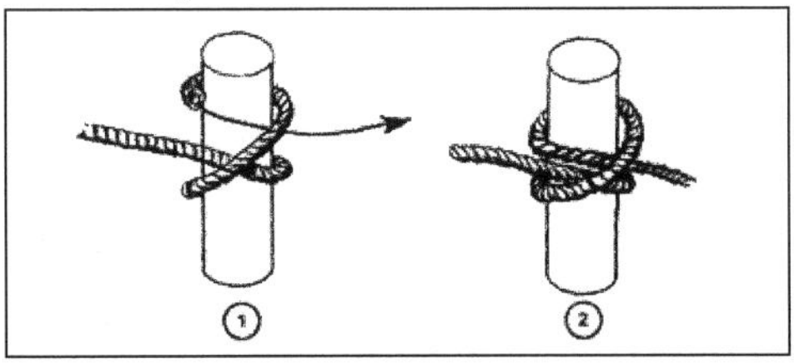

d. **Middle of the Rope Clove Hitch**. This knot (Figure 10-8) secures the middle of a rope to an anchor. The knot forms two turns around the anchor. A locking bar runs diagonally from one side to the other. Leave no more than one rope width between turns. Ensure the locking bar is opposite the direction of pull.

Figure 10-8. MIDDLE-OF-THE-ROPE CLOVE HITCH

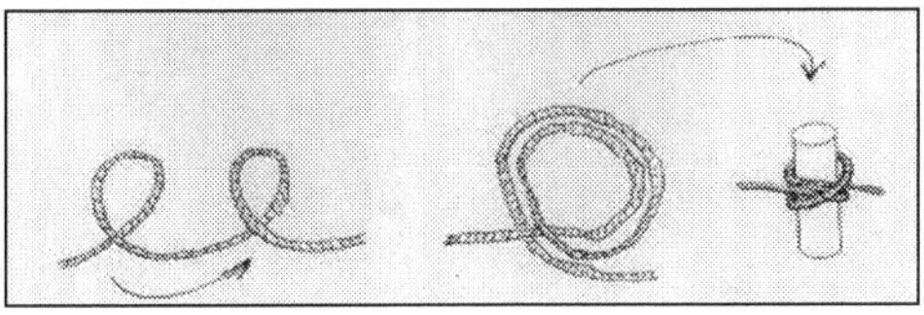

e. **Rappel Seat**. The rappel seat (Figure 10-9) is a rope harness used in rappelling and climbing. It can be tied for use with the left or right hand. Leg straps do not cross, and are centered on buttocks and tight. Leg straps form locking half hitches on rope around waist. A square knot is tied on the right hip and finished with two overhand knots. Tails must be even, no longer than 6 inches. A carabiner is inserted around all ropes, with the gate opening up and away. The carabiner must not contact the square or overhand knot. The rappel seat must be tight enough to prevent a fist from fitting between the rappeller's body and the harness.

Figure 10-9. RAPPEL SEAT

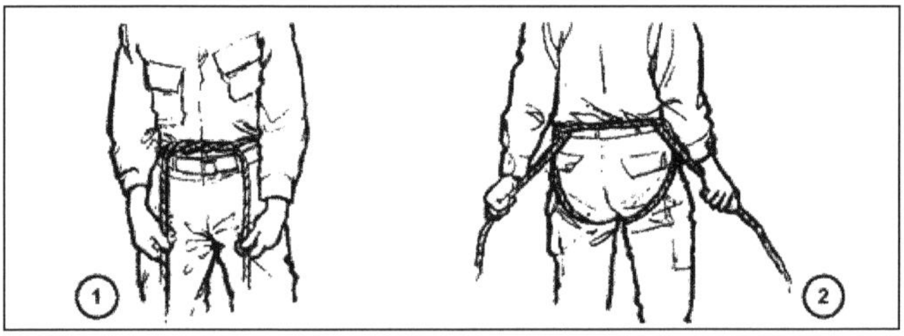

Figure 10-9. RAPPEL SEAT (CONTINUED)

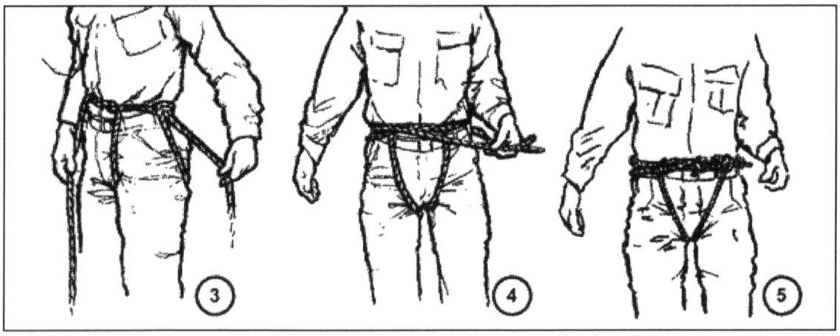

f. **Double Figure Eight**. Use a Figure Eight loop knot (Figure 10-10) to form a fixed loop in the end of the rope. It can be tied at the end of the rope or anywhere along the length of the rope. Figure Eight loop knots are formed by two ropes parallel to each other in the shape of a Figure Eight, no twists are in the Figure Eight. Fixed loops are large enough to insert a carabiner. When you dress the knot, leave at least a 4-inch tail on the working end.

Figure 10-10. DOUBLE FIGURE EIGHT-LOOP KNOT

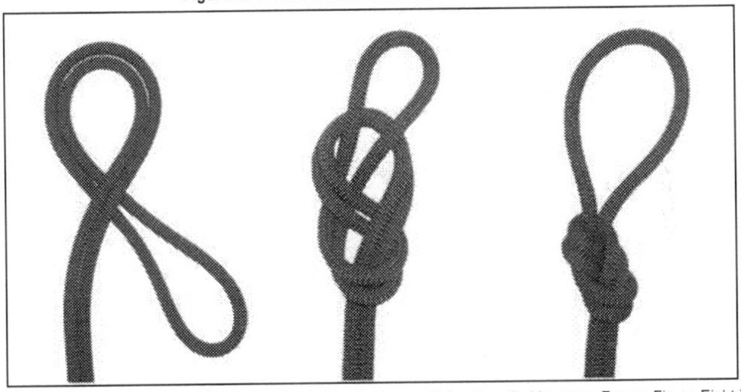

g. **Rerouted Figure Eight Knot**. This anchor knot also attaches a climber to a climbing rope. Form a Figure Eight in the rope, and pass the working end around an anchor. Reroute the end back through to form a double Figure Eight (Figure 10-11). Tie the knot with no twists. When you dress the knot, leave at least a 4-inch tail on the working end.

Figure 10-11. REROUTED FIGURE EIGHT KNOT

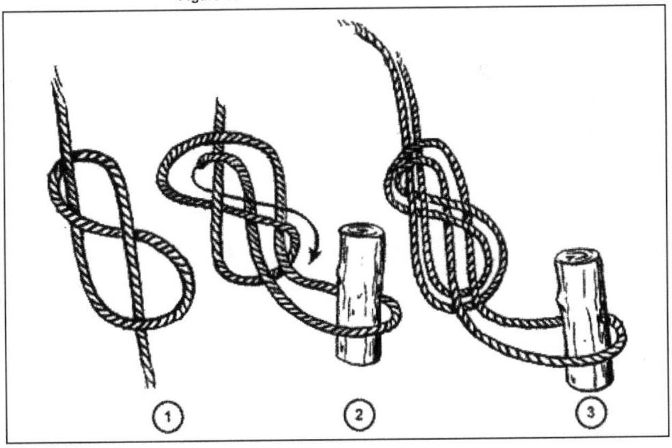

h. **Figure Eight Slip Knot**. The Figure Eight slip knot is used to form an adjustable bight in the middle of a rope. Knot is in the shape of a Figure Eight. Both ropes of the bight pass through the same loop of the Figure Eight. The bight is adjustable by means of a sliding section (Figure 10-12).

Figure 10-12. FIGURE EIGHT SLIP KNOT

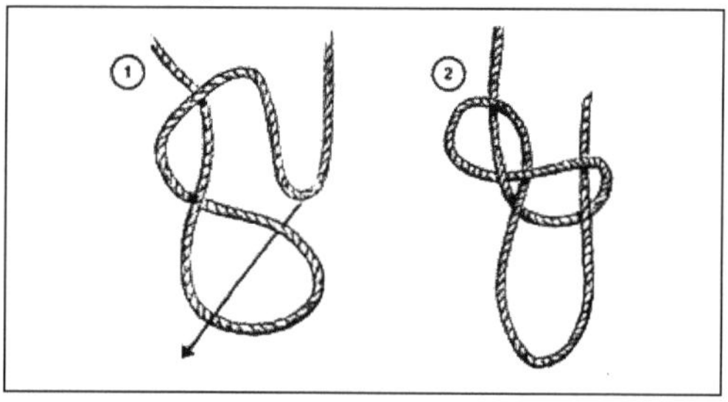

i. **End-of-the-Rope Prusik**. The End-of-the-Rope Prusik (Figure 10-13) attaches a movable rope to a fixed rope. The knot has two round turns, with a locking bar perpendicular to the standing end of the rope. Tie a bowline within 6 inches of the locking bar. When you dress the knot, leave at least a 4-inch tail on the working end.

Figure 10-13. END-OF-THE-ROPE PRUSIK

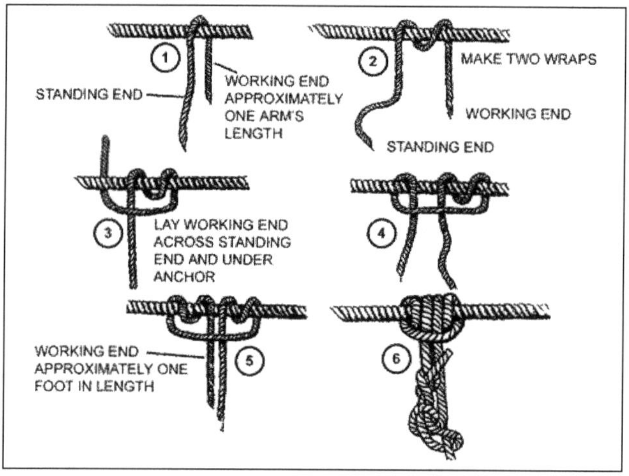

j. **Middle-of-the-Rope Prusik**. The middle-of-the-rope Prusik (Figure 10-14) attaches a movable rope to a fixed rope, anywhere along the length of the fixed rope. Make two round turns with a locking bar perpendicular to the standing end. Ensure the wraps do not cross and that the overhand knot is within 6 inches from the horizontal locking bar. Ensure the knot does not move freely on fixed rope.

Figure 10-14. MIDDLE-OF-THE-ROPE PRUSIK

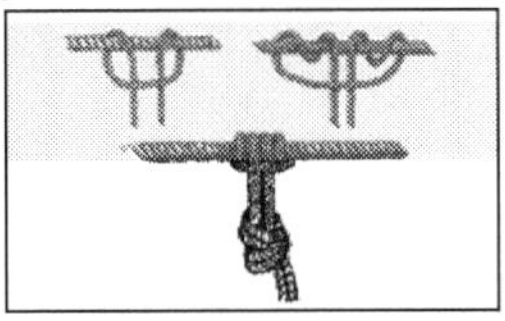

k. **Bowline on a Coil**. This knot (Figure 10-15) secures a climber to the end of a climbing rope. Make at least four parallel wraps around the body, between the hipbone and lower set of ribs. Keep all coils free of clothing, touch and are tight enough to ensure that a fist cannot be inserted between the wraps and the body. Three distinct coils show through the bight of the bowline. The rope coming off the bottom of the coils is on the right side, forward of the hip and forms the bight and the overhand knot. The rope coming off the top of the coils is on the left side, forward of the hips and forms the third and final coil showing through the bight of the bowline. Bowline is centered on the gig line and secured with an overhand knot with a minimum of 4 inch tail remaining after the knot is dressed.

Figure 10-15. BOWLINE ON A COIL

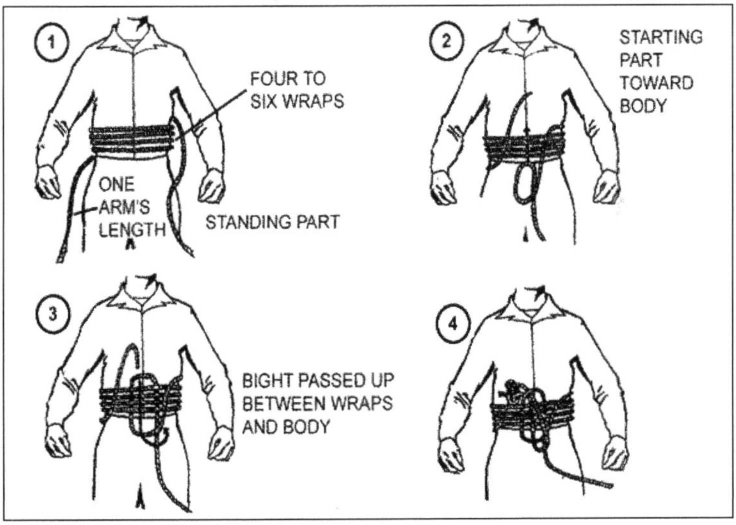

10-6. **BELAYS**. Belaying is any action taken to stop a climber's fall or to control the rate a load descends. The belayer also helps manage the rope for a climber by controlling the amount of rope that is taken out or in, or by controlling a rappeller's rate of descent. The belayer must be stably anchored to prevent him from being pulled out of his position, and losing control of the rope. The two types of belays are body and mechanical belays.

a. **Body Belay**. This belay (Figure 10-16) uses the belayer's body to apply friction by routing the rope around the belayer's body. Caution must be used when belaying from the body since the entire weight of the load may be placed on the belayer's body.

Figure 10-16. BODY BELAY

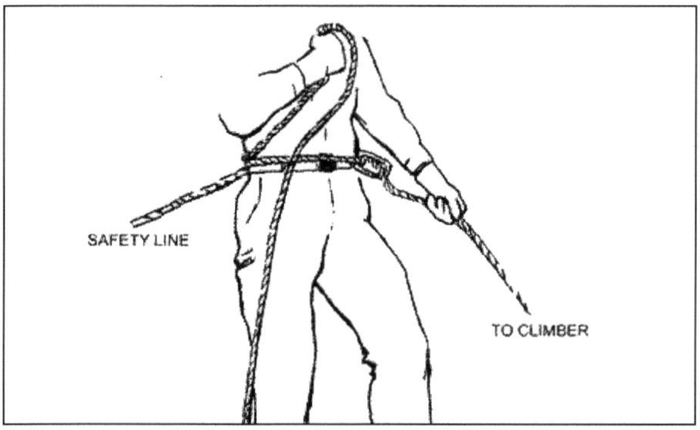

b. **Mechanical Belay**. This belay (Figure 10-17) uses mechanical devices to help the belayer control the rope, as in rappelling. A variety of devices are used to construct a mechanical belay for mountaineering.

Figure 10-17. MECHANICAL BELAY

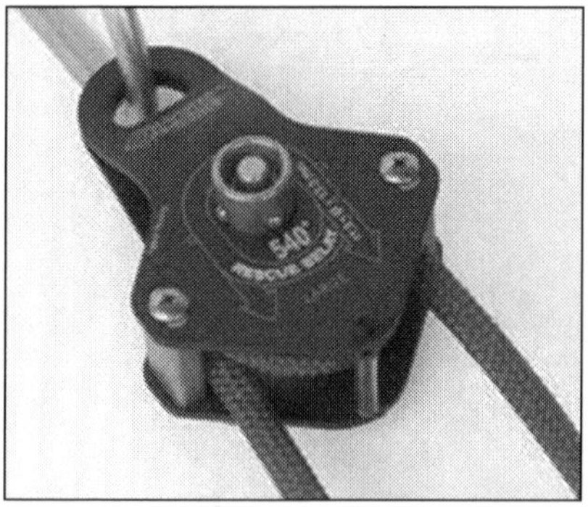

(1) ***Munter Hitch***. One of the most often used belays, the munter hitch (Figure 10-18), requires very little equipment. The rope is routed through a locking, pear-shaped carabiner, then back on itself. The belayer controls the rate of descent by manipulating the working end back on itself with the brake hand.

Figure 10-18. MUNTER HITCH

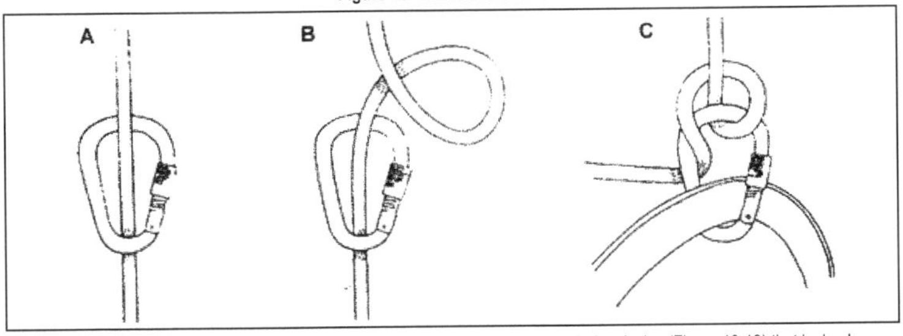

(2) ***Air Traffic Controller***. The ATC (air traffic controller) is a mechanical belay device (Figure 10-19) that locks down on itself when tension is applied in opposite directions. The belayer need apply very little force with his brake hand to control descent or arrest a fall.

Figure 10-19. AIR TRAFFIC CONTROLLER

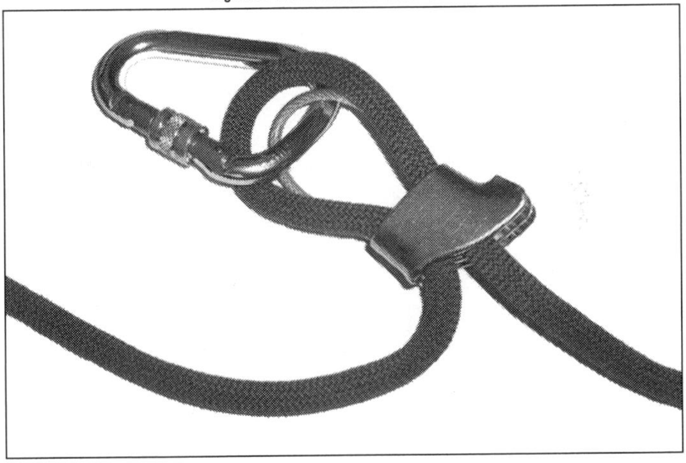

10-7. **CLIMBING COMMANDS**. Table 10-1 shows the sequence of commands used by climber and belayer.

Table 10-1. SEQUENCE OF CLIMBING COMMANDS

SEQUENCE	COMMAND	GIVEN BY	MEANING
1.	BELAY ON, CLIMB	Belayer	Alerts climber that belay is on and climber may climb.
2.	CLIMBING	Climber	Alerts belayer that climber is climbing.
3.	UP-ROPE	Climber	Belayer, remove excess slack in the rope.
4.	BRAKE	Climber	Alerts belayer to immediately apply brake.
5.	FALLING	Climber	Alerts belayer that climber is falling and that the belayer should immediately apply the brake and prepare to arrest the fall.
6.	TENSION	Climber	Alerts belayer to remove all slack from climbing rope until rope is tight, then apply brake and hold position.
7.	SLACK	Climber	Alerts belayer to pull slack into the climbing rope (Belayer may have to assist)
8.	ROCK	Anyone	Alerts everyone that an object is about to fall near them. Belayer immediately applies brake.
9.	POINT	Climber	Alerts belayer that the direction of pull on the climbing rope has changed in the event of a fall.
10.	STAND-BY	Climber or Belayer	Alerts the other to hold position, stand by, I am not ready.
11.	DO YOU HAVE ME?	Climber	Alerts belayer to prepare for a fall or to prepare to lower the climber.
12.	I HAVE YOU	Belay	Alerts climber that the brake is on and the belayer is prepared for the climber to fall, or to lower him.
13.	OFF-BELAY	Climber	Alerts belayer that climber is safetied in or it is safe to come off belay.
14.	3-METERS	Belayer	Alerts climber to the amount of rope between climber and belayer (May be given in feet or meters)
15.	BELAY-OFF	Belayer	Alerts climber that belayer is off belay

10-8. **ROPE INSTALLATIONS**. Rope installations may be constructed by teams to assist units in negotiating natural and manmade obstacles. installation teams consist of a squad sized element, with 2 to 4 trained mountaineers. Installation teams deploy early and prepare the AO for safe, rapid movement by constructing various types of mountaineering installations. Following construction of an installation, the squad, or part of it, remains on site to secure and monitor the system, assist with the control of forces across it, and make adjustments or repairs during its use. After passage of the unit, the installation team may then disassemble the system and deploy to another area as needed.

a. **Fixed Rope Installations**. A fixed rope is anchored in place to help Rangers move over difficult terrain. Its simplest form is a rope tied off on the top of steep terrain. As terrain becomes steeper or more difficult, fixed rope systems may require intermediate anchors along the route. Planning considerations are as follows:

- Does the installation allow you to bypass the obstacle?
- (Tactical considerations) Can obstacle be secured from construction through negotiation, to disassembly?
- Is it in a safe and suitable location? Is it easy to negotiate? Does it avoid obstacles?
- Are natural and artificial anchors available?
- Is the area safe from falling rock and ice?

b. **Vertical Hauling Line**. This installation (Figure 10-20) is used to haul men and equipment up vertical or near vertical slopes. It is often used in conjunction with the fixed rope.

Figure 10-20. VERTICAL HAULING LINE

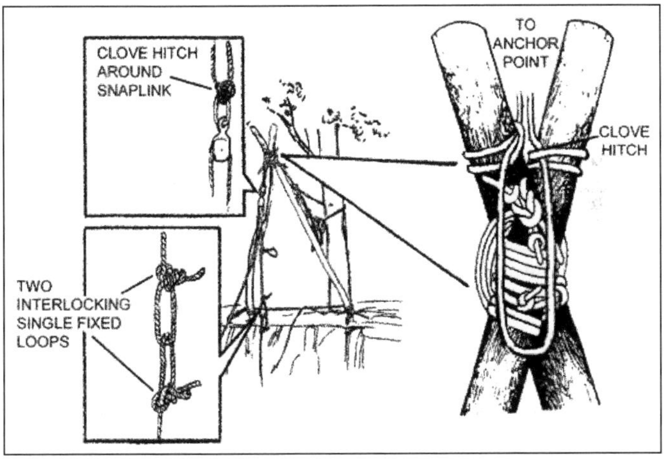

(1) ***Planning Considerations***.
- Does the installation allow you to bypass the obstacle?
- (Tactical considerations) Can you secure the installation from construction through negotiation, to disassembly?
- Does it have good loading and off-loading platforms? Are the platforms natural and easily accessible? Do they provide a safe working area?
- Does it allow sufficient clearance for load? Is there enough space between the slope and the apex of the A-frame to allow easy on-loading and off-loading of troops and equipment?
- Does it have an A-Frame for artificial height?
- Does it allow you to haul line in order to move personnel and equipment up and down slope?
- Does the A frame have a pulley or locking carabiner to ease friction on hauling line?
- Does it have a knotted hand line to help Rangers up the installation?
- Does it allow for Rangers top and bottom to monitor safe operation?

(2) ***Equipment***.
- Three 120-foot (50m) static ropes.
- Three 15-foot sling ropes for constructing A-frame.
- Two A-frame poles, 7 to 9 feet long, 4 to 6 inches in diameter (load dependent).
- Nine carabiners.
- One pulley with steel locking carabiner.

c. **Bridges**. Rope bridges are employed in mountainous terrain to bridge linear obstacles such as streams or rivers where the force of flowing water may be too great or temperatures are too cold to conduct a wet crossing.

(1) ***Construction***. The rope bridge is constructed using a static ropes. The max span that can be bridged is 50 percent of the ropes entire length for a dry crossing, 75 percent for a wet crossing. The ropes are anchored with an anchor knot on the far side

of the obstacle, and are tied off at the near end with a transport tightening system. Rope Bridge Planning Considerations are as follows:

- Does the installation allow you to bypass the obstacle?
- (Tactical considerations) Can you secure the installation from construction through negotiation, to disassembly?
- Is it in the most suitable location such as a bend in the river? Is it easily secured?
- Does it have near and far-side anchors?
- Does it have good loading and off-loading platforms?

(2) ***Equipment (1-Rope Bridge)***.

- One sling rope per-man.
- One steel locking carabineer.
- Two steel ovals.
- Two 120-foot static ropes.

(3) ***Construction Steps***. The first Ranger swims the rope to the far side and ties a tensionless anchor (Fig 10-4) between knee and chest level, with at least 6 to 8 wraps. The BTC ties a transport tightening system (Fig 10-21) to the near side anchor point. Then, he ties a Figure Eight slipknot and incorporates a locking half hitch around the adjustable bight. Insert 2 steel oval carabineers into the bight so the gates are opposite and opposed. The rope is then routed around the near side anchor point at waist level and dropped into the steel oval carabineers.

- A three-Ranger pulling team moves forward from the PLT. No more than three men should tighten the rope. Using more can overtighten the rope, bringing it near failure.
- Once the rope bridge is tight enough, the bridge team secures the transport-tightening system using two half hitches, without losing more than 4 inches of tension.
- Personnel cross using the commando crawl, rappel seat, or monkey crawl method (Figures 10-22 thru 24).

Figure 10-21. TRANSPORT TIGHTENING SYSTEM

Figure 10-22. COMMANDO CRAWL METHOD

Figure 10-23. RAPPEL SEAT (TYROLEAN TRAVERSE) METHOD

Figure 10-24. MONKEY CRAWL METHOD

(4) ***Bridge Recovery***. Once all but two Rangers have crossed the rope bridge, the bridge team commander (BTC) chooses either the wet or dry method to dismantle it. If he chooses the dry method, he should he should first anchor his tightening system with the transport knot.

- The BTC back-stacks all of the slack coming out of the transport knot, then ties a fixed loop and places a carabiner into the fixed loop.
- The next to last Ranger to cross attaches the carabiner to his rappel seat or harness, and then moves across the bridge using the Tyrolean traverse method.
- The BTC then removes all knots from the system. The far side remains anchored. The rope should now only pass around the near side anchor.
- A three-man pull team, assembled on the far side, takes the end brought across by the next to last Ranger and pulls the rope tight again and holds it.
- The BTC then attaches himself to the rope bridge and moves across.
- Once across, the BTC breaks down the far side anchor, removes the knots, and then pulls the rope across. If the BTC chooses a wet crossing, any method can be used to anchor the tightening system.
- All personnel cross except the BTC or the strongest swimmer.
- The BTC then removes all knots from the system.
- The BTC ties a fixed loop, inserts a carabiner, and attaches it to his rappel seat or harness. He then manages the rope as the slack is pulled to the far side.
- The BTC then moves across the obstacle while being belayed from the far side.

d. **Suspension Traverse**. The suspension traverse is used to move personnel and equipment over rivers, ravines, chasms, and up or down a vertical obstacle. By combining the transport tightening system used during the Rope bridge, an A-Frame used for the Vertical haul Line (Figure 10-25), and belay techniques device, Units can construct a suspension traverse (Figures 10-26 and 10-27). Installation of a suspension traverse can be time-consuming and equipment-intensive. All personnel must be well trained and rehearsed in the procedures.

(1) ***Construction***. The suspension traverse must be constructed using a static ropes. The max span that can be bridged is generally 75 percent length of the shortest rope. Planning considerations are the same as rope bridge and vertical haul line combined.

(2) ***Equipment***.

- Three static installation ropes.
- Seven sling Ropes.
- Nine carabiners.
- One heavy duty double pulley.
- One locking carabiner.
- One canvas pad.

Figure 10-25. ANCHORING THE TRAVERSE ROPE TO THE A-FRAME

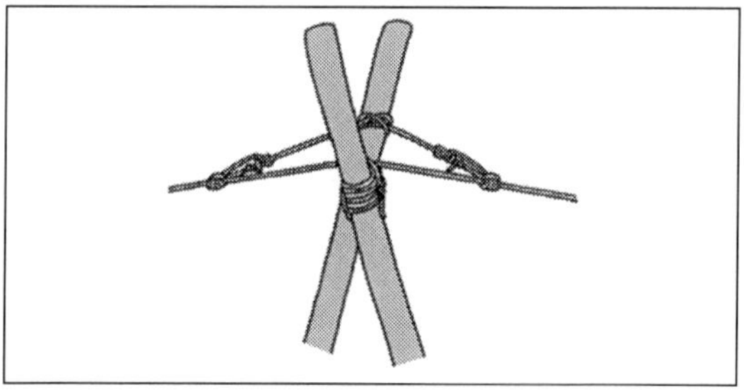

Figure 10-26. CARRYING ROPE FOR USE ON A SUSPENSION TRAVERSE

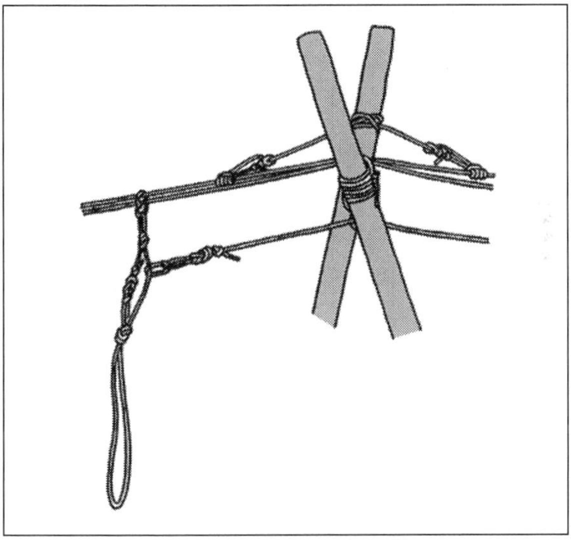

Figure 10-27. SUSPENSION TRAVERSE

10-9. **RAPPELLING**. Rappelling is a quick method of descent but can be extremely dangerous. These dangers include anchor failure, equipment failure, and individual error. Anchors in a mountainous environment should be selected carefully. Great care must be taken to load the anchor slowly and to ensure that no excessive stress is placed on the anchor. The best way to ensure this is to prohibit bounding rappels, and to use only walk-down rappels.

a. **Hasty and Body Rappels**. Hasty (Figure 10-28) and body rappels (Figure 10-29) are easier and faster than other methods, but should only be used on moderate pitches--never on vertical or overhanging terrain. Wear gloves to prevent rope burns.

Figure 10-28. Hasty Rappel

Figure 10-29. Body Rappel

b. **Seat-Hip Rappel**. The seat-hip rappel uses a mechanical rappel device that is inserted in a sling rope seat and is fastened to the rappeller to provide the necessary friction to remain in control. This method provides a faster and more controlled descent than any other method (Figure 10-30). There are several mechanical rappel devices (Figure 10-31 and 10-32) that can be used for the seat-hip rappel.

Figure 10-30. FIGURE EIGHT DESCENDER

Figure 10-30. FIGURE EIGHT DESCENDER (CONTINUED)

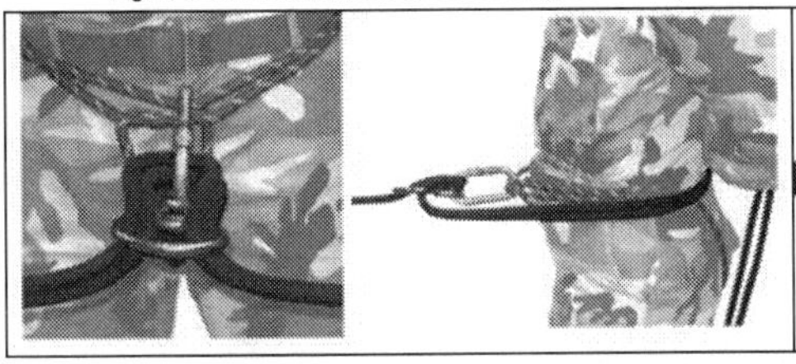

Figure 10-31. CARABINER WRAP

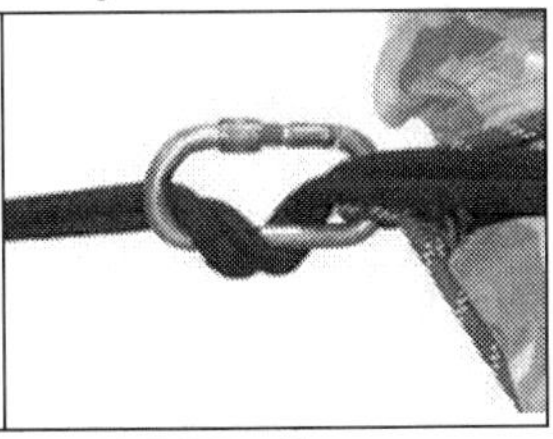

Figure 10-32. SEAT HIP RAPPEL WITH CARABINER WRAP DESCENDER

c. **Site Selection**. The selection of the rappel point depends on factors such as mission, cover, route, anchor points, and edge composition (loose or jagged rocks). There must be good anchors (primary and secondary). The anchor point should be above the rappeller's departure point. Suitable loading and off-loading platforms should be available.

- Each rappel point has primary and secondary anchors.
- Rappel point has equal tension between all anchor points.
- Double rope is used when possible.
- Ropes must reach the off loading platform.
- Site has suitable on and off loading platforms.
- Personnel working near the edge are tied in.
- Select a smooth route free of loose rock and debris.

Chapter 11

EVASION AND SURVIVAL

This chapter will cover basic material dealing with evading the enemy and survival techniques. Field expedient methods of acquiring food/ water, land navigation, constructing shelters, and making fires will be presented.

11-1. **EVASION**. When you become isolated or separated in a hostile area, either as an individual or as a group, your evasion and survival skills will determine whether or not you return to friendly lines.

a. When unable to continue the mission or unable to rejoin your unit, leave the immediate area, and move to your last rally point.

b. Observe activity in the area and form a plan.

c. Traveling alone offers the least possibility of detection, but traveling in groups of two to three is more desirable.

d. Plan a primary and alternate route. Consider distance, cover, food, and water. The easiest and shortest route may not be the best.

e. Food and water are daily requirements. You can do without food for several days; water, however, is essential.

f. Move at night. Use the daylight to observe, plan, and rest in a hide position.

g. Linkup only during daylight hours. Place friendly lines under observation.

h. Attempt to identify the unit you approach, note their movements and routine.

i. After carefully considering your approach route, make voice contact with the unit as soon as possible.

11-2. **SURVIVAL**.

With training, equipment, and the ***will to survive***, you can overcome *any* obstacle you may face. ***You will survive.*** Understand the emotional states associated with survival. "Knowing thyself" is extremely important in a survival situation. It bears directly on how well you cope with serious stresses, anxiety, pain, injury, illness; cold, heat, thirst, hunger, fatigue, sleep deprivation, boredom, loneliness and isolation.

a. **Memory Aid**. You can overcome and reduce the shock of being isolated behind enemy lines if you keep the key word S-U-R-V-I-V-A-L foremost in your mind. Its letters can help guide you in your actions.

(1) S - Size up the situation, the surroundings, your physical condition, and your equipment.

(2) U - Undue haste makes waste; don't be too eager to move. Plan your moves.

(3) R - Remember where you are relative to friendly and enemy units and controlled areas; water sources (most important in the desert); and good cover and concealment. This information will help you make intelligent decisions.

(4) V - Vanquish fear and panic.

(5) I – Improvise/Imagine. You can improve your situation. Learn to adapt what is available for different uses. Use your imagination.

(6) V - Value living. Remember your goal - to get out alive. Remain stubborn. Refuse to give in to problems and obstacles. This will give you the mental and physical strength to endure.

(7) A - Act like the natives; watch their daily routines and determine when, where, and how they get their food and water.

(8) L - Live by your wits. Learn basic skills.

b. **Navigation**. In a survival situation, you might find yourself without a compass. The ability to determine direction can enable you to navigate back to your unit or to a friendly sanctuary. In sunlight, there are two simple ways to determine direction: the shadow-tip and the watch.

(1) ***Shadow Tip***. Use the sun to find approximate true North. Use this in light bright enough to cast shadows. Find a fairly straight stick about 3 feet long, and follow the diagram below (Figure 11-l).

Figure 11-1. SHADOW-TIP METHOD

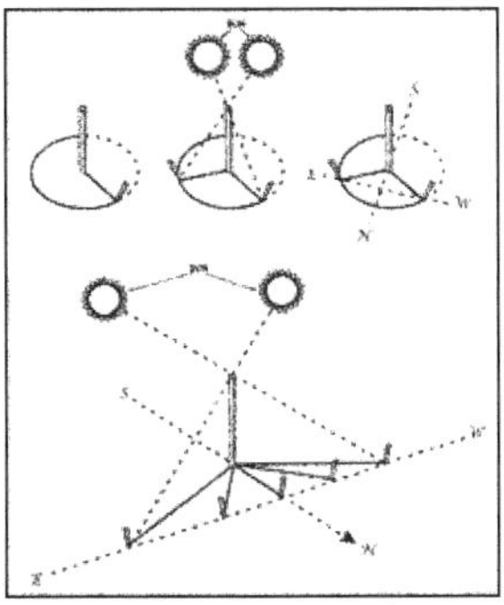

(2) ***Watch Method***. You can also determine direction using a watch (Figure 11-2). The steps you take will depend on whether you are in the northern Temperate Zone or in the southern Temperate Zone. The northern temperate zone is located between 23.4 north and 26.6 north. The southern Temperate Zone is located between 23.4 south and 66.6 south.

Figure 11-2. WATCH METHOD

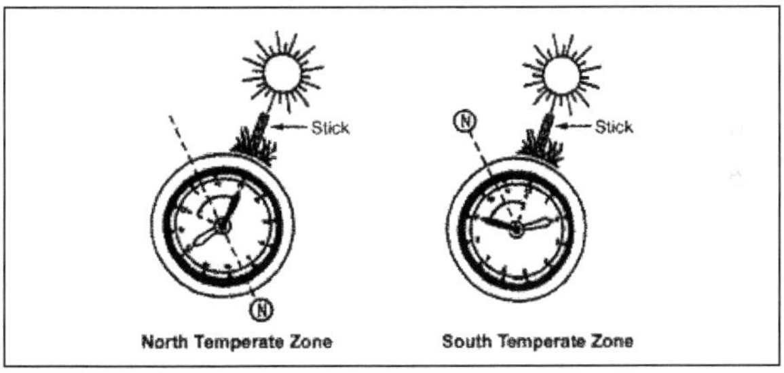

c. **Northern Temperate Zone**. Procedures in the northern temperate zone using a conventional watch are as follows:

(1) Place a small stick in the ground so that it casts a definite shadow.

(2) Place your watch on the ground so that the hour hand points toward and along the shadow of the stick.

(3) Find the point on the watch midway between the hour hand and 12 o'clock and draw an imaginary line from that point through and beyond the center of the watch. This imaginary line is a north-south line. You can then tell the other directions.

Note: If your watch is set on daylight savings time, then use the midway point between the hour hand and 1 o'clock to draw your imaginary line.

d. **Southern Temperate Zone**. Procedures in the southern temperate zone using a conventional watch are as follows:

(1) Place a small stick in the ground so that it casts a definite shadow.

(2) Place your watch on the ground so that 2 o'clock points to and along the shadow.

(3) Find the midway point between the hour and 12 o'clock and draw an imaginary line from the point through and beyond the center of the watch. This is a north-south line.

e. A hasty shortcut using a conventional watch is simply to point the hour hand at the sun in the northern temperate zone (or point the 12 at the sun in the southern temperate zone) and then follow the last step of the watch method above to find your directions. This shortcut, of course, is not as accurate as the regular method but quicker. Your situation will dictate which method to use.

11-3. **WATER**. Water is one of your most urgently needed resources in a survival situation. You can't live long without it, especially in hot areas where you lose so much through sweating. Even in cold areas, you need a minimum of 2 quarts of water a day to maintain efficiency. More than three-fourths of your body is composed of fluids. Your body loses fluid as a result of heat, cold, stress, and exertion. The fluid your body loses must be replaced for you to function effectively. So, one of your first objectives is to obtain an adequate supply of water.

a. **Purification**. Purify all water before drinking. Either--

(1) Boil it for at least one minute (plus 1 more minute for each additional 1,000 feet above sea level) or for a maximum of 10 minutes anywhere.

(2) Use water purification tablets.

(3) Add eight drops of 2-1/2 percent iodine solution to a quart (canteen full) of water. Let it stand for 10 minutes before drinking.

(4) Collect rain water directly in clean containers or on plants. This is generally safe to drink without purifying. Never drink urine or sea water; the salt content is too high. Use old, bluish sea ice, but newer, grayer ice may be salty. Glacier ice is safe to melt and drink.

b. **Desert Environment**. In a desert environment, water has a huge effect on Rangers. If a unit fails to plan properly for water, and resupply is unavailable, then they can run out of water. In the desert, look for four signs of water: animal trails, vegetation, birds, and civilization. Adequate water is critical in a hot environment if a unit is to survive and maintain the physical condition necessary to accomplish the mission. Unit leaders must enforce water discipline and plan for water resupply. The leader can use the following planning considerations for water resupply:

(1) Units average water consumption.

(2) Drop sites.

(3) Aviation support.

(4) DZ and LZ parties.

(5) Caches.

(6) Targets of opportunity (enemy).

c. **Survival Water Still**. Dig a belowground still (Figure 11-3). Select a site where you believe the soil will contain moisture (such as a dry stream bed or a spot where rain water has collected). The soil should be easy to dig, and be in sunlight most of the day:

(1) Dig a bowl-shaped hole about 3 feet across and 2 feet deep.

(2) Dig a sump in center of the hole. The depth and the perimeter of the sump will depend on the size of the container that you have to set inside of it. The bottom of the sump should allow the container to stand upright.

(3) Anchor the tubing to the bottom of the container by forming a loose overhand knot in the tubing.

(4) Place the container upright in the sump.

(5) Extend the unanchored end of the tubing up, over, and beyond the lip of the hole.

(6) Place plastic sheeting over the hole covering the edge with soil to hold it in place.

(7) Place a rock in the center of the plastic.

(8) Allow the plastic to lower into the hole until it is about 15 inches below ground level. The plastic now forms an inverted cone with the rock at its apex. Make sure that the apex of the cone is directly over your container. Also, make sure the plastic cone does not touch the sides of the hole, because the earth will absorb the condensed water.

(9) Put more soil on the edges of the plastic to hold it securely in place and to prevent loss of moisture.

(10) Plug the tube when not being used so that moisture will not evaporate.

d. You can drink water without disturbing the still by using the tube as a straw. You may want to use plants in the hole as a moisture source. If so, when you dig the hole you should dig out additional soil from the sides of the hole to form a slope on which to place the plants. Then proceed as above.

Figure 11-3. SURVIVAL WATER STILL

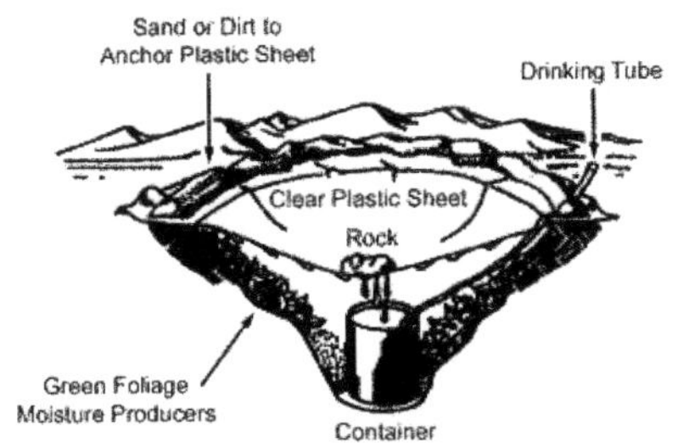

11-4. **PLANT FOOD**. There are many plants throughout the world. Tasting or swallowing even a small portion of some can cause severe discomfort, extreme internal disorders, or death. Therefore, if you have the slightest doubt as to the edibility of a plant, apply the universal edibility test described below before eating any part of it.

a. **Universal Edibility Test**. Before testing a plant for edibility, make sure there are a sufficient number of plants to make testing worth your time and effort. You need more than 24 hours to apply the edibility test outlined below:

(1) Test only one part of a potential food plant at a time.

(2) Break the plant into its basic components, leaves, stems, roots, buds, and flowers.

(3) Smell the food for strong or acid odors. Keep in mind that smell alone does not indicate a plant is edible.

(4) Do not eat for 8 hours before starting the test.

(5) During the 8 hours you are abstaining from eating, test for contact poisoning by placing a piece of the plant you are testing on the inside of your elbow or wrist. Usually 15 minutes is enough time to allow for reaction.

(6) During the test period, take nothing by mouth except purified water and the plant part being tested.

(7) Select a small portion and prepare it the way you plan to eat it.

(8) Before putting the prepared plant part in your mouth, touch a small portion (a pinch) to the outer surface of the lip to test for burning or itching.

(9) If after 3 minutes there is no reaction on your lip, place the plant part on your tongue, holding there for 15 minutes.

(10) If there is no reaction, thoroughly chew a pinch and hold it in your mouth for 15 minutes. *Do not swallow.*

(11) If no burning, itching, numbing, stinging, or other irritation occurs during the 15 minutes, swallow the food.

(12) Wait 8 hours. If any ill effects occur during this period, induce vomiting and drink a lot of water.

(13) If no ill effects occur, eat 1/2 cup of the same plant part prepared the same way. Wait another 8 hours. If no ill effects occur, the plant part as prepared is safe for eating.

b. **Poisonous Plants**. ***Do not*** eat unknown plants that have the below characteristics:

(1) Have a milky sap or a sap that turns black when exposed to air.

(2) Are mushroom like.

(3) Resemble onion or garlic.

(4) Resemble parsley, parsnip, or dill.

(5) Have carrot-like leaves, roots, or tubers.

11-5. **ANIMAL FOOD**. Animal food contains the most food value per pound. Anything that creeps, crawls, swims, or flies is a possible source of food, however you must first catch, kill and butcher it before this is possible. There are numerous methods for catching fish and animals in a survival situation. You can catch fish by using a net across a small stream, (Figure 11-4) or by making fish traps and baskets. Improvise fish hooks and spears as indicated in Figure 11-5, and use them for conventional fishing, spearing and digging.

Figure 11-4. SETTING A GILL NET IN THE STREAM

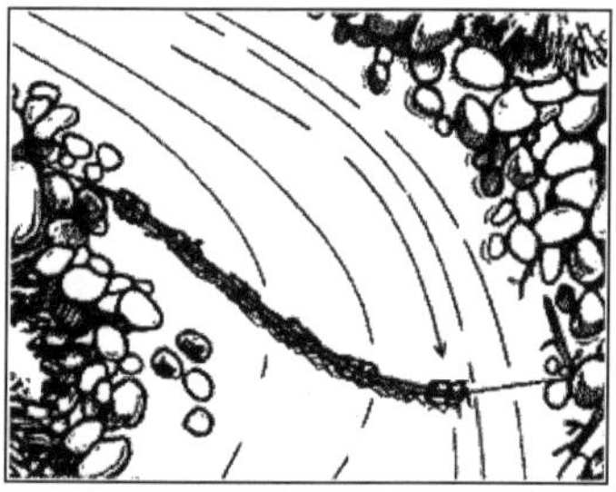

Figure 11-5. SPEARS AND FISH HOOKS

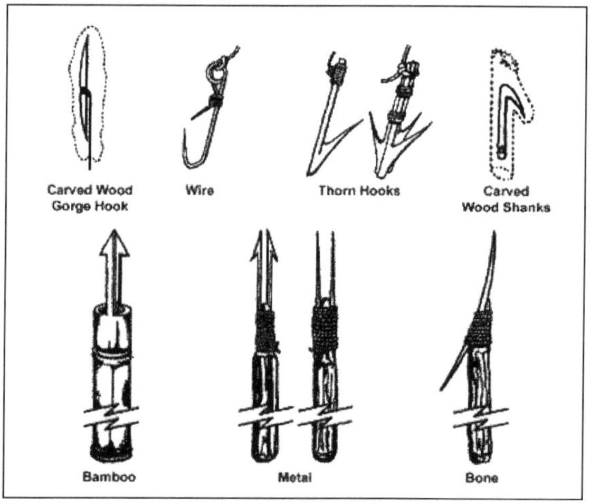

11-6. **TRAPS AND SNARES**. For an unarmed survivor or evader, or when the sound of a rifle shot could be a problem, trapping or snaring wild game is a good alternative. Several well-placed traps have the potential to catch much more game than a Ranger with a rifle is likely to shoot.

a. To be effective with any type of trap or snare, you must---
   (1) Know the species of animal you intend to catch.
   (2) Know how to construct a proper trap.
   (3) Avoid alarming the prey with signs of your presence.

b. There are no catchall traps you can set for all animals. You must determine what species are in a given area and set your traps specifically with those animals in mind. Look for the following:

(1) Runs and trails.

(2) Tracks.

(3) Droppings.

(4) Chewed or rubbed vegetation.

(5) Nesting or roosting sites.

(6) Feeding and watering areas.

c. Position your traps and snares where there is proof that animals pass through. You must determine if it is a "run" or a "trail." A trail will show signs of use by several species and will be rather distinct. A run is usually smaller and less distinct and will only contain signs of one species. You may construct a perfect snare, but it will not catch anything if haphazardly placed in the woods. Animals have bedding areas, waterholes, and feeding areas with trails leading from one to another. You must place snares and traps around these areas to be effective.

d. An evader in a hostile environment must conceal traps and snares. It is equally important, however, to avoid making a disturbance that will alarm the animal and cause it to avoid the trap. Therefore, if you must dig, remove all fresh dirt from the area. Most animals will instinctively avoid a pitfall-type trap. Prepare the various parts of a trap or snare away from the site, carry them in, and set them up. Such actions make it easier to avoid disturbing the local vegetation, thereby alerting the prey. Do not use freshly cut, live vegetation to construct a trap or snare. Freshly cut vegetation will "bleed" sap that has an odor the prey will be able to smell. It is an alarm signal to the animal.

e. You must remove or mask the human scent on and around the trap you set. Although birds do not have a developed sense of smell, nearly all mammals depend on smell even more than on sight. Even the slightest human scent on a trap will alarm the prey and cause it to avoid the area. Removing the scent from a trap is difficult but masking it is relatively easy. Use the fluid from the gall and urine bladders of previous kills. Do not use human urine. Mud, particularly from an area with plenty of rotting vegetation, is also good. Use it to coat your hands when handling the trap and to coat the trap when setting it. In nearly all parts of the world, animals know the smell of burned vegetation and smoke. It is only when a fire is actually burning that they become alarmed. Therefore, smoking the trap parts is an effective means to mask your scent. If one of the above techniques is not practical, and if time permits, allow a trap to weather for a few days and then set it. Do not handle a trap while it is weathering. When you position the trap, camouflage it as naturally as possible to prevent detection by the enemy and to avoid alarming the prey.

f. Traps or snares placed on a trail or run should use canalization. To build a channel, construct a funnel-shaped barrier extending from the sides of the trail toward the trap, with the narrowest part nearest the trap. Canalization should be inconspicuous to avoid alerting the prey. As the animal gets to the trap, it cannot turn left or right and continues into the trap. Few wild animals will back up, preferring to face the direction of travel. Canalization does not have to be an impassable barrier. You only have to make it inconvenient for the animal to go over or through the barrier. For best effect, the canalization should reduce the trail's width to just slightly wider than the targeted animal's body. Maintain this constriction at least as far back from the trap as the animal's body length, then begin the widening toward the mouth of the funnel.

(1) Use a treadle snare against small game on a trail (Figure 11-6). Dig a shallow hole in the trail. Then drive a forked stick (fork down) into the ground on each side of the hole on the same side of the trail. Select two fairly straight sticks that span the two forks. Position these two sticks so that their ends engage the forks. Place several sticks over the hole in the trail by positioning one end over the lower horizontal stick and the other on the ground on the other side of the hole. Cover the hole with enough sticks so that the prey must step on at least one of them to set off the snare. Tie one end of a piece of cordage to a twitch-up or to a weight suspended over a tree limb. Bend the twitch-up or raise the suspended weight to determine where you will tie a 5 centimeter or so long trigger. Form a noose with the other end of the cordage.

(2) Route and spread the noose over the top of the sticks over the hole. Place the trigger stick against the horizontal sticks and route the cordage behind the sticks so that the tension of the power source will hold it in place. Adjust the bottom horizontal stick so that it will barely hold against the trigger. As the animal places its foot on a stick across the hole, the bottom horizontal stick moves down, releasing the trigger and allowing the noose to catch the animal by the foot. Because of the disturbance on the trail, an animal will be wary. You must therefore use canalization.

Figure 11-6. TREADLE SNARE

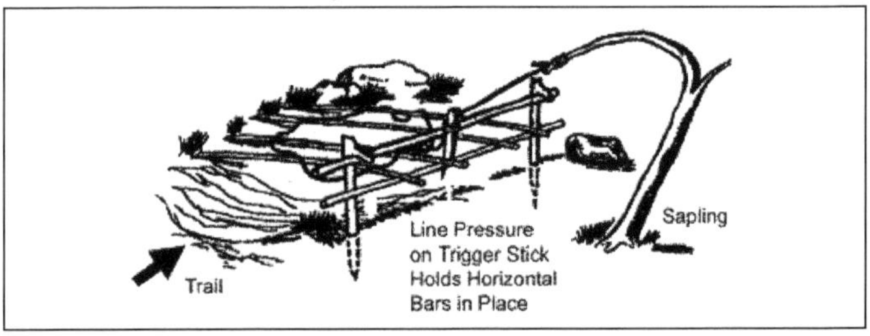

g. Trapping game can be accomplished through the use of snares, traps, or deadfalls. A snare is a noose that will slip and strangle or hold any animal caught in it. You can use inner core strands of parachute suspension lines, wire, bark of small hardwood saplings as well as hide strips from previously caught animals to make snares.

(1) The drag noose snare, Figure 11-7, is usually the most desirable in that it allows you to move away from the site, plus it is one of the easiest to make and fastest to set. It is especially suitable for catching rabbits. To make the drag noose snare, make a loop in the string using a bowline or wireman's knot. (When using wire, secure the loop by intertwining the end of the wire with the wire at the top of the loop). Pull the other end of the string (or wire) through the loop to form a noose that is large enough for the animal's head but too small for its body; tie the string (or attach the wire) to a sturdy branch. The branch should be large enough to span the trail and rest on the bush or support (two short forked sticks) you have selected. A snared animal will dislodge the drag stick, pulling it until it becomes entangled in the brush. Any attempt to escape tightens the noose, strangling or holding the animal.

Figure 11-7. DRAG NOOSE SNARE

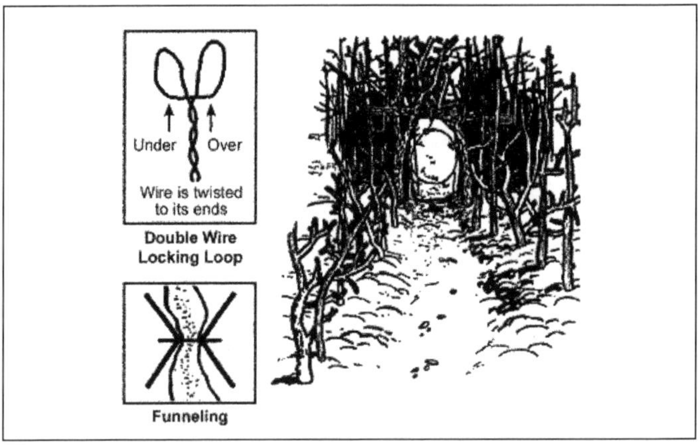

(2) Another type snare is the locking type snare loop (Figure 11-8) that will lock when pulled tight, ensuring the snared animal cannot escape. Use lightweight wire to make this snare, such as, trip wire or the wire from a vehicle or aircraft electrical system. To construct this snare, cut a piece of wire twice the length of the desired snare wire. Double the wire and attach the running ends to a securely placed object, such as the branch of a tree. Place a stick about 1/2 inch in diameter through the loop end of the wire; holding the wire taut, turn the stick in a winding motion so that the wire is twisted together. You should have four to five twists per inch. Detach the wire from the branch and then remove the loop from the stick; make a Figure Eight in the l/2-inch loop by twisting the loop over itself then fold the Figure Eight so the small loops are almost overlapping; run the loose wire ends through these loops. This forms a stiff noose that is strong. Tie the loose end to the stick (for a drag noose square) or branch you are using to complete the snare. This is an excellent snare for catching large animals.

Figure 11-8. FORMING A LOCKING-TYPE SNARE LOOP

(3) Another means of obtaining game is the use of the deadfall trap (Figure 11-9 and Figure 11-10).

Figure 11-9. TRIGGER WITH DEADFALL

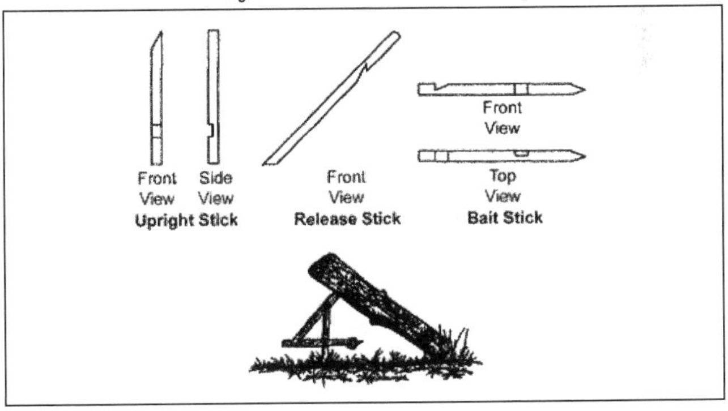

Figure 11-10. TRIP-STRING DEADFALL TRAP

h. Once you have obtained your fish or game, you must clean/butcher and cook/store it. Improper cleaning storing can result in inedible fish and game.

(1) ***Fish***. You must know how to tell if fish are free of bacterial decomposition that makes the fish dangerous to eat. Although cooking may destroy the toxin from bacterial decomposition, do not eat fish that appear spoiled.

(a) Signs of spoilage are:

- A peculiar odor.
- A suspicious color. (Gills should be red or pink. Scales should be a pronounced-not faded shade of gray).
- A dent remaining after pressing the thumb against the flesh.
- A slimy rather than moist or wet body.
- A sharp or peppery taste.

(b) Eating spoiled or poisoned fish may cause diarrhea, nausea, cramps, vomiting, itching; paralysis, or a metallic taste in the mouth. These symptoms appear suddenly 1 to 6 hours after eating. If you are near the sea, drink sea water immediately upon on set of such symptoms and force yourself to vomit.

(c) Fish spoil quickly after death, especially on a hot day, so prepare fish for eating as soon as possible after you catch them.

(d) Cut out the gills and large blood vessels that lie next to the backbone. (You can leave the head if you plan to cook the fish on a spit).

(e) Gut fish that are more than 4 inches long cut along the abdomen and scrape out the intestines.

(f) Scale or skin the fish.

(g) You can impale a whole fish on a stick and cook it over an "open fire." However, boiling the fish with the skin on is the best way to get the most food value. The fats and oil are under the skin, and by boiling the fish, you can save the juices for broth. Any of the methods used for cooking plant food can be used for cooking fish. Fish is done when the meat flakes off.

(h) To dry fish in the sun, hang them from branches or spread them on hot rocks. When the meat has dried, splash it with sea water, if available, to salt the outside. Keep seafood only if it is well dried or salted.

(2) ***Snakes***. All poisonous and nonpoisonous fresh water and land snakes are edible.
To prepare snakes for eating use the following steps (Figure 11-11):

(a) Grip the snake firmly behind the head and cut off the head with a knife.

(b) Slit the belly and remove the innards. (You can use the innards for baiting traps and snares).

(c) Skin the snake. (You can use the skin for improvising, belts, straps, or similar items).

CAUTION

Take extreme care in securing snakes--the bite of some poisonous snakes can be fatal. Even after a snake's head is cut off, its reflex action can cause it to bite, injecting poison. The best time to capture snakes is in the early morning or late evening when temperatures are low and they move slow. Kill or use a long stick to pin down its head and capture it. To pick up a snake, place the index finger on the top rear of its head with your thumb and middle finger on either side of the head behind the jaws. Keep your index finger on top of snake's head to prevent it from turning inside its skin and biting you.

Figure 11-11. CLEANING A SNAKE

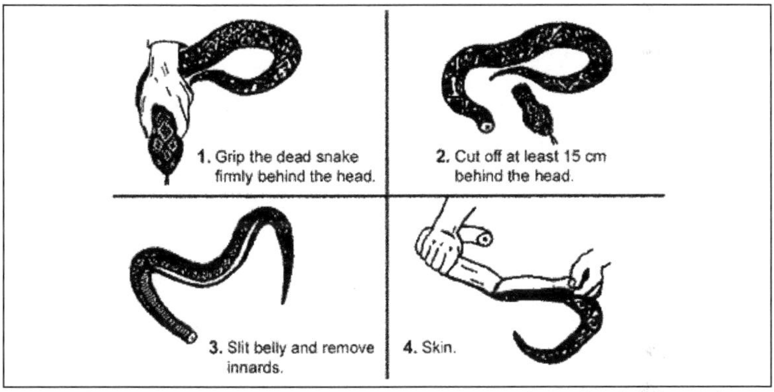

(3) ***Fowl***. Your first step after killing a fowl for eating or preserving is to pluck its feathers. If plucking is impractical, you can skin the fowl. Keep in mind, however, that a fowl cooked with the skin on retains more food value. Waterfowl are easier to pluck while dry, but other fowl are easier to pluck after scalding. After you pluck the fowl--

(a) Cut off its neck close to the body.

(b) Cut an incision in the abdominal cavity and clean out the insides. Save the neck, liver, and heart for stew. Thoroughly clean and dry the entrails to use for cordage.

(c) Wash out the abdominal cavity with fresh clean water. You can boil fowl or cook it on a spit over a fire. You should boil scavenger birds such as vultures and buzzards for at least 20 minutes to kill any parasites. Use the feathers from fowl for insulating your shoes clothing, or bedding. You can also use feathers for fish lures.

(4) **Medium-Sized Mammals**. The game you trap or snare will generally be alive when you find it and therefore dangerous. Be careful when you approach a trapped animal. Use a spear or club to kill it so you can keep a safe distance from it. After you kill an animal, immediately bleed it by cutting its throat. If you must drag the carcass any distance, do so before you cut off the hide so that the carcass is protected from dirt and debris that might contaminate it. Clean the animal near a stream if possible so that you can wash and cool the carcass and edible parts. Fleas and parasites will leave a cooled body so if the situation allows, wait until the animal cools before cleaning and dressing the carcass. To skin and dress the animal (Figure 11-12 and 11-13).

Figure 11-12. SKINNING AND BUTCHERING LARGE GAME

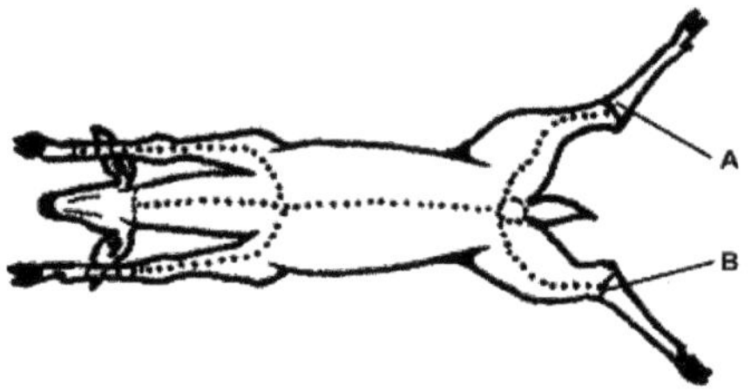

Figure 11-13. SKINNING SMALL GAME

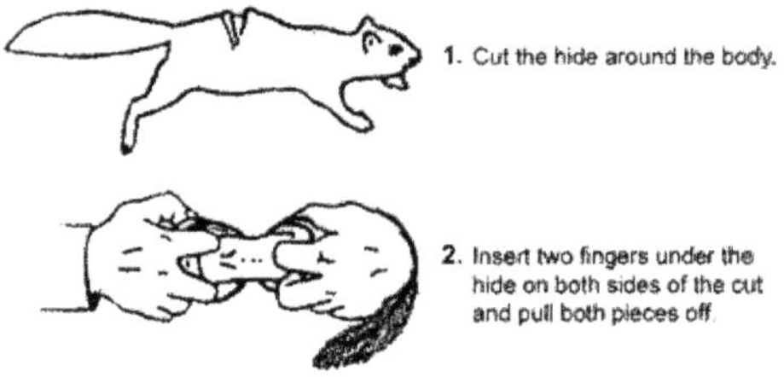

(a) Place carcass, belly up, on a slope if available. You can use rocks or brush to support it.
(b) Remove genitals or udder.
(c) Remove musk glands to avoid tainting meat.
(d) Split hide from tail to throat. Make the cut shallow so that you do not pierce the stomach.
(e) Insert your knife under the skin, taking care not to cut into the body cavity. Peel the hide back several inches on each side to keep hair out of the meat.
(f) Open the chest cavity by splitting the sternum. You can do this by cutting to one side of the sternum where the ribs join.
(g) Reach inside and cut the windpipe and gullet as close to the base of the skull as possible.
(h) With the forward end of the intestinal tract free, work your way to the rear lifting out internal organs and intestines. Cut only where necessary to free them.
(i) Carefully cut the bladder away from the carcass so that you do not puncture the bladder (urine can contaminate meat). Pinch the urethra tightly and cut it beyond the point you are pinching.
(j) Remove the bladder.
(k) From the outside of the carcass, cut a circle around the anus.
(l) Pull the anus into the body cavity and out of the carcass.
(m) Lift or roll the carcass to drain all blood.

Note: Try to save as much blood as you can as it is a vital source of food and salt. Boil the blood.

(n) Remove the hide, and make cuts along the inside of the legs to just above the hoof or paw. Peel the skin back, using your knife in a slicing motion to cut the membrane between the skin and meat. Continue this until the entire skin is removed.

(o) Most of the entrails are usable. The heart, liver, and kidneys are edible. Cut open the heart and remove the blood from its chambers. Slice the kidneys and if enough water is available, soak or rinse them. In all animals except those of the deer family, the gall bladder (a small, dark-colored, clear-textured sac) is attached to the liver.

(p) Sometimes, the sac looks like a blister on the liver. To remove the sac, hold the top portion of it and cut the liver around and behind the sac. If the gall bladder breaks and gall gets on the meat, wash it off immediately so the meat will not become tainted. Dispose of the gall.

(q) Clean blood splattered on the meat will glaze over and help preserve the meat for a short time. However, if an animal is not bled properly, the blood will settle in the lowest part of its body and will spoil in a short time. Cut out any meat that becomes contaminated.

(r) When temperatures are below 40 degrees, you can leave meat hanging for several days without danger of spoilage. If maggots get on the meat, remove the maggots and cut out the discolored meat. The remaining meat is edible. Maggots, which are the larvae of insects, are also edible.

(s) Blood, which contains salts and nutrients, is a good base for soups.

(t) Thoroughly clean the intestines and use them for storing or smoking food or lashings for general use. Make sure they are completely dry to preclude rotting.

(u) The head of most animals contains a lot of meat, which is relatively easy to get. Skin the head, saving the skin for leather. Clean the mouth thoroughly and cut out the tongue. Remove the outer skin from the tongue after cooking. Cut or scrape the meat from the head. If you prefer, you can roast the head over an open fire before cutting off the meat. Eyes are edible. Cook them but discard the retina (this is a plastic like disc). The brain is also edible; in fact, some people consider it a delicacy. The brain is also used to tan leather, the theory being that the brain of an animal is adequate to tan its hide.

(v) Use the tendons and ligaments of the body of large animals for lashings.

(w) The marrow in bones is a rich-food source. Crack the bones and scrap out the marrow, and use bones to make weapons.

(x) If the situation and time allow, you should preserve the extra meat for later use. If the air is cold enough, you can freeze the meat. In warmer climates however, you will need to use a drying or smoking process to preserve it. One night of heavy smoking will make meat edible for about 1 week. Two nights will make it remain edible for 2 to 4 weeks. To prepare meat for drying or smoking, cut it with the grain in quarter inch strips. To air dry the meat, hang it in the wind and hot sun out the reach of animals; cover it so that blow flies cannot land on it.

(y) To smoke meat, you will need an enclosed area – for instance, a teepee (Figure 11-14) or a pit. You will also need wood from deciduous trees, preferably green. Do not use conifer trees such as pines, firs, spruces, or cedars, as the smoke from these trees give the meat a disagreeable taste.

(z) When using the para-teepee or other enclosed area with a vent at the top, set the fire in the center and let it burn down to coals, then stoke it with green wood. Place the strips of meat on a grate or hang them from the top of the enclosure so that they are about 2 feet above the smoking coals. To use the pit method of smoking meat dig, a hole about 3 feet (1 meter) deep and 1 1/2 feet (1/2 meter) in diameter. Make a fire at the bottom of the hole. After it starts burning well, add chipped green wood or small branches of green wood to make it smoke. Place a wooden grate about 1 1/2 feet (1/2 meter) above the fire and lay the strips of meat on the grate. Cover the pit with poles, boughs, leaves, or other material.

11-7. **SHELTERS**. A shelter can protect you from the sun, insects, wind, rain, snow, hot or cold temperatures, and enemy observation. In some areas your need for shelter may take precedence over your need for food, possibly even your need for water. After determining your shelter site, you should keep in mind the type of shelter (protection) you need. You need to know how to make different types of shelters. Only two are described in this handbook. Additional information is available in FM 21-76.

a. **Planning Considerations**.

- How much time and effort are needed to build the shelter?
- Will the shelter adequately protect you from the elements (rain, snow, wind, sun, and so on)?
- Do you have tools to build it? If not, can you improvise tools from materials in the area?
- Do you have the type and amount of manmade materials needed to build it? If not, are there sufficient natural materials in the area?

(f) If you plan to use the lean-to for more than one night, or if you expect rain, make a center support to the lean-to. You can do this by stretching a rope between two upright poles or trees that are in line with the center of the poncho.

(g) Tie another rope to the poncho hood; pull it upward so that it lifts the center of the poncho, and tie it firmly to the rope stretched between the two uprights.

(h) Another method is to cut a stick to place upright under the center of the lean-to. This method, however, will restrict your space and movements in the shelter.

(i) To give additional protection from wind and rain, place boughs, brush, your rucksack, or other equipment at the sides of the lean-to.

(j) To reduce heat loss to the ground, place some type of insulating material, such as leaves or pine needles, inside your lean-to.

Note: When at rest, as much as 80 percent of your body heat can be lost to the ground.

(l) To increase your security from enemy observation, lower the silhouette of the lean-to by making two modifications.

- Secure the support lines to the trees knee-high rather than waist-high.
- Use two knee-high sticks in the two center grommets (sides of lean-to), and angle the poncho to the ground, securing it with sharpened sticks as above.

(2) ***Field Expedient Lean-To***. If you are in a wooded area and have sufficient natural materials, you can make an expedient lean-to (Figure 11-16) without the aid of tools or with only a knife. You need more time to make it than the shelter previously mentioned, but it will protect you from most environmental elements. You will need two trees, (or two upright poles), about 6 feet apart; one pole about 7 feet long and 1 inch in diameter. Five to eight poles about 10 feet long and 1 inch in diameter for beams, cord or vines for securing, the horizontal support to the trees and other poles, saplings, or vines to crisscross the beams. To make this lean-to:

Figure 11-16. FIELD-EXPEDIENT LEAN-TO

(a) Tie the 7-foot pole to the two trees at point about waist to chest high. This is your horizontal support. If there is a fork in the tree, you can rest the pole in it instead of tying the pole in place. If a standing tree is not available, construct a bipod using an Y-shaped sticks or two tripods.

(b) Place one end of the beams (10-foot poles) one side of the horizontal support. As with all lean-to type shelters, make sure the backside of the lean-to is placed into the wind.

(c) Crisscross sapling or vines on the beams.

(d) Cover the framework with brush, leaves, pine needles, or grass, starting at the bottom and working your way up like shingling.

(f) If you plan to use the lean-to for more than one night, or if you expect rain, make a center support to the lean-to. You can do this by stretching a rope between two upright poles or trees that are in line with the center of the poncho.

(g) Tie another rope to the poncho hood; pull it upward so that it lifts the center of the poncho, and tie it firmly to the rope stretched between the two uprights.

(h) Another method is to cut a stick to place upright under the center of the lean-to. This method, however, will restrict your space and movements in the shelter.

(i) To give additional protection from wind and rain, place boughs, brush, your rucksack, or other equipment at the sides of the lean-to.

(j) To reduce heat loss to the ground, place some type of insulating material, such as leaves or pine needles, inside your lean-to.

Note: When at rest, as much as 80 percent of your body heat can be lost to the ground.

(l) To increase your security from enemy observation, lower the silhouette of the lean-to by making two modifications.

- Secure the support lines to the trees knee-high rather than waist-high.
- Use two knee-high sticks in the two center grommets (sides of lean-to), and angle the poncho to the ground, securing it with sharpened sticks as above.

(2) ***Field Expedient Lean-To***. If you are in a wooded area and have sufficient natural materials, you can make an expedient lean-to (Figure 11-16) without the aid of tools or with only a knife. You need more time to make it than the shelter previously mentioned, but it will protect you from most environmental elements. You will need two trees, (or two upright poles), about 6 feet apart; one pole about 7 feet long and 1 inch in diameter. Five to eight poles about 10 feet long and 1 inch in diameter for beams, cord or vines for securing, the horizontal support to the trees and other poles, saplings, or vines to crisscross the beams. To make this lean-to:

Figure 11-16. FIELD-EXPEDIENT LEAN-TO

(a) Tie the 7-foot pole to the two trees at point about waist to chest high. This is your horizontal support. If there is a fork in the tree, you can rest the pole in it instead of tying the pole in place. If a standing tree is not available, construct a bipod using an Y-shaped sticks or two tripods.

(b) Place one end of the beams (10-foot poles) one side of the horizontal support. As with all lean-to type shelters, make sure the backside of the lean-to is placed into the wind.

(c) Crisscross sapling or vines on the beams.

(d) Cover the framework with brush, leaves, pine needles, or grass, starting at the bottom and working your way up like shingling.

(e) Place straw, leaves, pine needles, or grass inside the shelter for bedding.

(f) In cold weather, you can add to the comfort of your lean-to by building a fire-reflector wall (Figure 11-16). Drive four stakes about 4 feet long into the ground to support the wall. Stack green logs on top of one another between the support stales. Bind the top of the support stakes so the green logs will stay in place. Fill in the spaces between the logs with twigs or small branches. With just a little more effort, you can have a drying rack. Cut a few 3/4-inch diameter poles. The length depends on distance between the lean-to support and the top of the fire-reflector wall. Lay one end of the poles on the lean-to horizontal support and the other ends on top of the reflector wall. Place and tie into place smaller sticks across these poles. You now have a place to dry clothes, meat, or fish.

11-8. **FIRES**. A fire can full fill several needs. It can keep you warm, it can keep you dry: you can use it to cook food, to purify water, and to signal. It can also cause you problems when you are in enemy territory: it creates smoke, which can be smelled and seen from a long distance: It causes light which can be seen day or night and it leaves signs of your presence. Remember you should always weigh your need for a fire against your need to avoid enemy protection. When operating in remote areas you should always take a supply of matches in a waterproof case and keep them on your person.

a. **Selection**. When selecting a site to build a fire, you should consider the following:

(1) The area (terrain and climate) in which you are operating.
(2) The material and tools available.
(3) How much time you have.
(4) Why you need a fire.
(5) The nearness of the enemy.

b. **Preparation**. If you are in a wooded or brush-covered area, clear brush away, and scrape the surface soil from the spot you selected. The cleared circle should be at least 3 feet (1 meter) in diameter so that there is little chance of the fire spreading. To prepare the site for a fire, ensure that it is dry and that it--look for a dry spot that--

(1) Offers protection from the wind.
(2) Is suitably placed in relation to your shelter (if any).
(3) Concentrates the heat in the direction you desire.
(4) Has a supply of wood or other fire burning material.

c. **Dakota Fire Hole**. In some situations, an underground fireplace will best meet your needs. It conceals the fire to some extent and serves well for cooking food. To make an underground fireplace or Dakota fire hole (Figure 11-17):

Figure 11-17. DAKOTA FIRE HOLE

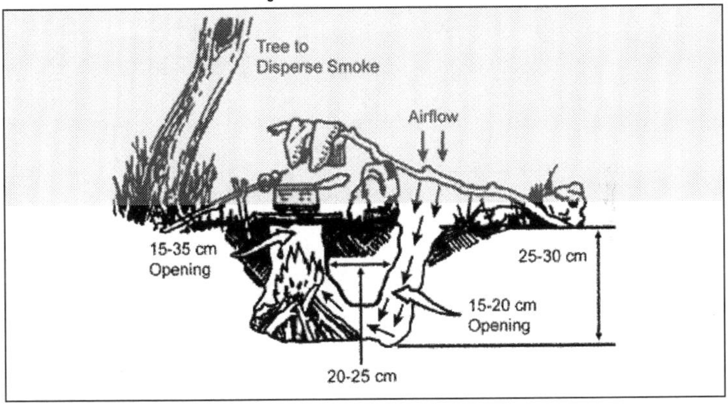

(1) Dig a hole in the ground.

(2) On the upwind side of this hole, poke one large connecting hole for ventilation.

d. **Aboveground fire**. If you are in a snow covered or wet area, you can use green logs to make a dry base for your fire (Figure 11-18). Trees with wrist-size trunks are easily broken in extreme cold. Cut or break several green logs and lay them side by side on top of the snow. Add one or two more layers, laying the top layer logs in a direction opposite those of the layer below it.

Figure 11-18. BASE FOR FIRE IN SNOW COVERED AREA

e. **Methods**. There are several methods for laying a fire for quick fire making. Three easy methods are Tepee, lean-to, and cross-ditch. Tepee (Figure 11-19). Arrange tinder and a few sticks of kindling in the shape of a cone. Fire the center. As the cone burns away, the outside logs will fall inward, feeding the heart of the fire. This type of fire burns well even with wet wood.

(1) ***Lean-To*** (Figure 11-19). Push a green stick into the ground at a 30 degree angle. Point the end of the stick in the direction of the wind. Place some tinder (at least a handful) deep inside this lean-to stick. Light the tinder. As the kindling catches fire from the tinder, add more kindling.

Figure 11-19. METHODS FOR LAYING A FIRE

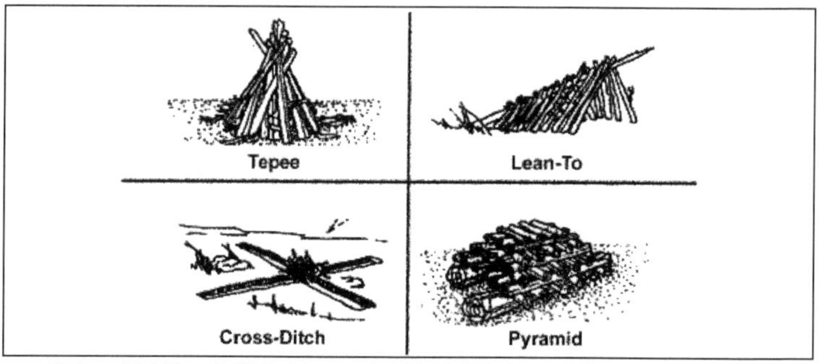

(2) ***Cross-Ditch*** (Figure 11-19). Scratch a cross about 1 foot in size in the ground. Dig the cross 3 inches deep. Put a large wad of tinder in the middle of the cross. Build a kindling pyramid above the tinder. The shallow ditch allows air to sweep under the fire to provide a draft.

NOTES

Chapter 12
FIRST AID

Patrolling, more than some other types of missions, puts Rangers in harms way. CASEVAC planning is vital. Also, because trained medical personnel might be unavailable at the initial point of injury, everyone must know how to diagnose and treat injuries, wounds, and common illnesses. The unit should also have a plan for handling KIAs.

12-1. **LIFESAVING STEPS**. Whatever the injury, (1) open the airway and restore breathing; (2) stop the bleeding and protect the wound; (3) check, treat and monitor for shock; and (4) MEDEVAC the casualty.

12-2. **CARE UNDER FIRE**. When still under fire, (1) maintain situational awareness; (2) return fire; (3) protect the casualty; (4) Identify and control severe bleeding with bandage or tourniquet; (5) and move the casualty to cover.

12-3. **PRIMARY SURVEY**. Use the ABC's to help you remember how to identify and deal with life-threatening injuries, airway complications, breathing problems, or uncontrolled hemorrhaging (bleeding).

A	Airway. Open airway by patient position or with airway adjuncts.
B	Breathing. Seal open chest wounds with occlusive dressing.
C	Circulation. Identify uncontrolled bleeding and control with pressure or tourniquet. Start IV if needed.
D	Disability. Determine Level of consciousness.
E	Exposure. Fully expose patient. (Environment dependent)

12-4. **AIRWAY MANAGEMENT**. The *airway* is most often obstructed at the base of the tongue.

a. If this happens, open the airway using the chin lift (nontrauma, Figure 12-1) or jaw thrust (trauma, Figure 12-2).

Figure 12-1. CHIN LIFT

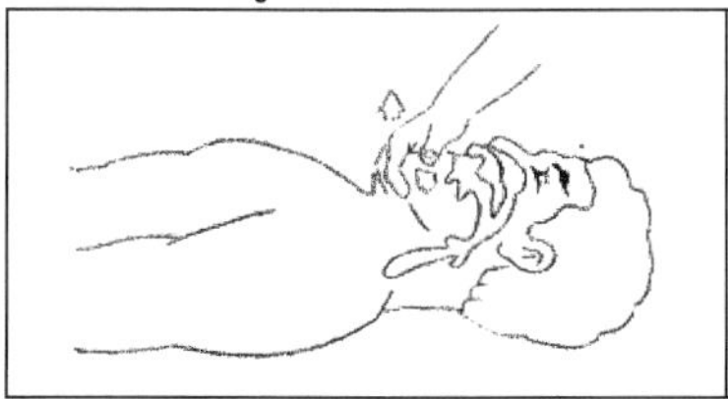

Figure 12-2. JAW THRUST

b. **Remove** debris (teeth, blood clots, bone) from the oral cavity; use suction if you have it; and place airway adjuncts to allow the victim to breathe through their nose (Figure 12-3) or mouth (Figure 12-4).

Figure 12-3. NASAL AIRWAY

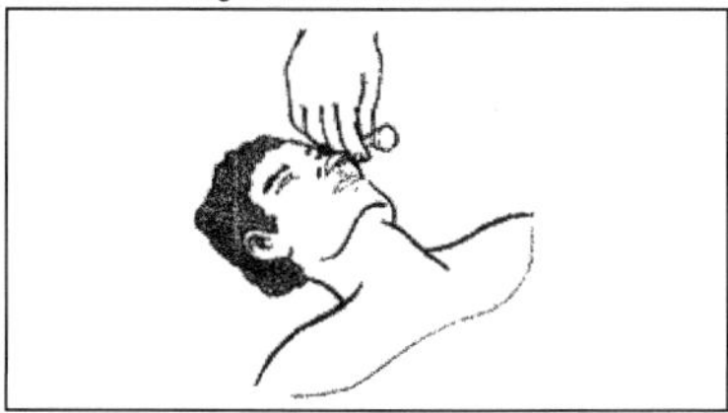

Figure 12-4. MOUTH AIRWAY

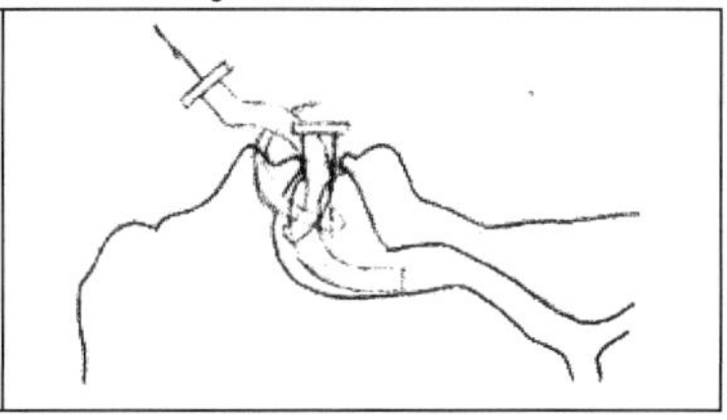

12-5. **BREATHING**. If the patient is having difficulty breathing, (1) expose the chest and identify open chest injuries; (2) place occlusive dressing (plastic and tape, or a COTS chest dressing) over open entry and exit chest wounds; and (3) place the patient on his injured side, or position him where he can breathe most comfortably.

12-6. **BLEEDING**. Quickly identify and control bleeding. (1) Apply a standard or pressure dressing. (2) If this does not control the bleeding, immediately apply a tourniquet. (3) Reassess dressings frequently to ensure bleeding has been controlled.

12-7. **SHOCK**. Shock causes an inadequate flow of oxygen to the tissues. Shock can cause bleeding (hemorrhagic) or not (nonhemorrhagic). Signs and symptoms of shock can include increased pulse, increased respiration; and decreased level of consciousness. (1) Open the victim's airway; (2) restore his breathing; (3) control the bleeding; (4) initiate an IV or a saline lock; and (5) monitor his condition.

12-8. **EXTREMITY INJURIES**. Identify and control bleeding. If you suspect a fracture, splint it as it lies. Do not reposition injured extremity.

12-9. **ABDOMINAL INJURIES**. Identify and control bleeding. Treat for shock. (1) If internal organs are exposed, cover them with dry, sterile dressing. Do not place organs back in wound area. Do not handle internal organs. (2) Place patient in position of comfort. Flex knees to relax abdomen. (3) Do not give anything by mouth to the patient.

12-10. **BURNS**. Remove patient from burn source. Then (1) remove all clothing and jewelry from the area; (2) cover with dry, sterile dressings, and ensure fingers and toes have dressings between them before covering entire area; and (3) Evacuate immediately any casualties with burns of the face, neck, hands, genitalia, or over 20 percent of his body surface.

12-11. **WEATHER (HEAT AND COLD) INJURIES**. Tables 12-1, 12-2, and 12-3 show first aid for (respectively) heat injuries, cold injuries, and environmental injuries.

Table 12-1. HEAT INJURIES

INJURY	*SIGNS/SYMPTOMS*	*FIRST AID*
Heat Cramps	Casualty experiences muscle cramps in arms, legs and/or stomach, may also have wet skin and extreme thirst.	1. Move the casualty to a shaded area and loosen clothing. 2. Allow casualty to drink 1 quart of cool water slowly per hour. 3. Monitor casualty and provide water as needed. 4. Seek medical attention if cramps persist.
Heat Exhaustion	Casualty experiences loss of appetite, headache, excessive sweating, weakness or faintness, dizziness, nausea, muscle cramps. The skin is moist, pale, and clammy.	1. Move the casualty to a cool, shaded area and loosen clothing. 2. Pour water on casualty and fan to increase cooling effect of evaporation. 3. Provide at least one quart of water to replace lost fluids. 4. Elevate legs. 5. Seek medical aid if symptoms continue.
Heatstroke (Sunstroke)	Casualty stops sweating (hot, dry skin), may experience headache, dizziness, nausea, vomiting, rapid pulse and respiration, seizures, mental confusion. Casualty may suddenly collapse and lose consciousness. THIS IS A MEDICAL EMERGENCY!	1. Move casualty to a cool, shaded area, loosen clothing, and remove outer clothing if the situation permits. 2. Immerse in cool water. If cool bath is not available, massage arms and legs with cool water. Fan casualty to increase the cooling effect of evaporation. 3. If conscious, slowly consume one quart of water. 4. SEEK MEDICAL AID AND EVACUATE AS SOON AS POSSIBLE. Perform any lifesaving measures.

Table 12-2. COLD INJURIES

INJURY	***SIGNS/SYMPTOMS***	***FIRST AID***
Chilblain	Red, swollen, hot, tender, itching skin. Continued exposure may lead to infected (ulcerated bleeding) skin lesions.	1. Area usually responds to locally applied warming (body heat). 2. Do Not rub or massage area. 3. Seek medical treatment.
Immersion (trench) foot	Affected parts are cold, numb, and painless. As parts warm, they may be hot, with burning and shooting pains. Advanced stage: skin pale with bluish cast: pulse decreases, blistering, swelling, heat hemorrhages, and gangrene may follow.	1. Gradual warming by exposure to warm air. 2. DO NOT massage or moisten skin. 3. Protect affected parts from trauma. 4. Dry feet thoroughly: avoid walking. 5. Seek medical treatment.
Frostbite	***SUPERFICIAL*** Redness, blisters in 24-36 hours and sloughing of the skin. ***DEEP*** Preceded by superficial frostbite; skin is painless, pale- yellowish, waxy, "wooden" or solid to touch, blisters form in 12-36 hours	***SUPERFICIAL*** 1. Keep casualty warm; gently warm affected parts. 2. Decrease constricting clothing, increase exercise and insulation. ***DEEP*** 1. Protect the part from additional injury. 2. Seek medical treatment as fast as possible.
Snow Blindness	Red, scratchy, or watery eyes; headache; increased pain in eyes with exposure to light.	1. Cover the eyes with a dark cloth. 2. Seek medical treatment.
Dehydration	Similar to heat exhaustion.	1. Keep warm, loosen clothes. 2. Replace lost fluids, rest, and additional medical treatment.
Hypothermia	Casualty is cold, shivers uncontrollably until shivering stops. Rectal (core) temp less 95° F can affect consciousness. Uncoordinated movements, shock, and coma may occur as body temperature drops.	***MILD HYPOTHERMIA*** 1. Warm body evenly and without delay. (Heat source must be provided.) 2. Keep dry, protect from elements. 3. Warm liquids may be given to conscious casualty only. 4. Be prepared to start CPR. 5. Seek medical treatment immediately. ***SEVERE HYPOTHERMIA*** 1. Quickly stabilize body temperature. 2. Attempt to prevent further heat loss. 3. Handle the casualty gently. 4. Evacuate to nearest medical treatment facility as soon as possible.

Table 12-3. ENVIRONMENTAL INJURIES

TYPE	*FIRST AID*
Snake bite	1. Get the casualty away from the snake. 2. Remove all rings and bracelets from the affected extremity. 3. Reassure the casualty and keep him quiet. 4. Apply constricting band(s) 1 to 2 finger widths close to the bite. You should be able to slip 1 finger between the band and skin. ***ARM OR LEG BITE*** - Place one band above and one band below the bite site. ***HAND OR FOOT BITE*** - Place one band above the wrist or ankle. 5. Immobilize the affected limb below the level of the heart. 6. Kill the snake, if possible, (without damaging its head or endangering yourself) and send it with the casualty. 7. Seek medical treatment immediately.
Brown recluse or black widow, spider bite	1. Keep the casualty calm. 2. Wash the area. 3. Apply ice or a freeze pack, if available. 4. Seek medical treatment.
Tarantula bite, scorpion sting, ant bite	1. Wash the area. 2. Apply ice or a freeze pack, if available. 3. Apply baking soda, calamine lotion, or meat tenderizer to the bite site to relieve pain and itching. 4. If site of bite(s) or sting(s) is on the face, neck (possible airway blockage), or genital area, or if reaction is severe, or if the sting is by the dangerous Southwestern scorpion, keep the casualty as quiet as possible, administer the epinephrine auto-injector (EAI) if needed, and then seek immediate medical aid.
Wasp or bee sting	1. If the stinger is present, remove by scraping with a knife or finger nail. DO NOT squeeze venom sack on stinger, more venom may be injected. 2. Wash the area. 3. Apply ice or freeze pack, if available. 4. If allergic signs or symptoms appear, be prepared to administer the EAI, and then seek medical assistance.
Human and other animal bites	1. Cleanse the wound thoroughly with soap or detergent solution. 2. Flush bite well with water. 3. Cover bite with a sterile dressing. 4. Immobilize injured extremity. 5. Transport casualty to a medical treatment facility. 6. For animal bites, without endangering yourself or damaging the animal's head, kill the animal and send its head with the casualty. 7. For human bites, try to extract some of the attacker's saliva from the wound and send that in a sealed, identified container with the casualty.

Table 12-3. ENVIRONMENTAL INJURIES (CONTINUED)

Poison Ivy, Oak, Sumac	1. Gently clean affected area. Clean areas 2-3 times daily. Wash clothing. 2. Apply topical anti-itch lotion or ointment as needed, and cover. 3. Avoid scratching or itching the area. 4. Observe for signs of infection (increasing redness, tenderness, warm to touch). 5. Seek medical attention if rash persists or signs of infection develop.

12-12. **POISONOUS PLANT IDENTIFICATION**

Poison plants include, among others, poison ivy, oak, and sumac, as well as a few more such as stinging nettles which are not discussed here.

a. **Poison Ivy**. Poison ivy is grows in two forms, a vine or shrub. The compound leaves of poison ivy consist of three pointed leaflets; the middle leaflet has a much longer stalk than the two side ones. The leaflet edges can be smooth or toothed but are rarely lobed. The leaves vary greatly in size, from 8 to 55 mm (0.31" to 2.16") in length. They are reddish when they emerge in the spring, turn green during the summer, and become various shades of yellow, orange, or red in the autumn. Small greenish flowers grow in bunches attached to the main stem close to where each leaf joins it. Later in the season, clusters of poisonous, berrylike drupes form. They are whitish, with a waxy look.

b. **Poison Oak**. Poison oak is a widespread deciduous shrub throughout mountains and valleys of North America, generally below 5,000 feet elevation. It commonly grows as a climbing vine with aerial (adventitious) roots that adhere to the trunks of oaks and sycamores. Poison oak also forms dense thickets in chaparral and coastal sage scrub. Poison oak leave typically have three leaflets (sometimes five), the terminal one on a slender rachis (also called a stalk or petiolule). Eastern poison ivy often has a longer rachis and the leaflet margins tend to be less lobed and serrated (less "oak-like"). Like many members of the sumac family (Anacardiaceae) new foliage and autumn leaves often turn brilliant shades of pink and red.

c. **Poison Sumac**. Poison Sumac is a woody perennial shrub or small tree growing from 5 - 25 feet tall that grows in peat bogs and swamps. To identify Poison Sumac, look for the fruit that grows between the leaf and the branch. Leaves are arranged in an alternate pattern on the vine. There are about 7-13 leaflets forming a feather-like appearance. The foliage has brilliant orange or scarlet coloring in the fall.

Figure 12-5. POISONOUS PLANTS

12-13. **FOOT CARE.** Use moleskin to prevent blisters prior to movement or foot march. Drain large blisters. Clean area, puncture with needle, drain blister. Place moleskin over area. Observe for signs of infection. Keep feet as clean and dry as possible. Use foot powder and change socks. Let feet air dry as mission permits.

12-14. **CASUALTY RESCUE AND TRANSPORT SYSTEM.** Remove the hoistable CRTS litter from its pack, and place the litter on the ground. Unfasten the retainer strap, step on the foot end of the litter, and unroll the litter completely to the opposite end. Then--

a. Bend the CRTS litter in half and backroll it. Repeat with the opposite end of the litter. The litter will now lay flat. Lift the sides of the litter, and fasten the four cross-straps to the buckles directly opposite the straps.

b. Place the CRTS litter next to the casualty. Insure the head end of the stretcher is adjacent to the head of casualty. Place the cross-straps under the litter. Logroll the casualty and slide the litter as far under the casualty as possible. Gently roll the casualty down on the litter. Slide the casualty to the center of the litter. Be sure to keep his spinal column as straight as possible.

c. Pull the straps out from under the CRTS litter and fasten it to the buckles. Position the foot end of the litter at the head of the casualty. Have one rescuer straddle the litter and support the casualty's head, neck, and shoulders.

d. Grasp the foot straps of the litter, and slide the litter under the casualty. Center the casualty on the litter, and feed foot straps thru the unused grommets at the foot end of the litter. Fasten the straps to the buckles.

12-15. **HYDRATION AND ACCLIMATIZATION**

a. Table 12-4 shows strategies for minimizing dehydration and increasing acclimatization and good hydration practices.

Table 12-4. HYDRATION MANAGEMENT AND ACCLIMATIZATION

STRATEGY	***SUGGESTIONS FOR IMPLEMENTATION***
Start early	1. Start at least 1 month prior to School 2. Be flexible and patient: performance benefits take longer than the physiological benefits
Mimic the training environment climate	1. In warm climates, acclimatize in the heat of day. 2. In temperate climates workout in a warm room wearing sweats.
Ensure adequate heat stress	1. Induce sweating. 2. Work up to 100 minutes of continuous physical exercise in the heat. Be patient. The first few days, you may not be able to go 100 minutes without resting. 3. Once you can comfortably exercise for 100 minutes in the heat, then continue for at least 7-14 days with added exercise intensity (loads or training runs).
Teach yourself to drink and eat	1. Your thirst mechanism will improve as you become heat acclimatized, but you will still under-drink if relying on thirst sensation. 2. Heat acclimatization will increase your water requirements. 3. Dehydration will negate most benefits of physical fitness and heat acclimatization. 4. You will sweat out more electrolytes when not acclimatized, so add salt to your food, or drink electrolyte solutions during the first week of heat acclimatization. 5. A convenient way to learn how much water your body needs to replace is to weigh yourself before and after the 100 minutes of exercise in the heat. For each pound lost, you should drink about one-half quart of fluid. 6. Do not skip meals, as this is when your body replaces most of its water and salt losses.

12-16. **WORK, REST, AND WATER CONSUMPTION.** Table 12-5 shows a work, rest, and water consumption table. The guidance applies to the average size, heat acclimated Soldier wearing ACU, not hot weather gear, except as specified otherwise. The work and rest times and fluid replacement volumes shown will help the Soldier sustain his performance and hydration for at least 4 hours of work in the specified heat category. Fluid needs can vary based on individual differences (±1 quart per hour).

a. "NL" means that there is no limit to work time per hour. "Rest" means minimal physical activity such as sitting or standing, preferably in the shade.

b. Consume no more than 1.5 quarts of fluid per hour, and no more than 12 quarts per day.

c. If you are wearing body armor in a humid climate, then add 5° F to the WBGT. if wearing MOPP 4 clothing, add 10° F to the WBGT.

d. Work categories include easy, moderate, and hard.

1) ***Easy Work***. This includes, for example, maintaining weapons; walking on hard surfaces at 2.5 mph with a load weighing no more than 30 pounds; participating in marksmanship training; and participating in drills or ceremonies.

2) ***Moderate Work***. This includes, for example, walking in loose sand at 2.5 mph (maximum) or under with no load; walking on a hard surface at 3.5 mph (maximum) with a load weighing no more than 40 pounds; performing calisthenics; patrolling; or conducting individual movement techniques such as the low or high crawl.

3) ***Hard Work***. This includes, for example, walking on a hard surface at 3.5 mph with a load weighing 40 or more pounds; walking in loose sand at 2.5 mph while carrying a load; and conducting field assaults.

Table 12-5. WORK, REST, AND WATER CONSUMPTION TABLE

HEAT CATEGORY	WBGT INDEX, F°	*EASY WORK*		*MODERATE WORK*		*HARD WORK*	
		WORK/ REST	WATER INTAKE (QT/H)	WORK/ REST	WATER INTAKE (QT/H)	WORK/ REST	WATER INTAKE (QT/H)
1	78° to 81.9°	NL	0.50	NL	0.75	40/20	0.75
2 (Green)	82° to 84.9°	NL	0.50	50/10	0.75	30/30	1.00
3 (Yellow)	95° to 87.9°	NL	0.75	40/20	0.75	30/30	1.00
4 (Red)	88° to 89.9°	NL	0.75	30/30	0.75	20/40	1.00
5 (Black)	90° or more	50/10 min	1.00	20/40	1.00	10/50	1.00

NOTES

Chapter 13
DEMOLITIONS

This chapter introduces Rangers to the characteristics of explosives, to initiation systems, MDI components, detonation systems, safety considerations, expedient explosives, breaching charges, and timber cutting charges (FM 5-250). The two categories of explosives are low and high (Table 13-1).

- *Low explosives* have a detonating velocity up to 1,300 feet per second, which produces a pushing or shoving effect.
- *High explosives* have a detonating velocity of 3,280 to 27,888 feet per second, which produces a shattering effect.

Table 13-1. CHARACTERISTICS OF US DEMOLITIONS EXPLOSIVES

NAME	*APPLICATIONS*	*DETONATION VELOCITY*		*RE FACTOR**	*FUME TOXICITY*	*WATER RESISTANCE*
		MIN/SEC	*FT/SEC*			
AMMONIUM NITRATE	CRATERING CHG	2,700	8.800	0.42	DANGEROUS	POOR
PETN	DET CORD BLASTING CAPS DEMOLITION CHGS	8,300	27,200	1.66	SLIGHT	EXCELLENT
RDX	BLASTING CAPS COMPOSITION EXPLOSIVE	8,350	27,400	1.60	DANGEROUS	EXCELLENT
TRINITROTOLUENE (TNT)	DEMOLITION CHG COMPOSITION EXPLOSIVE	6,900	22,600	1.00	DANGEROUS	EXCELLENT
TETRYL	BOOSTER CHG COMPOSITION EXPLOSIVE	7,100	23,300	1.25	DANGEROUS	EXCELLENT
NITROGLYCERIN	COMMERCIAL DYNAMITE	7,700	25,200	1.50	DANGEROUS	GOOD
BLACK POWDER	TIME FUSE	400	1,300	0.55	DANGEROUS	POOR
AMATOL 80/20	BURSTING CHG	4,900	16,000	1.17	DANGEROUS	POOR
COMPOSITION A3	BOOSTER CHG BURSTING CHG	8,100	26,500	---	DANGEROUS	GOOD
COMPOSITION B	BURSTING CHG	7,800	25,600	1.35	DANGEROUS	EXCELLENT
COMPOSITION C4 (M112)	CUTTING CHG BREACHING CHG	8.040	26,400	1.34	SLIGHT	EXCELLENT
COMPOSITION H6	CRATERING CHG	7,190	23,600	1.33	DANGEROUS	EXCELLENT
TETRYTOL 75/25	DEMOLITION CHG	7,000	23,000	1.20	DANGEROUS	EXCELLENT
PENTOLITE 50/50	BOOSTER CHG BURSTING CHG	7,450	24,400	---	DANGEROUS	EXCELLENT

NAME	APPLICATIONS	DETONATION VELOCITY		RE FACTOR*	FUME TOXICITY	WATER RESISTANCE
		MIN/SEC	FT/SEC			
M1 DYNAMITE	DEMOLITION CHG	6,100	20,000	0.92	DANGEROUS	FAIR
DET CORD	PRIMING DEMOLITION CHG	6,100 TO 7,300	20,000 TO 24,000	---	SLIGHT	EXCELLENT
SHEET EXPLOSIVE M118 AND M186	CUTTING CHG	7,300	24,000	1.14	DANGEROUS	EXCELLENT
BANGALORE TORPEDO M1A2	DEMOLITION CHG	7,800	25,600	1.17	DANGEROUS	EXCELLENT
SHAPED CHARGES M2A3, M2A4, AND M3A1	CUTTING CHG	7,800	25,600	1.17	DANGEROUS	EXCELLENT
** TNT = 1.00 RELATIVE EFFECTIVENESS*						

13-1. **INITIATING (PRIMING) SYSTEMS**. The best way to prime demolition systems is with modernized demolition initiators (MDI). These consist of blasting caps attached to various lengths of time fuse or shock tube. MDI can be used with a fuse igniter and detonating cord to create many firing systems. In the absence of MDI, field expedient methods may be used.

a. **Shock Tube**.

(1) Thin, plastic tube of extruded polymer with a layer of special explosive material on the interior surface.

(2) Explosive material propagates a detonation wave which moves along the shock tube to a factory crimped and sealed blasting cap.

(3) Detonation is normally contained within the plastic tubing. However, burns may occur if the shock tube is held.

(4) Advantages of shock tube.

(a) The shock tube offers the instantaneous action of electric initiation without the risk of accidental initiation caused by radio transmitters, static electricity, and so on.

(b) Extremely reliable.

(c) May be extended using left over pieces from previous operations.

b. **Blasting Caps**. Five MDI blasting caps are available which replace the M6 electric and M7 Non-electric blasting cap. Three of these are high-strength caps and two are low-strength. High-strength blasting caps can be used to prime all standard military explosives (including detonating cord) or to initiate the shock tube of other MDI blasting caps.

(1) ***M11***.

- Factory crimped to 30 feet of shock tube.
- A movable "J" hook is attached for quick and easy attachment to detonating cord.
- A red flag is attached 1 meter from the blasting cap and a yellow flag 2 meters from the blasting cap.

(2) ***M14***.

- Factory crimped to 7.5 feet of time fuse.
- May be initiated using a fuse igniter or match.
- Burn-time for total length is about five minutes.
- Yellow bands indicate calibrated one-minute time intervals.

(3) ***M15***.

- Two blasting caps factory crimped to 70 feet of shock tube.
- Each blasting cap has delay elements to allow for staged detonations.
- Low-strength blasting caps. Used as a relay device to transmit a shock tube detonation impulse from an initiator to a high strength-blasting cap.

(4) ***M12***. Factory crimped to 500 feet of shock tube on a cardboard spool.

(5) ***M13***. Factory crimped to 1,000 feet of shock tube.

c. **Matches**. If fuse igniter is unavailable, light the time blasting fuse with a match. Split the fuse at the end (Figure 13-1), and place the head of an unlit match in the powder train. Light the inserted match head with a flaming match, or rub the abrasive on the match box against it. It may be necessary to use two match heads during windy conditions. Burn time will increase with altitude and colder temperatures.

d. **M81 Fuse Igniter**. This is used to ignite time blasting fuse or to initiate the shock tube of MDI blasting caps.

Figure 13-1. LIGHTING TIME FUSE WITH A MATCH

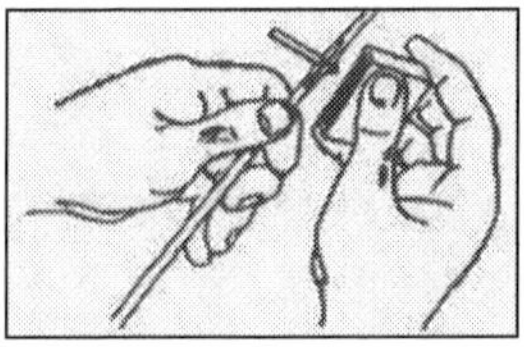

Notes: Low strength blasting caps cannot reliably set off explosives. They should only be used to set off additional shock tubes. The M60 fuse igniter may still be used to ignite time blasting fuse. However, it will not reliably initiate the shock tube.

13-2. **DETONATION (FIRING) SYSTEMS**. The two types of firing systems are MDI alone, or MDI plus detonating cord.

a. **MDI Alone**. An MDI firing system is one in which the initiation set, transmission and branch lines are constructed using MDI components and the explosive charges are primed with MDI blasting caps. Construct the charge in the following manner.

(1) Emplace and secure explosive charge (C4, TNT, cratering charge, and so on.) on target.

(2) Place a sandbag or other easily identifiable marker over the M11, M14, or M15 blasting cap to be used.

(3) Connect to an M12 or M13 transmission line if desired.

(4) Connect blasting cap with shock tube to an M14 cap with time fuse. Cut the time blasting fuse to desired delay time.

(5) Prime the explosive charge by inserting the blasting cap into the charge.

(6) Visually inspect firing system for possible misfire indicators such as cracks, bulges, or corrosion.

(7) Return to the firing point and secure a fuse igniter to the cut end of the time fuse.

(8) Remove the safety cotter pin from the igniter's body.

(9) Actuate the charge by grasping the igniter body with one hand while sharply pulling the pull ring.

b. **MDI and Detonating Cord**. Construct the charge using the above steps for MDI stand-alone system. Incorporate detonating cord branch lines into the system using the "J" hooks of the M11 shock tube. Taping the ends of the Detonation Cord reduces the effect of moisture on the system.

13-3. **SAFETY**. MDI is not recommended for below ground use, except in quarry operations with water-gel or slurry explosives. Use detonating cord when it is necessary to bury primed charges.

a. Do not handle misfires downrange until the required 30 minute waiting period for both primary and secondary initiation systems has elapsed and other safety precautions have been accomplished.

b. Never yank or pull hard on the shock tube. This may actuate the blasting cap.

c. Do not dispose of used shock tubes by burning because of potentially toxic fumes given off from the burning plastic.

d. Do not use M1 dynamite with the M15 blasting cap. The M15 delay-blasting cap should be used only with water-gel or slurry explosives.

e. Always use protective equipment when handling demolitions. Minimum protection consists of leather gloves, ballistic eye protection, and helmet.

13-4. **EXPEDIENT EXPLOSIVES--IMPROVISED SHAPED CHARGE**. An improvised shaped charge (Figure 13-2) concentrates the energy of the explosion released on a small area, making a tubular or linear fracture in the target.

a. The versatility and simplicity of these charges make them effective against targets, especially those made of concrete or those with armor plating.

(1) Bowls, funnels, cone shaped glasses, (champagne glasses with stem removed) used as cones. Champagne or cognac bottles are excellent.

(2) Charge characteristics.

(a) *Cavity Liners.* These are made of copper, tin, or zinc. If none is available, cut a cavity out of the plastic explosive.

(b) *Cavity Angle.* This will work with 30 to 60 degree angles. The cavity angle in most high-explosive antitank (HEAT) ammunition is 42 to 45 degrees.

(c) *Explosive Height (In Container).* 2 x the height of the cone measured from the base of the cone to the top of the explosive.

(d) *Standoff.* Normal standoff is one and one half the cone's diameter.

(e) *Detonation Point.* The exact top center of the charge is the detonation point. Cover the blasting cap with a small amount of C4 if any part of the blasting cap is exposed.

b. Remove the narrow neck of a bottle or the stem of a glass by wrapping it with a piece of soft, absorbent twine or by soaking the string in gasoline and lighting it. Place two bands of adhesive tape, one on each side of the twine, to hold the twine firmly in place. The bottle or stem must be turned continuously with the neck up, to heat the glass uniformly.

c. A narrow band of plastic explosive placed around the neck and burned gives the same result. After the twine or plastic has burned, submerge the neck of the bottle in water and tap it against some object to break it off. Tape the sharp edge of the bottle to prevent cutting hands while tamping the explosive in place.

d. Do not immerse the bottle in water before the plastic has been completely burned, or it could detonate.

Figure 13-2. IMPROVISED SHAPED CHARGE

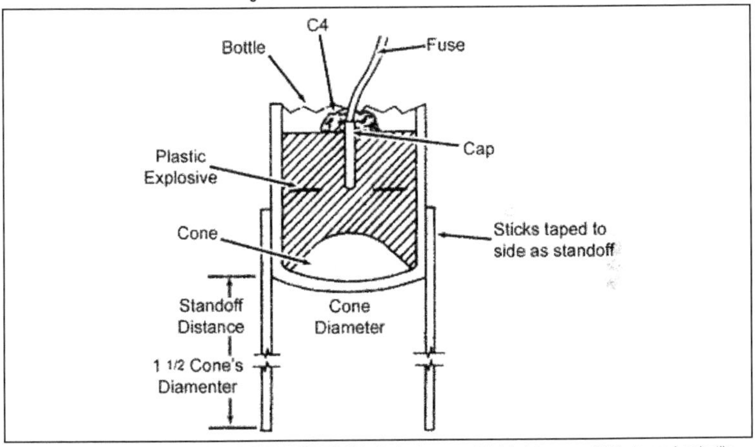

13-5. **EXPEDIENT EXPLOSIVES--PLATTER CHARGE**. This device turns a metal plate into a powerful blunt-nosed projectile (Figure 13-3). The plate should be steel, preferably round but square will work, and should weigh from 2 to 6 pounds.

a. The weight of the explosive should equal the weight of the platter.

b. Uniformly pack the explosive behind the platter. You will only need a container if the explosives fail to remain firmly against the platter. You can use tape to anchor the explosives, if needed.

c. Prime the charge at the exact, rear center of the charge. If any part of the blasting cap is exposed, cover the blasting cap with a small quantity of C4.

d. Aim charge at the direct center of the target, and ensure that the charge is on the opposite side of the platter from the target. Effective range is 35 yards for a small target. With practice, you might hit a 55-gallon drum at 25 yards 90 percent of the time. A gutted fuse igniter can serve as an expedient aiming device.

Figure 13-3. PLATTER CHARGE

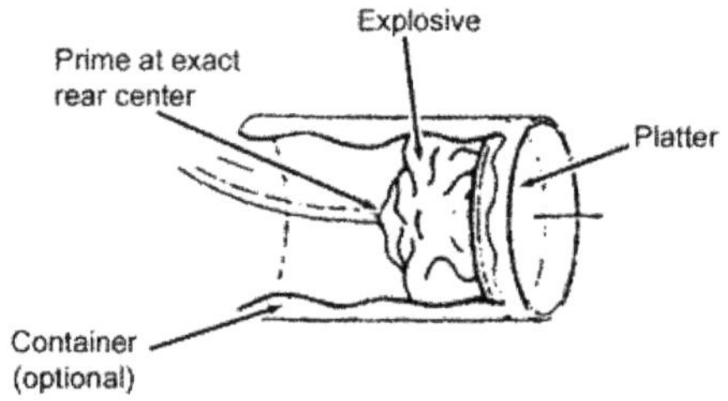

13-6. **EXPEDIENT EXPLOSIVES--GRAPESHOT CHARGE**. To use this antipersonnel fragmentation mine (Figure 13-4)--

a. **Hole**. Create a hole in the center, bottom of the container, for the blasting cap.

b. **Explosives**. Place explosives evenly on the bottom of the container. Remove all voids and air pockets by pressing the C4 into place using a non-sparking instrument.

c. **Buffer**. Place buffer material directly over top of the explosives.

d. **Projectiles**. Place projectiles over top of the buffer materials, then cover to prevent spilling from movement.

e. **Aim**. Aim at target from about 100 feet. Use a small amount of C4 on any exposed portion of the blasting cap.

Figure 13-4. GRAPESHOT CHARGE

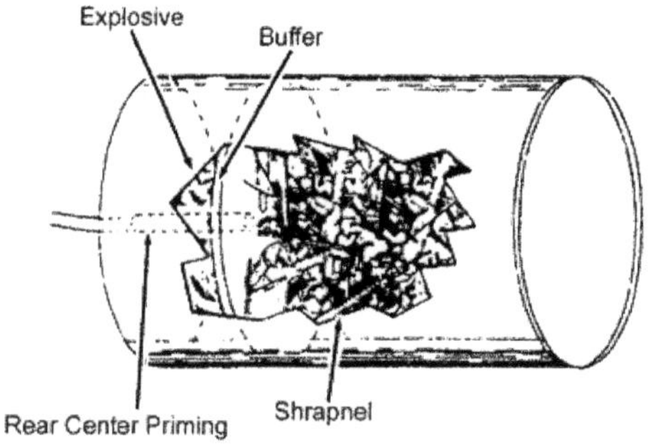

13-7. **DEMOLITION KNOTS.** Several knots are used in demolitions. Figures 13-5 and 13-6 show a few simple knots that can join demolitions to detonation cord.

Figure 13-5. VARIOUS JOINING KNOTS USED IN DEMOLITIONS

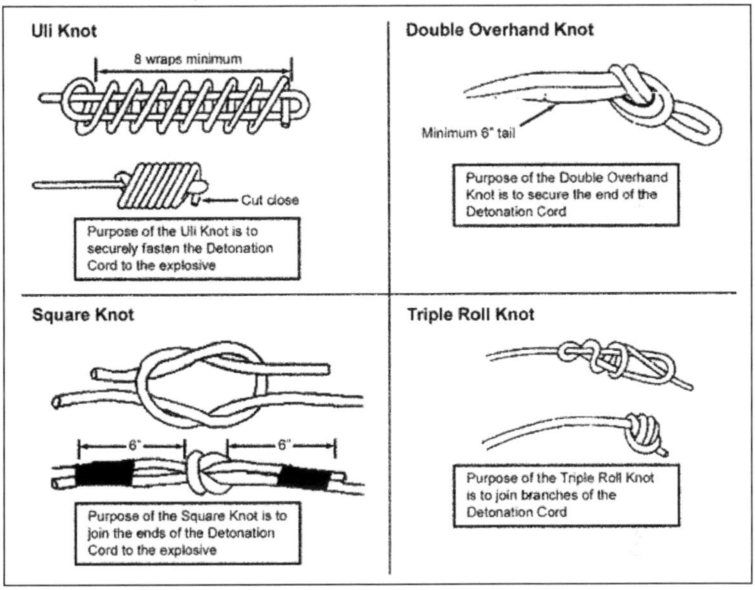

Figure 13-6. BRITISH JUNCTION

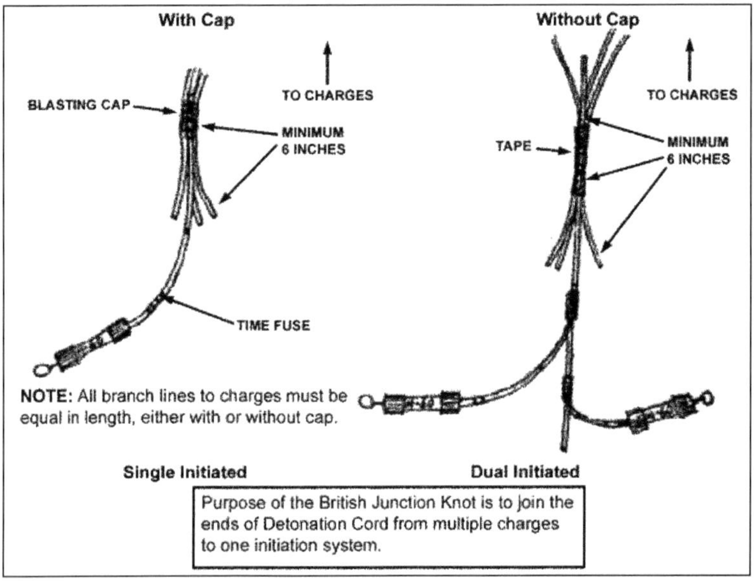

13-8. **MINIMUM SAFE DISTANCES**. Rangers must remain especially aware of their situation when using demolitions. (Table 13-2 shows minimum safe distances for employing demolitions.) For charges over 500 pounds, use the formulas shown in Figure 13-7.

Table 13-2. MINIMUM SAFE DISTANCE FOR PERSONNEL IN OPEN (BARE CHARGE)

EXPLOSIVE WEIGHT (LB)	SAFE DISTANCE		EXPLOSIVE WEIGHT (LB)	SAFE DISTANCE	
	FEET	METERS		FEET	METERS
27 OR LESS	985	300	175	1,838	560
30	1,021	311	200	1,920	585
35	1,073	327	225	1,999	609
40	1,123	342	250	2,067	630
45	1,168	356	275	2,136	651
50	1,211	369	300	2,199	670
60	1,287	392	325	2,258	688
70	1,355	413	350	2,313	705
80	1,415	431	375	2,369	722
90	1,474	449	400	2,418	737
100	1,526	465	425	2,461	750
125	1,641	500	500	2,625	800
150	1,752	534	---	---	---

Figure 13-7. MINIMUM SAFE DISTANCES FOR CHARGES OVER 500 POUNDS

$$\textit{safe distance (meters)} = 100\sqrt[3]{\textit{pounds of explosive}}$$

$$\textit{safe distance (feet)} = 300\sqrt[3]{\textit{pounds of explosive}}$$

Notes: For charges on targets, the minimum radius of danger is 1,000 meters. Minimum safe distance when in a missile-proof shelter from the point of detonation is 100 meters.

13-9. **BREACHING CHARGES**. For Table 13-3, the left column represents the thickness of reinforced concrete wall. The remaining 7 columns show the number of packages of C4 required to breach the wall using the charge placements shown in the drawings above the columns.

a. Use Table 13-3, 13-4 and 13-5 for breaching charges.

b. Use the formula in Figure 13-8 to calculate the charges.

c. Multiply number of packages of C4 from Table 13-3 by conversion factor from Table 13-4 for materials other than reinforced concrete.

Table 13-3. BREACHING CHARGES FOR REINFORCED CONCRETE

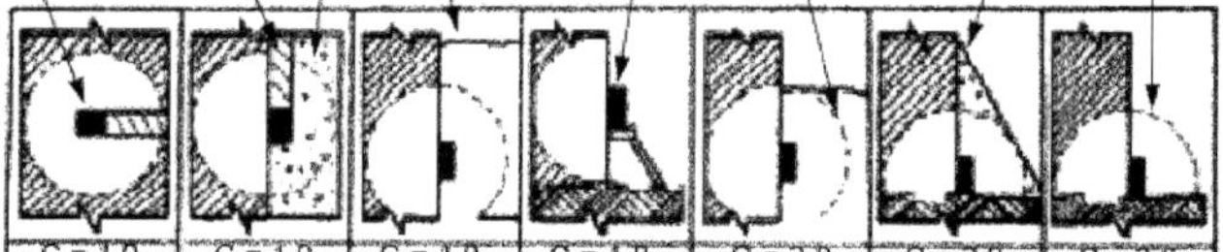

Rein-forced-Concrete Thick-ness (ft)	Placement Methods: Placed in center of mass, C = 1.0	Tamped or stemmed, C = 1.0	Deep water, C = 1.0	Elevated untamped, C = 1.8	Shallow water, C = 2.0	Earth tamping, C = 2.0	Ground placed, untamped, C = 3.6
	Packages of M112 (C4)						
2.0	1	5	5	9	10	10	17
2.5	2	9	9	17	18	18	33
3.0	2	13	13	24	26	26	47
3.5	4	21	21	37	41	41	74
4.0	5	31	31	56	62	62	111
4.5	7	44	44	79	88	88	157
5.0	9	48	48	85	95	95	170
5.5	12	63	63	113	126	126	226
6.0	13	82	82	147	163	163	293
6.5	17	104	104	186	207	207	372
7.0	21	111	111	200	222	222	399
7.5	26	137	137	245	273	273	490
8.0	31	166	166	298	331	331	595

NOTE: The results of all calculations for this table have been rounded UP to the next whole package.

Table 13-4. CONVERSION FACTORS FOR MATERIALS OTHER THAN REINFORCED CONCRETE

MATERIAL	*CONVERSION FACTOR*
• EARTH	0.1
• ORDINARY MASONRY • HARDPAN • SHALE • ORDINARY CONCRETE • ROCK • GOOD TIMBER • EARTH CONSTRUCTION	0.5
• DENSE CONCRETE • FIRST-CLASS MASONRY	0.7

Table 13-5. MATERIAL FACTOR (K) FOR BREACHING CHARGES

Material	R	K
Earth	All values	0.07
Poor masonry, shale, hardpan, good timber, and earth construction	Less than 1.5 m (5 ft)	0.32
	1.5 m (5 ft) or more	0.29
Good masonry, concrete block, and rock	0.3 m (1 ft) or less	0.88
	Over 0.3 m (1 ft) to less than 0.9 m (3 ft)	0.48
	0.9 m (3 ft) to less than 1.5 m (5 ft)	0.40
	1.5 m (5 ft) to less than 2.1 m (7 ft)	0.32
	2.1 m (7 ft) or more	0.27
Dense concrete and first-class masonry	0.3 m (1 ft) or less	1.14
	Over 0.3 m (1 ft) to less than 0.9 m (3 ft)	0.62
	0.9 m (3 ft) to less than 1.5 m (5 ft)	0.52
	1.5 m (5 ft) to less than 2.1 m (7 ft)	0.41
	2.1 m (7 ft) or more	0.35
Reinforced concrete (factor does not consider cutting steel)	0.3 m (1 ft) or less	1.76
	Over 0.3 m (1 ft) to less than 0.9 m (3 ft)	0.96
	0.9 m (3 ft) to less than 1.5 m (5 ft)	0.80
	1.5 m (5 ft) to less than 2.1 m (7 ft)	0.63
	2.1 m (7 ft) or more	0.54

Figure 13-8. FORMULA FOR COMPUTING SIZES OF CHARGES TO BREACH CONCRETE, MASONRY, AND ROCK

$$P = R^3KC$$

where—

P = *TNT required, in pounds*

R = *breaching radius, in feet*

K = *material factor, which reflects the strength, hardness, and mass of the material to be demolished (Table 3-5)*

C = *tamping factor, which depends on the location and tamping of the charge (Figure 3-15)*

13-10. **TIMBER CUTTING CHARGES**. Table 13-6 shows timber-cutting charge sizes. Figures 13-9 through 13-15 show the types of charges and the formulas to use with each.

Table 13-6. TIMBER-CUTTING CHARGE SIZE

	Packages of C4 Required (1.25-lb Packages) by Timber Diameter (in)											
Charge Type	6	8	10	12	15	18	21	24	27	30	33	36
Internal	1	1	1	1	1	1	2	2	2	3	3	4
External	1	1	2	3	4	5	7	9	11	14	17	20
Abatis	---	---	---	---	---	---	---	7	9	11	14	16
NOTE: **Packages required are rounded UP to the next whole package.**												

Figure 13-9. ABATIS

Figure 13-10. FORMULA FOR FALLEN TREE OBSTACLES OR TEST SHOT

$$P = \frac{D^2}{50} = P = 0.02D^2$$

where—

P = TNT required per tree, in pounds

D = diameter or least dimension of dimensioned timber, in inches

Figure 13-11. TIMBER-CUTTING RING CHARGE

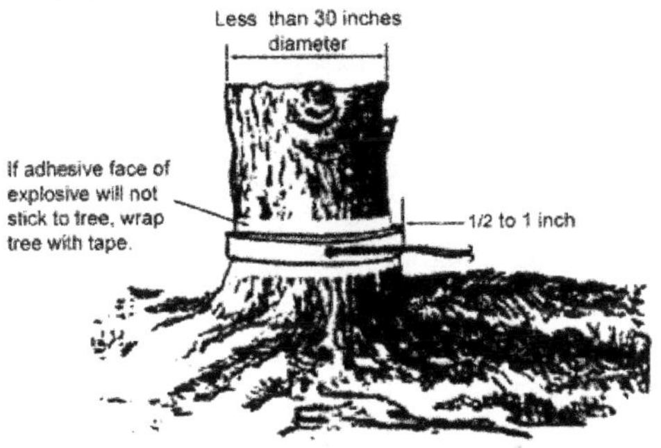

Figure 13-12. TIMBER-CUTTING CHARGE (EXTERNAL)

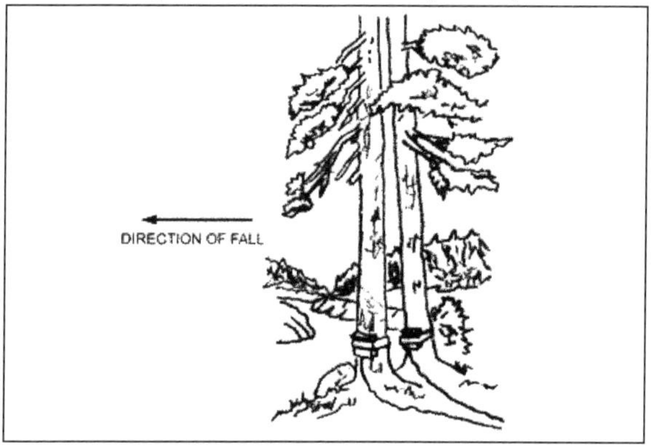

Figure 13-13. FORMULA FOR EXTERNAL TIMBER-CUTTING CHARGE

$$P = \frac{D^2}{40} \text{ or } P = 0.025D^2$$

where—

P = TNT required per target, in pounds

D = diameter or least dimension of dimensioned timber, in inches

Figure 13-14. TIMBER-CUTTING CHARGE (INTERNAL)

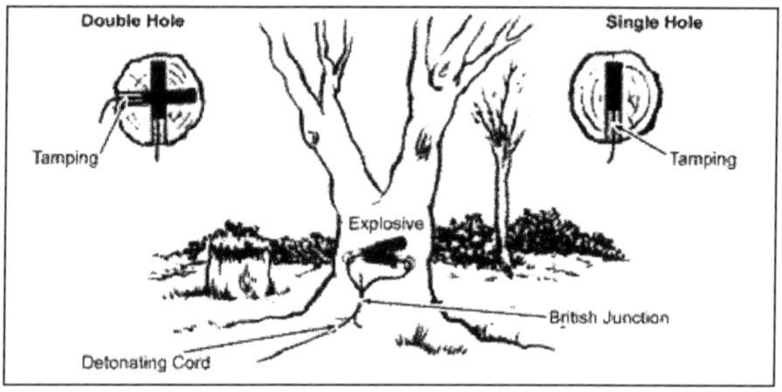

Figure 13-15. FORMULA FOR INTERNAL TIMBER-CUTTING CHARGE

$$P = \frac{D^2}{250} \text{ or } P = 0.004D^2$$

where—

P = *TNT required per tree, in pounds*

D = *diameter or least dimension of dressed timber, in inches*

NOTES

Chapter 14

RANGER URBAN OPERATIONS

Urban operations are planned and conducted in an area of operations (AO) that includes one or more urban areas. An urban area is a topographical complex where man-made construction or high population density is the dominant feature (FM 30-06.11). Urban terrain is likely to be one of the most significant future areas of operations for American forces throughout the world. Expanding urban development affects military operations as the terrain is altered. The increasing focus on stability and support operations, urban terrorism, and civil disorder emphasizes that combat in urbanized areas is unavoidable. Urban areas are the power centers, the centers of gravity, and thus the future battlefield. For further study, see FM 30-06.11, FM 90-10, FM 90-1, FM 7-8, 75th Ranger Regiment Advanced UO SOP, and Ranger Training Circular 350-1-2.

14-1. **TYPES**. Infantry units must be trained to conduct urban combat under high-intensity conditions. High-intensity urban combat requires the employment of combat power of the joint combined arms team. An Infantry unit's mission is normally to recon, isolate, penetrate, systematically clear, defend the urban area, and engage and defeat the enemy with decisive combat power. Although the changing world situation may have made urban combat under high-intensity conditions less likely for US forces, it represents the high end of the combat spectrum, and units must be trained for it. High-intensity urban operations can be casualty-intensive for both sides. With the integrated firepower of the joint, combined arms team, leaders must make every attempt to limit unnecessary destruction of critical infrastructure and casualties among noncombatants.

a. **Precision Conditions**. Infantry units train to defeat an enemy that is mixed with non-combatants in precision urban combat. Leaders plan to limit civilian casualties and collateral damage through the establishment of strict rules of engagement (ROE) and the employment of precision weapons and munitions. The ROE provides the focus for the use and restraint of combat power. The ROE may be significantly more restrictive than under high-intensity conditions.

b. **Surgical Conditions**. Operations conducted under surgical conditions include special-purpose raids, small precision strikes, or small-scale personnel seizure or recovery operations in an urban environment (for example, hostage rescue). Joint special operation forces usually conduct these operations. They may closely resemble US police operations performed by Special Weapons and Tactics (SWAT) teams. They may even involve cooperation between US forces and host nation police. Though regular units may not usually be involved in the actual surgical operation, they may support it by isolating the area, by providing security or crowd control, or providing search and rescue teams.

c. **Transitions**. Leaders must always be prepared to transition rapidly from one type of urban combat to another, and back. Real-world combat shows us that urban operations can deteriorate rapidly and without warning. A force involved in stability and support operations can suddenly find itself in a high-intensity combat situation.

14-2. **PRINCIPLES**

a. **Surprise**. Strike the enemy at a time or place or in a manner for which he is unprepared. Key to success: gives the assaulting element the advantage.

b. **Security**. Never permit the enemy to acquire unexpected advantage.

(1) Maintain during all phases of the operation.

(2) Four-dimensional battlefield (height, depth, width, subterranean).

(3) Always maintain 360 degree security (include elevated and subterranean areas).

(4) Mission is never complete as long as you remain in the urban environment. The status of actors in the urban environment does not afford the sense of security offered by "open" terrain. The key to survivability is a constant state of situational awareness.

c. **Simplicity**. Prepare clear, uncomplicated plans, and provide subordinates with concise orders to ensure thorough understanding.

(1) Always keep plans simple.

(2) Ensure everyone understands the mission and the commander's intent.

(3) Plan and prepare for the worst.

d. **Speed**. Rate of military action.

(1) Act as security.

(2) Move in a careful hurry.

(3) Smooth is fast and fast is smooth.

(4) Never move faster than you can accurately engage targets.

(5) Exercise tactical patience.

e. **Violence of Action**. Eliminate the enemy with sudden, explosive force.

(1) Combined with speed gives surprise.

(2) Prevents enemy reaction.

(3) Both physical and mental.

14-3. **METT-TC**. To effectively plan combat operations in urban environments, leaders must use Troop Leading Procedures and conduct a thorough analysis utilizing METT-TC factors. The following lists specific planning guidance that must be incorporated when planning for urban operations. For more specifics on mission planning, refer to Chapter 2, Ranger Handbook.

a. **Mission**. Know correct Task Organization to accomplish the mission (Offense, Defense, or Stability and Support Operations).

b. **Enemy**.

(1) ***Disposition***. Analyze the arrayal of enemy forces in and around your objective, known and suspected. Example: Known or suspected locations of minefields, obstacles, and strong points.

(2) ***Composition and Strength***. Analyze the enemy's task organization, troops available, suspected strength, and amount of support from local civilian populace based on intelligence estimates. Is the enemy a conventional or unconventional force?

(3) ***Morale***. Analyze the enemy's current operational status based on friendly intelligence estimates. For example, is the enemy well supplied; have they had recent success against friendly forces, taken many casualties; and what is the current weather?

(4) ***Capabilities***. Determine what the enemy can employ against your forces. Example: Enemy's weapons, artillery assets, engineer assets, air defense assets, NBC threats, thermal/NVG capabilities, close air support, armor threat, and so on.

(5) ***Probable Course(s) of Action***. Based on friendly intelligence estimates, determine how the enemy will fight within his area of operation (in and around your area of operation).

c. **Terrain**. Leaders conduct a detailed terrain analysis of each urban setting, considering the types of built-up areas and composition of existing structures. They use OCOKA when analyzing terrain, in and around the area of operation.

(1) ***Observation and Fields of Fire***. Always be prepared to conduct urban operations under limited visibility conditions. This includes the effects of reduced illumination, as well as natural and manmade obscuration. Leaders should ensure that Rangers are equipped with adequate resources, which allow them to successfully operate in the urban environment under these types of conditions. (2) ***Cover and Concealment***. Leaders must perform a thorough analysis of peripheral as well as intra-urban areas. Leaders should identify routes to objectives, which afford assault forces with the best possible cover and concealment. Additionally, leaders should take advantage of limited visibility conditions, which would allow forces to move undetected to their final assault / breaching positions. When in the final assault position, forces should move as rapidly as tactically possible to access structures, which afford additional cover and concealment. Leaders must learn to properly employ obscurants and exercise "tactical patience" to fully take advantage of these effects. Finally, all members of the urban force must practice noise and light discipline. Soldiers must avoid unnecessary voice communications, learn the proper use of white light, and limit contact with surfaces that may alert the enemy of their presence.

(3) ***Obstacles***. There are many manmade and natural obstacles on the periphery, as well as within the urban environment. Leaders should conduct a detailed reconnaissance of routes and objectives (this must include subterranean complexes), taking into consideration route adjustments and special equipment needs.

(4) ***Key Terrain***. Analyze which buildings, intersections, bridges, LZ/PZ, airports, and elevated areas that provide a tactical advantage to you or the enemy. Additionally, the leader must identify critical infrastructure within his area of operations, which would provide the enemy with a tactical advantage on the battlefield. These may include, but are not limited to, communication centers, medical facilities, governmental facilities, and facilities that are of psychological significance.

(5) ***Avenues of Approach***. Consider roads, intersections, inland waterways, and subterranean constructions (subways, sewers, and basements). Leaders should classify areas as go, slow go, or no-go based on the navigability of the approach.

Note: Military maps may not provide enough detail for urban terrain analysis or reflect the underground sewer systems, subways, water systems, or mass transit routes.

(6) ***Troops***. Analyze your forces utilizing their disposition, composition, strength, morale, capabilities, and so on. Leaders must also consider the type and size of the objective to plan effective use of troops available.

(7) ***Time***. Operations in an urban environment have a slower pace and tempo. Leaders must consider the amount of time required to secure, clear, or seize the urban objective and stress and fatigue Rangers will encounter. Additional time must also be allowed for area analysis efforts; these may include, but are not limited to--

- Maps and urban plans Recon and analysis.
- Hydrological data analysis.
- Line-of-sight surveys.
- Long Range Surveillance and Scout reconnaissance.
- Artillery.
- Armor.
- Aviation.
- Armor.
- Engineers.

(8) ***Civilians***. Authorities such as the National Command will establish the Rules of Engagement. Commanders at all levels may provide further guidance regarding civilians occupying the area of operations (AO). Leaders must daily reiterate the ROE to subordinates, and immediately inform them of any changes to the ROE. Rangers must have the discipline to identify the enemy from noncombatants and ensure civilians understand and follow all directed commands.

Note: Civilians may not speak English, may be hiding (especially small children), or dazed from a breach. Civilians must not be given the means to resist. Rehearse how clearing/search teams will react to these variables. Never compromise the safety of your Rangers.

14-4. **CLOSE QUARTERS COMBAT.** Due to the very nature of a CQC encounter, engagements will be very close (within 10 meters) and very fast (targets exposed for only a few seconds). Most close quarter's engagements are won by who hits first and puts the enemy down. It is more important to knock a Ranger down as soon as possible than it is to kill him. In order to win a close quarters engagement, Rangers must make quick, accurate shots by mere reflex. This is accomplished by reflexive fire training. Remember, no matter how proficient you are, always fire until the enemy goes down. All reflexive fire training is conducted with the eyes open.

Note: Research has determined that, on average, only three out of ten people actually fire their weapons when confronted by an enemy during room clearing operations. Close quarters combat success for the Ranger begins with the Ranger being psychologically prepared for the close quarter's battle. The foundation for this preparedness begins with the Ranger's proficiency in basic rifle marksmanship. Survival in the urban environment does not depend on advanced skills and technologies. Rangers must be proficient in the basics.

14-5. **REHEARSALS**. Similar to the conduct of other military operations, leaders need to designate time for rehearsals. Urban operations require a variety of individual, collective, and special tasks, which are not associated with operations on less complex terrain. These task require additional rehearsal time for clearing, breaching, obstacle reduction, casualty evacuation, and support teams. Additionally, rehearsal time must be identified for rehearsals with combined arms elements. These may include, but are not limited to--

a. **Stance**. Feet are shoulder width apart, toes pointed straight to the front (direction of movement). The firing side foot is slightly staggered to the rear of the non-firing foot. Knees are slightly bent and the upper body leans slightly forward. Shoulders are not rolled or slouched. Weapon is held with the butt stock in the pocket of the shoulder. The firing side elbow is kept in against the body. The stance should be modified to ensure the Ranger maintains a comfortable boxer stance.

(1) ***Low-Carry Technique***. The butt stock of the weapon is placed in the pocket of the shoulder. The barrel is pointed down so the front sight post and day optic is just out of the field of vision. The head is always up identifying targets. This technique is safest and is recommended for use by the clearing team once inside the room.

(2) ***High-Carry Technique***. The butt stock of the weapon is held in the armpit. The barrel is pointed slightly up with the front sight post in the peripheral vision of the individual. Push out on the pistol grip and thrust the weapon forward and pull straight back into the pocket of the shoulder to assume the proper firing position. This technique is best suited for the line-up outside the door. Exercise caution with this technique always maintaining situational awareness, particularly in a multi-floored building.

Note: Muzzle awareness is critical to the successful execution of close quarter's operations. Rangers must never point their weapons or cross the bodies of their fellow Rangers at any time. Additionally Rangers should always avoid exposing the muzzle of their weapons around corners; this is referred to as "flagging."

b. **Malfunction**. If a Ranger has a malfunction with his weapon during any CQC training, he will take a knee to conduct immediate action. Once the malfunction is cleared, there is no need to immediately stand up to engage targets. Rangers can save precious seconds by continuing to engage from one knee. Whenever other members of the team see a Ranger down, they must automatically clear his sector of fire. Before rising to his feet, the Ranger warns his team members of his movement and only rises after they acknowledge him. If a malfunction occurs once committed to a doorway, the Ranger must enter the room far enough to allow those following him to enter and move away from the door. This drill must be continually practiced until it is second nature.

c. **Approach to a Building or Breach Point**. One of the trademarks of ranger operations is the use of limited visibility conditions. Whenever possible, breaching and entry operations should be executed during hours and conditions of limited visibility. Rangers should always take advantage of all available cover and concealment when approaching breach and entry points. When natural or manmade cover and concealment is not available, Rangers should employ obscurants to conceal their approach. There are times when Rangers will want to employ obscurants to enhance existing cover and concealment. Members of the breach / entry team should be numbered for identification, communication, and control purposes.

(1) The # 1 Ranger should always be the most experienced and mature member of the team, usually the team leader. The # 1 Ranger is responsible for frontal and entry and breach point security. The team leader is responsible for initiating all voice and physical commands. The team leader must exercise situational awareness at all time with respect to the task, friendly force, and enemy activity.

(2) The # 2 Ranger is directly behind the # 1 Ranger in the order of movement, and he moves through the breach point in the opposite direction from the #1 Ranger.

(3) The # 3 Ranger will simply go opposite the # 2 Ranger inside the room at least 1 meter from the door.

(4) The # 4 Ranger moves the opposite the # 3 Ranger and is responsible for rear security (and is normally the last Ranger into the room). An additional duty of the # 4 Ranger is breaching.

Note: Consider how much firepower each Ranger delivers. Where do you put the SAW gunner in the order? You must weigh the benefits of firepower against those of quick, accurate shots. If the # 4 Ranger has breaching responsibilities, it should not be the SAW gunner, because this would reduce your firepower.

d. **Actions Outside the Point of Entry**. Entry point position and individual weapon positions are important. The clearing team members should stand as close to the entry point as possible, ready to enter. Weapons are oriented in such a manner that the team provides itself with 360 degree security at all times. Team members must signal to one another that they are ready at the point of entry. This is best accomplished by sending up a "squeeze." If a tap method is used, an inadvertent bump may be misunderstood as a tap.

e. **Enter Building/Clear Room**. See Battle Drill 5, Chapter 6.

f. **Locking Down the Room**.

(1) Control the situation within the room.

(2) Use clear, concise arm and had signals. Voice commands should be kept to a minimum to reduce the amount of confusion and to prevent the enemy--who might be in the next room--from discerning what is going on. This enhances the opportunity for surprise and allows the assault force the opportunity to detect any approaching force.

(3) Physically and psychologically dominate.

(4) Establish security and report status.

(5) Cursory search of the room to include the ceiling (3 Dimensional Fight).

(6) Identify the dead using reflexive response techniques (Eye thump method).

(7) Search the room for PIR, precious cargo as per the mission and time available.

(8) Evacuate personnel.

(9) Mark room clear using chemical lights, engineer tape, chalk, paint, VS-17 panels, and so on.

14-6. **TTPS FOR MARKING BUILDINGS AND ROOMS**. Units have long identified a need to mark specific buildings and rooms during UO. Sometimes rooms need to be marked as having been cleared, or buildings need to be marked as containing friendly forces. The US Army Infantry School is currently testing a remote marking device that can be used to mark doors from as far away as across a wide street. In the past, units have tried several different field-expedient marking devices; some with more success than others. Chalk has been the most common. It is light and easily obtained but not as visible as other markings. Some of the other techniques have been to use spray paint, and paintball guns.

a. **NATO Standard Marking SOP**. The North Atlantic Treaty Organization (NATO) has developed a standard marking SOP for use during urban combat. It uses a combination of colors, shapes, and symbols. These markings can be fabricated from any material available. (Figure 14-1 shows examples.)

Figure 14-1. EXAMPLE NATO STANDARD MARKINGS

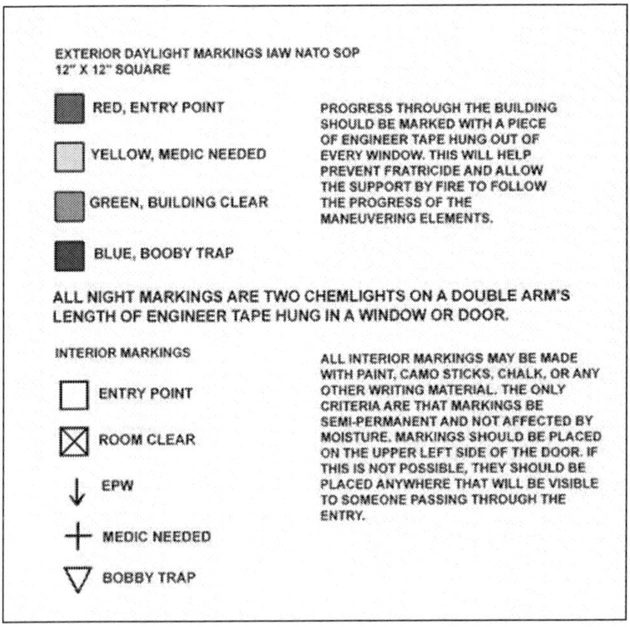

b. **Spray Paint**. Canned spray paint is easily obtained and comes in a wide assortment of colors including florescent shades that are highly visible in daylight. It cannot be removed once used. Cans of spray paint are bulky and hard to carry with other combat equipment. Paint is not visible during darkness nor does it show up well through thermal sights.

c. **Paintball Guns**. Commercial paintball guns have been purchased by some units and issued to small unit leaders. Some models can be carried in standard military holsters. They can mark a building or door from about 30 meters. The ammunition and propellant gas is not easily obtainable. The ammunition is fragile and often jams the gun if it gets wet. The available colors are not very bright, and just like spray paint, cannot be seen at night or through thermal sights.

d. **Wolf Tail**. A simple, effective, easy-to-make, lightweight device called a "wolf tail" can be fabricated to mark buildings, doorways, and windows (Figure 14-1). A unit has changed its tactical SOP to require that each Infantryman carry one of these devices in his ACU cargo pocket. Wolf tails, when used IAW a simple signaling plan understood by all members of the unit, can aid in command and control, reduce the chances of fratricide, and speed up casualty collection during urban combat.

(1) The wolf tail marking device is simple to make and versatile. It can be used together with the NATO marking scheme. Rolled up, it makes a small, easily accessible package that can be carried in the cargo pocket of the ACUs. It can be recovered easily and used again if the situation changes. All its components can be easily obtained through unit supply. It combines a variety of visual signals (colored strapping and one or more chemical lights of varying colors) with a distinctive heat signature that is easily identified through a thermal weapon sight. An infrared chemlight can be used either as a substitute for the colored chemlight(s) or in addition to them.

(2) Constructing the wolf-tail marking device requires the following material:

- A 2-foot length of nylon strap (the type used for cargo tie-downs) (engineer tape can be substituted).
- About 5 feet of 550 cord.
- A small weight such as a bolt or similar object.
- Duct tape.
- Chemical lights (colored or IR).
- Two 9-volt batteries.

(3) Assemble the items by tying or taping the cord to the small weight. Attach the other end of the cord to the nylon strapping, securing it with duct tape. Attach the 9-volt batteries in pairs to the lower end of the strapping with several wraps of duct tape, making sure that the negative terminals are opposite the positive, but not actually touching. Use more duct tape to attach the chemical lights, about 2 inches above the batteries, to the strapping.

(4) When you want to mark your position, push the batteries together firmly until the male and female plugs lock. This shorts out the battery, causing it to heat up rapidly. The hot battery is easily identified through the thermal sights of tanks or BFVs. The batteries stay visible for about 45 minutes. Activating the chemical lights provides an easily identified light source visible to the naked eye. You can use infrared chemical lights if you want them to be seen through night vision devices but not with the naked eye.

Figure 14-2. EXAMPLE OF A WOLF-TAIL MARKING DEVICE

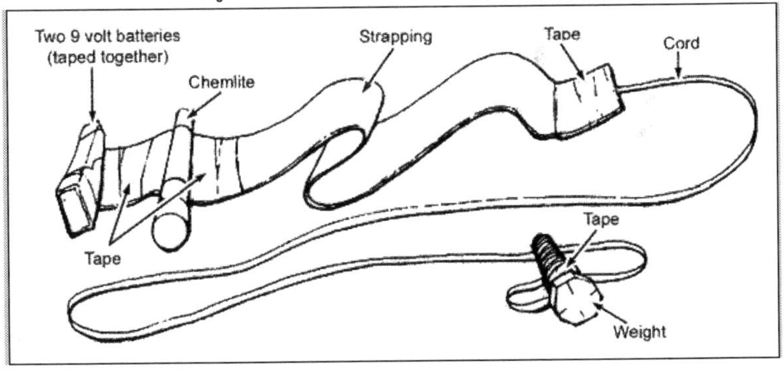

Note: One option is to place chemical lights and batteries at both ends of the wolf tail to mark the inside and outside of a building or room.

(5) Use the cord and the small weight to hold the wolf tail in position by tying or draping it out a window or hanging it on a door, wherever it is best seen by other friendly troops. Squads or platoons can vary the numbers and colors of chemical lights, or use multiple battery sets to identify precisely what unit is in which building.

(6) Medics and combat lifesavers can carry a standardized variation that can be used to clearly identify a building as containing wounded personnel needing evacuation. This could be a white strap with multiple red chemical lights, or any other easily identified combination.

14-7. **URBAN ASSAULT BREACHES**. Understanding how to employ and incorporate breaching as part of a leader's planning process is an important part of urban operations. It is imperative that elements of your patrol be skilled in the art of breaching. Whether infiltrating or exfiltrating from an objective, leaders must plan for either option. One constant disadvantage with the employment of explosives is that placement requires Rangers to expose themselves to possible enemy fires. Breach teams need to be supported by fires or obscurants and breaching operations should be performed during hours of limited visibility whenever possible. Breaching classes follow:

a. **Mechanical**. Mechanical breaching should be an important part of a leader's breaching training program because it is usually an option. Mechanical breaching is best described as gaining access by the use of tools or saws. Although most tools and saws used are recognizable and self explanatory to the individual Ranger, one must practice on various techniques to increase speed an effectiveness. This reduces fatigue and expedites the actual assault.

(1) **Tools**.

(a) Hooligan tools..........................Doors/windows of all types.
(b) Sledge hammer.......................Heavy duty doors, locks, and window frames.
(c) Picket pounder........................Doors of all types, light walls.
(d) Bolt Cutters.............................Chain link fence, locks, and wire obstacles.
(e) Pick Ax....................................Lightweight doors and locks.
(f) Saws.......................................Fences, light doors, locks.

b. **Ballistic**. Ballistic breaching is defined as a forced entry or exit by the use of weapons. Whether using shotguns, M16A2/M4, M249 SAW, specific considerations must be addressed:

(a) Type of round and ricochet factor.
(b) Composition of the breaching point.
(c) Composition of the floor beyond the door.
(d) Personnel behind the door (friendly/enemy).
(e) Always shoot at a 45-degree angle.

c. **Explosive**. Explosive breaching is the most viable because it is the most effective. When employing explosives during breaching operations, leaders must consider three major factors.

(1) ***Overpressure***. The amount of PSI released from the concussion of the blast.

(2) ***Missile Hazard***. Fragmentation or projectiles sent at tremendous speed from the explosion area. This occurs from either the charge or target being breached.

(3) ***Minimum Safe Distance Requirements***. Use of explosives in the urban environment must consider the presence of noncombatants and friendly forces. Additionally, the are many hazardous materials located in the urban environment, these may include chemicals as well as construction materials. There is always a risk of secondary explosions and fires, when employing explosive breaching techniques.

(4) ***Charges***. Various charges can be used for explosive breaching. Leaders must conduct extensive training on the use of the charges to get proper target feedback. Figure 14-3 shows examples of charges used for explosive breaching.

Figure 14-3. EXAMPLE CHARGES FOR EXPLOSIVE BREACHING

Charge	Target
Water Impulse................	Steel/wood doors
Flexible linear.................	Wood doors
Ranger wall breach........	Masonry/brick walls
Chain link ladder............	Chain link fence
E-Type silhouette...........	Wood doors
Brashier breacher..........	Concertina wire

Chapter 15
VEHICLE CONVOY OPERATIONS

This chapter outlines a technique for conducting vehicle convoy operations. Convoy operations present a challenge to the Ranger leader. Trucks and other combat vehicles produce a large signature on the battlefield and increase your unit's value as a target. Vehicle movement is restricted to roads and terrain that they can traverse; therefore, a sound plan must be implemented to minimize the possibility of compromise.

15-1. **PLANNING CONSIDERATIONS**. This paragraph explains how to plan a vehicle convoy IAW the eight steps of the TLP.

a. **Receive the Mission**. (PL will extract the following information from Company OPORD).
   (1) Truck Support (number and type of Truck, ACL).
   (2) Weather. Road conditions.
   (3) Truck pickup and drop-off location / marking.
   (4) Truck movement Timeline (Pick-up time, roll time, H-hour).
   (5) Truck Routes (primary and alternate, Check points).
   (6) Drop-off locations (primary and alternate).
   (7) IED/Contact compromise and contingencies.

b. **Issue the Warning Order**. During the WARNO the PL gives a basic overview of the Truck Movement and gives specific Task to Maneuver Units.
   (1) Manifest Personnel (PSG).
   (2) Pick up point security and set-up (PSG).
   (3) Primary and alternate routes (Security SL).

c. **Make a tentative Plan**. PL considers METT-TC.
   (1) ***Mission***. Develop your Ground Tactical Plan and a then develop a truck plan that best supports it.
   (2) ***Enemy***. Consider enemy patrols and activity in AO.
   (3) ***Terrain***. Evaluate routes and dropoff and pickup points using OCOKA.
   (4) ***Troops***.
   (a) Truck ACL.
   (b) Chalks and chalk leaders identified.
   (c) Chalks planed with tactical integrity & self sufficiency.
   (d) Key leaders and weapons are cross-loaded.
   (e) Pickup point marking team identified.
   (f) Pickup point security teams identified.
   (5) ***Time***. PL considers the following when evaluating time available.
   (a) Movement to Pickup point.
   (b) Recon of Pickup point.
   (c) Emplacement of security.
   (d) Marking the Pickup point.
   (e) Pickup point Posture time.

d. **Initiate Necessary Movement**. PL decides whether to move the platoon to the pickup location or to conduct a map reconnaissance and complete the plan.

e. **Conduct Reconnaissance**. (Is not necessary if pickup point is secured and established by Higher HQ's).
   (1) Leader's Recon consists of PL, RTO, Chalk Leaders and Pickup point Security element.
   (2) PL determines suitability of Pickup point.
   (3) PL determines security plan (either overwatch or strong point).

f. **Complete the Plan**. Complete Truck Movement Annex on Page ??? the Ranger Handbook.

g. **Issue the OPORD**.

h. **Supervise**.
   (1) Back brief from chalk leaders.
   (2) Rock drill of truck movement and contingency plans.

15-2. **FIVE PHASES OF TRUCK MOVEMENT**. Each phase must support the ground tactical plan, which specifies actions in the objective area to accomplish the commander's intent for the assigned mission, be it a raid, ambush, recon or other follow-on missions.

a. **Staging Plan**.
   (1) Secure the pickup point (strongpoint or overwatch).
   (2) Mark pickup point (day/night).

b. **Loading Plan**. Each Ranger is assigned a truck.

c. **Movement Plan**.
   (1) Troops awake and alert during movement.
   (2) Platoon leader and chalk leaders tracking route progress.
   (3) Compromise and contingency plan.
   (a) React to IED.
   (b) React to ambush.
   (c) Truck breakdown.

d. **Dropoff Plan**.
   (1) Establish security of dropoff point (overwatch or strongpoint).
   (2) Dismount vehicles.
   (3) Recon and secure assembly area.
   (4) Adjust perimeter as chalks arrive.
   (5) PSG clears the trucks when last chalk departs.

e. **Ground Tactical Plan**. Execute after platoon is reconsolidated or minimum force required is assembled.

15-3. **CONVOY TECHNIQUES**. The following convoy techniques have been included for your convenience:

- Convoy warning order
- Convoy brief
- React to Ambush Near/Far
- Convoy forced to stop (Methods 1 and 2)
- Break contact
- Recovery/CASEVAC Operations

Convoy Warning Order

1. SITUATION: A brief statement of the enemy and friendly situation. Who. What, Where
2. MISSION: Who, What, When, Where, Why
3. TASK ORGANIZATION: Based on tasking from higher WARNO.
4. INITIAL TIME SCHEDULE:

When	Who	What

5. SPECIAL INSTRUCTIONS: PCC/PCI Guidance, Rehearsals, additional tasks to be accomplished.
6. SERVICE SUPPORT:

Class I: (Rations & Water)

Class III: (POL)

Class V: (Ammunition /Pyrotechnics)

Weapon System	Rounds	Type

Pyrotechnic Device	Number	Location

Class VIII: (Medical)

7. UNIFORM AND EQUIPMENT COMMON TO ALL.

CONVOY BRIEF **(Modified from FM 55-30)**

Movement Order No. ____

References: A. ______________________________ *(Maps, tables and relevant documents)*

B. ______________________________

TASK ORGANIZATION: (Internal organization for convoy – Manifest)

1. **<u>SITUATION</u>:**

a. **Enemy Forces:**

(1) Weather. General forecast.

(2) Light Data (EENT, % Illumination, MR, MS, BMNT)

(3) Discuss Enemy.

Identification of enemy (If known).

Composition / capabilities / strength / equipment

Location

(Hot Spots highlighted on map)

b. **Friendly Forces:**

(1) Operational support provided by higher headquarters.

Helicopter / Gunships

Quick Reaction Forces (QRFs)

MP Escorts / Rat Patrols

Fire Support elements

c. **Attachments:** (From outside the organization)

2. **<u>MISSION</u>: (WHO, WHAT, WHEN, WHERE, & WHY)**

3. **Employment Tactics, Techniques and Procedures**

a. Readiness Level
- i. Ride / Scanning (Observation)
- ii. At the Ready

b. Scanning / Sector of Fire
- i. Driver
  1. Sector of Scan is 9 – 1 clock position
  2. Observation with Mirrors
  3. Sector of Fire is 9 – 11 o'clock
- ii. TC
  4. Sector of Scan is 11 – 3 clock position
  5. Sector of Fire is 1 – 3 clock position
- iii. Other Systems are based off vehicle type and load

b. Target ID
- i. Communicate / Signal

c. Body Positioning
- i. Engage as you Train (Right or Left handed Firing)
- ii. Firing-side Shoulder Down
- iii. Maintain your Body Position
- iv. Weapon to Head

d. Rules of Engagement Concerns – Employ the appropriate systems based off the threat.

e. Point of Aim
- i. Moving Platform – Stationary Target requires aim to the rear (trail) and low dependant upon speed.

CONVOY BRIEF **(Modified from FM 55-30)** **(Continued)**

3. Employment Tactics, Techniques and Procedures

e. Point of Aim
   ii. Moving Platform – Moving Target requires aim directly on and low.
   iii. Stationary Platform – Moving Target requires aim to the front (lead) and low dependent upon speed.

f. Rate of Fire
   i. Burst Mode
   ii. Steady Suppression (ROE)

g. Magazine Awareness
   i. Serviceability / maintenance
   ii. Tracer Mix
   iii. Magazine Storage / Placement
   iv. Mounted, Reload when time is available

Dismounted, Seek cover prior to the need to Reload

REACT TO AMBUSH (NEAR)

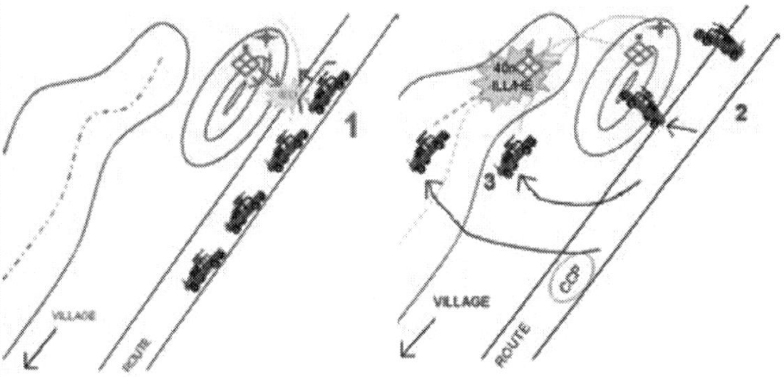

(Enemy Initiates Weapons Fire / RPG / IED)

1. Lead element of convoy reacts to contact and immediately returns fire. Lead fire team fires 40mm HE (at night also white star if NVGs drown out by vehicle headlights – leader calls it. FYI, most travel at night will be with headlights on).
2. Lead elements assualt through ambush – fire team dismounts while gun trucks use M249, MK-19, and .50 cal in close support.
3. Trail fire team and trail gun truck move to block most likely enemy escape route. Fire team uses 40mm HE/Illumination if needed. (Use of 40mm becomes less likely as units get closer to urban areas due to collateral damage risk.) Establish CCP upon action completion or enemy neutralized.

REACT TO AMBUSH (FAR)

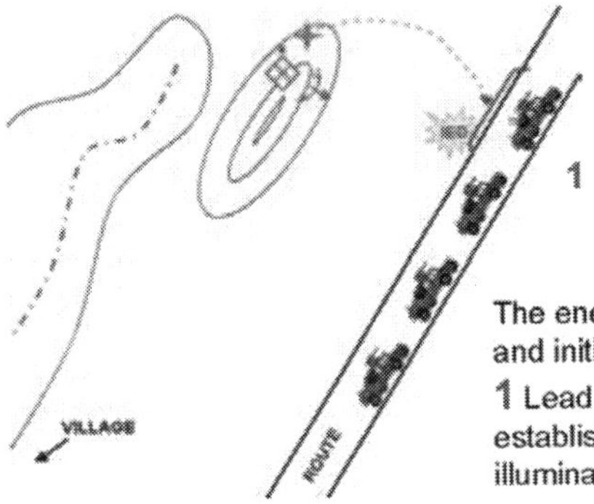

The enemy is outside 100 meters and initiates with IED/RPG

1 Lead gun truck and lead fire team establish SBF—fire team fires 40mm illumination

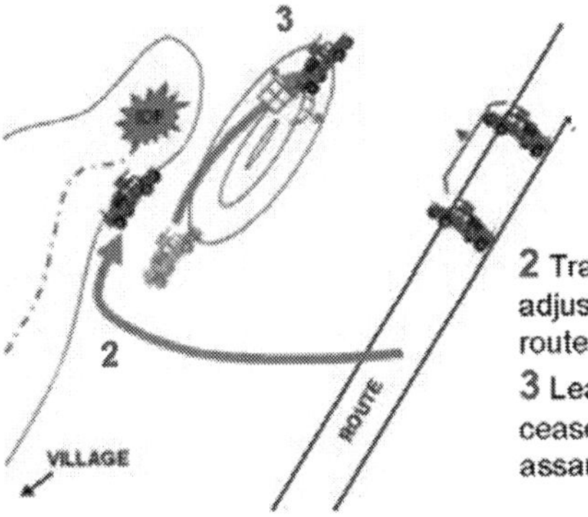

2 Trail element flanks—SL/PL adjusts IDF into enemy escape route(if in range)

3 Lead element shifts fire—arty cease loading—trail element assaults across—enemy destroyed

CONVOY FORCED TO STOP
(Weapons Fire / RPG / IED / Indirect)

1. Vehicles forced to stop. Activate vehicle turn signal to indicate direction of contact.
2. Vehicle(s) / Personnel not in direct contact, report on internal communication, identifying truck number, type of contact, clock direction and eight digit grid if available.

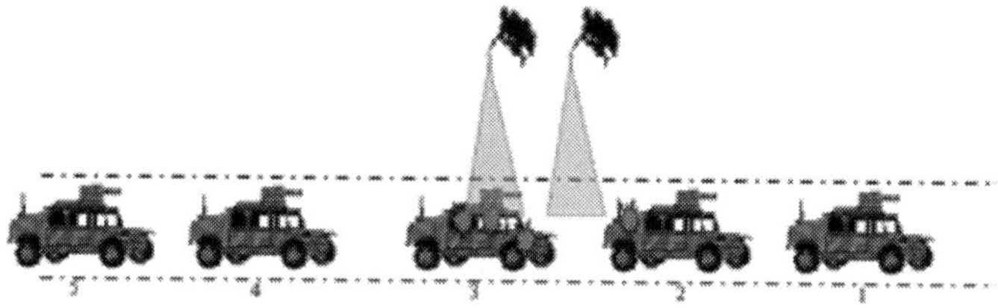

3. Personnel on vehicle(s) forced to stop dismount on the non-contact side, assume covered position(s) and provide initial base of fire.
4. The <u>entire</u> convoy halts; personnel will dismount vehicle(s) on the non-contact side and provide additional base of fire on the enemy if in range. Vehicle(s) not in contact will reposition to attack the attackers. M16's should now be on <u>Semi-Automatic</u> in order to conserve ammunition.
5. PL/CC/SC will move to better assess the situation and position the Gun Truck(s) in order to best suppress the enemy while maintaining <u>standoff</u>. Gun trucks will close with and engage the enemy.
6. Once the PL/CC/SC determines the convoy has either gained fire superiority or defeated the enemy contact, Recovery / CASEVAC operations will begin.
7. If the PL/CC/SC determines the convoy cannot gain fire superiority, leader will then conduct Break Contact procedures

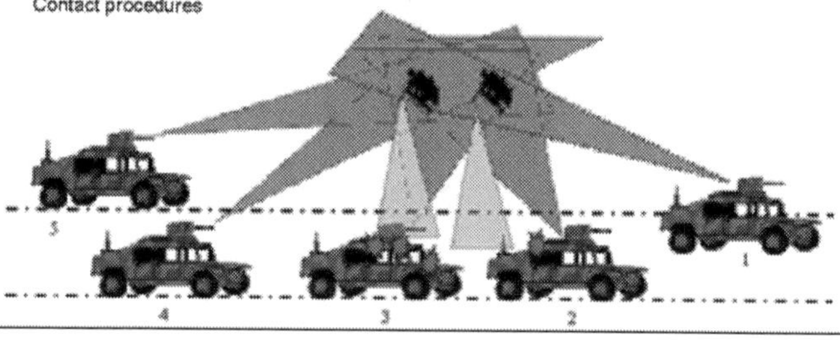

CONVOY FORCED TO STOP - METHOD 2

(Weapons Fire /RPG / IED / Indirect)

1. Vehicle(s) forced to stop. Activate vehicle turn signal to indicate direction of contact.

*** ALL PERSONNEL STAY IN VEHICLES ***

2. All vehicles immediately drive forward out of the kill zone.

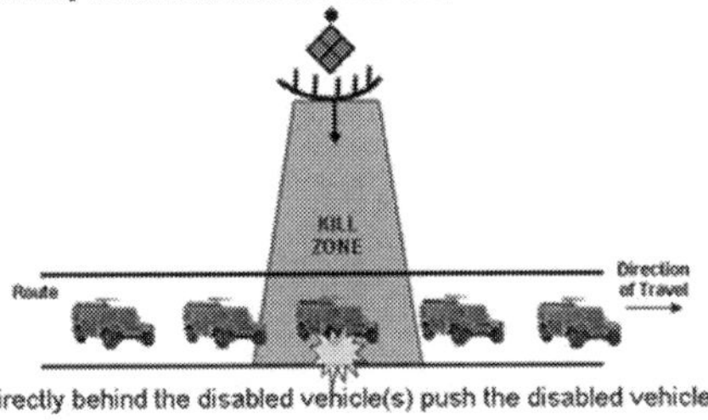

3. The vehicle(s) directly behind the disabled vehicle(s) push the disabled vehicle(s) out of the kill zone.
4. The vehicle(s) that are not disabled will establish a base of fire towards the suspected/known contact.
5. If fire superiority can be gained the PL will use the minimum force necessary to destroy the enemy.
6. If the PL determines the convoy cannot gain fire superiority, the leader will conduct break contact.

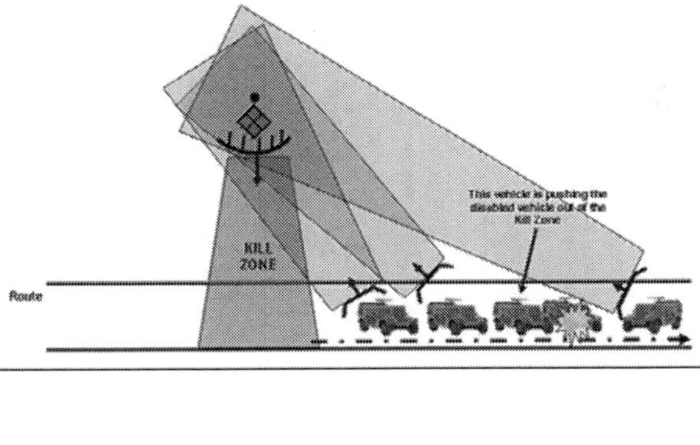

BREAK CONTACT

Always try to close with and destroy the enemy first. This way he cannot come back later on to attack you or another convoy again. However – if you must...

1. The PL/CC/SC determines the convoy cannot gain fire superiority and the decision has been made to Break Contact.
2. The PL/CC/SC designates that either Rally Point "Rear" or "Forward" will be used. If necessary, both Rally Points may be used. Communication systems and appropriate pyrotechnic signals will be used to communicate Break Contact and Rally Point.
3. Personnel will deploy obscuration measures if available. Utilizing cover and concealment, Aid & Litter team(s) will evacuate all casualty(s) under support of Gun Truck and other protective fire(s).
4. Personnel will maintain position and suppression in contact zone and assist Aid & Litter team(s) as necessary.
5. Disabled vehicle(s) will be hauled back or destroyed as directed by leaders. (Thermite or explosives)

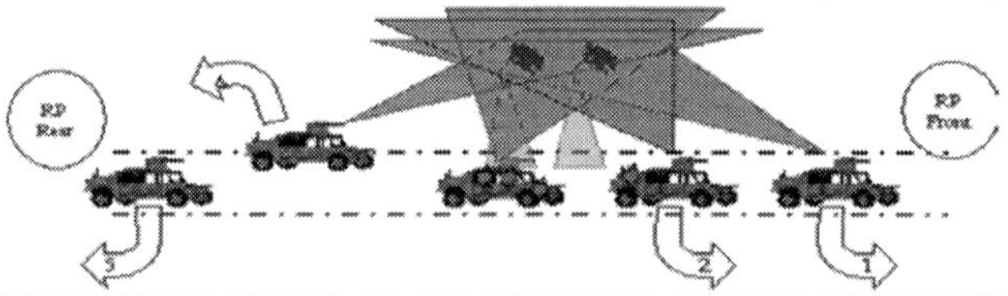

6. Vehicles will displace either backwards or forwards through the convoy lines under control of leaders. The most forward vehicle in the contact zone moves first, then the next most forward vehicle moves second. Vehicles will continue to displace. As vehicles displace, Gun Truck repositions as necessary until contact is broken.

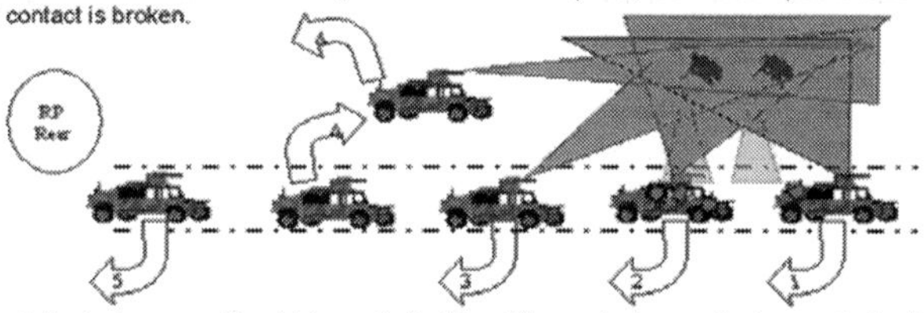

7. If Break Contact occurs with vehicles on both sides of the contact zone, displacement of vehicles will occur using an alternating displacement technique.
8. Upon occupation of the Rally Point, leaders will immediately position vehicles, setting 360 degree security, and conduct Consolidation and Reorganization.
9. If the convoy vehicles get separated when not in contact with the enemy, personnel and vehicles stay together and move to the closest Rally Point or Check Point.

RECOVERY / CASEVAC OPERATIONS

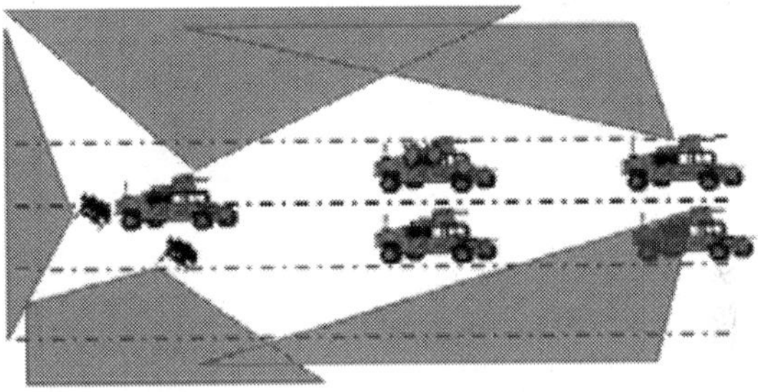

1. Once the leader assesses the enemy threat to be destroyed, neutralized, defeated and the area secured – Recovery / CASEVAC operations will begin. This keeps soldiers focused first on defeating and destroying the threat.

2. CASEVAC:

   a. Aid & Litter team will position on the safe side of the vehicle and extract casualty(s) and personnel.

   b. Treatment of casualty(s) will occur once they are safely removed from the contact area.

3. Vehicle Recovery Procedures:

   a. Recovery team will position on the safe side of the disabled vehicle.

   b. TC will dismount and assess the disabled vehicle.

   c. If determined the vehicle can be safely recovered, TC guide the recovery vehicle into position and conduct a hasty hook-up. TC will operate the disabled vehicle is necessary.

   d. Upon exiting the contact area, complete and correct hook-up procedures will occur.

   e. If assessment results in outside support necessary for recovery, leader will contact higher for guidance.

4. Once recovery operations are complete, the team will displace and conduct link-up with the convoy at the Rally Point.

5. Disabled vehicle(s) will be abandoned or destroyed as directed by leaders. (Thermite or explosives)

GLOSSARY

1SG	first sergeant
5 W's	Who, What, Where, When, and Why?
5-S Rule	Search, Silence, Segregate, Safeguard, and Speed to rear
AA	avenue(s) of approach
AAR	after-action review
AATF	air assault task force
ABCDE	method of identification and response to life-threatening conditions: Airway, Breathing, Circulation, Disability, Exposure
ACE	ammunition, casualties, and equipment
ACL	allowable combat load
ACP	aerial checkpoint
ACU	Army combat uniform
ADA	air defense artillery
AG	assistant gunner
ATC	air traffic controller; a mechanical belay device that locks down on itself when tension is applied in opposite directions
ALT	alternate
AMC	air mission commander
ammo	ammunition
ANCD	automated net-control device
AO	area of operations
AOO	actions on the objective
AR	automatic rifleman
ABF	attack by fire
ATL	Alpha team leader
ATM	Alpha Team
bangalore torpedo	A manually emplaced, 1.5-meter long explosive-filled tube used to breach wire and detonate simple, pressure-activated antipersonnel mines. Ten tubes clear a 1- by 15-meter lane.
belay	any action taken to stop a climber's fall or to control the rate a load descends
binos	binoculars
BMNT	begin morning nautical twilight
BN	battalion
body belay	belay that uses the belayer's body to apply friction by routing the rope around the his body
bowline on a coil	knot used to secure a climber to the end of a climbing rope
BP	battle position
British junction knot	knot used to join the ends of detonation cords from multiple charges to one initiation system
BTC	bridge team commander
BTL	Bravo Team leader
BTM	Bravo Team

C2	command and control
CAS	close air support
CASEVAC	casualty evacuation
CCA	close combat attack
CCIR	commander's critical information requirements
CCP	casualty collection point
CDR	commander
CDS	Camp Darby Special (map)
CLS	combat lifesaver
CO	company
COA	course of action
commo	communications
COMSEC	communications security
cordelette	short section of static rope or static cord. Also called "sling rope"
COTS	commercial off the shelf
CP	command post
CPR	cardiopulmonary resuscitation
CQC	close quarters combat
CS	combat support
CSS	*obsolete: now referred to as* sustainment
CTT	common task test
DAR	designated area of recovery
DOL	direction of landing
double figure eight knot	knot used to form a fixed loop in the end of the rope; loops are large enough to insert a carabiner
double overhand knot	knot used to secure the end of detonation cord
DP	duty position; decision point (depending on context)
DST	distance
DTG	date-time group
dynamic ropes	one of two classifications of kernmantle rope; used for climbing; *see also* static rope
DZ	drop zone
EA	engagement area
EEFI	essential elements of friendly information
EENT	end evening nautical twilight
end-of-the-rope clove hitch	intermediate anchor knot that requires constant tension
end-of-the-rope Prusik	knot used to attach a movable rope to a fixed rope; *see also* middle-of-the-rope Prusik
ENY	enemy
EPW	enemy prisoner of war
FA	field artillery
FDC	fire direction center
FFIR	friendly force information requirements

figure 8 slip knot	knot used to form an adjustable bight in the middle of a rope
FLIR	forward looking infrared
FLOT	forward line of own troops
FO	forward observer
FPF	final protective fires
FRAGO	fragmentary order
FSC	fire support coordinator
FSO	fire support officer
GOTWA	G - Where leader is GOING O - Others he is taking with him T - Time he plans to go W - WHAT to do if the leader does not return in time A - The unit's and the leader's ACTIONS on chance contact while the leader is gone
GPS	global positioning system
GRN	grenadier
HDG	heading
HE	high explosive
H-Hour	hit hour (the time the unit plans to accomplish the mission)
HI	high temperature (weather)
HPT	high-payoff target
HQ	headquarters
IAW	in accordance with
ICM	improved conventional munitions
ID	identification
IP	initial point
IR	intelligence requirements
ERRP	en route to release point
JAAT	joint air attack team
JD	Julian date
KIA	killed in action
LAW	light antiarmor weapon
LBV	load-bearing vest
LD	line of departure
LDA	linear danger area
LO	low temperature (weather)
LOGSTAT	logistical status
LP	listening post
LZ	landing zone
material factor	the strength, hardness, and mass of the material to be demolished
mb	millibar (a metric unit used to measure air pressure)

MDI	modernized demolition initiator
ME	main effort
mechanical belay	a belay that uses mechanical devices to help the belayer control the rope, as in rappelling
MEDEVAC	medical evacuation
METL	mission-essential task list
METT-TC	mission, enemy, terrain (and weather), troops (and support) available, time available, and civil considerations
middle of the rope clove hitch	knot that secures the middle of a rope to an anchor
middle-of-the-rope Prusik	knot that attaches a movable rope to a fixed rope, anywhere along the length of the fixed rope; *see also* end-of-the-rope Prusik
MG	machine gun/ner
MOPP	mission-oriented protective posture
MR	moonrise
MRE	meal, ready to eat
MS	moonset
MSD	minimum safe distance
MSL	mean sea level
munter hitch	commonly used belay that requires little equipment
NATO	North Atlantic Treaty Organization
NAV	navigation
NFA	no-fire area
NLT	no later than
NVD	night-vision device
NVG	night vision goggles
NVS	night vision system
OAKOC	observation and fields of fire, avenues of approach, key terrain, observation, and cover and concealment
OBJ	objective
OD	olive drab
OOM	order of movement
OP	observation post
OPORD	operations order
OPSKED	operational schedule
ORP	objective rally point
OT	observer-target
PB	patrol base
PCC	precombat checks
PCI	precombat inspection
PDF	principal direction of fire

PI	probability of incapacitation
PIR	priority intelligence requirements
PL	platoon leader
PLD	probable line of deployment
PLOT-CR	purpose, location, observer, trigger, communication method, resources (a format for planning fire support)
PLT	platoon
POL	petroleum, oils, and lubricants
PRI	primary
protection	a piece of equipment, natural or artificial, that is used to construct an anchor
PSG	platoon sergeant
PSI	pounds per square inch
PZ	pickup zone
R	rifleman
R&S	reconnaissance and surveillance
RACO	rear area combat operations
RAP	rocket-assisted projectile
rappel seat	a rope harness used in rappelling and climbing
RED	risk-estimate distance
REQ	required
rerouted figure 8 knot	anchor knot that also attaches a climber to a climbing rope
RFA	restrictive fire area
RFL	restrictive fire line
RFLM	rifleman
RHB	Ranger Handbook
ROE	rules of engagement
round-turn with two half hitches	a constant tension anchor knot
RP	release point
RTO	radio operator
S-2	intelligence staff officer
S-3	operations staff (and training) officer
SALUTE	Size, Activity, Location, Unit/Uniform, Time, and Equipment
SAW	squad automatic weapon
SBF	support by fire (position)
SDT	self-development test
SE	supporting effort
SEAD	suppression of enemy air defenses
SITREP	Situation Report
SITTEMP	situational template

SL	squad leader
sling rope	short section of static rope or static cord. Also called "cordelette"
SLLS	Stop, Look, Listen, Smell
SOC	succession of command
SOI	signal operating instructions
SOP	standing operating procedures
SP	start point
square knot	knot used to join two ropes of equal diameter; used to join the ends of the detonation cord to the explosive
SR	sunrise
SS	sunset
STANO	surveillance, target acquisition, and night observation
static ropes	one of two classifications of kernmantle rope; used where rope stretch is undesired, and when the rope is subjected to heavy static weight. *See also* dynamic rope
SURVIVAL	S Size up the situation, your surroundings, your physical condition, and your equipment. U Undue haste makes waste; don't be too eager to move. Plan your moves. R Remember where you are in relation to important friendly and locations and critical resources V Vanquish fear and panic. I Improvise. You can improve your situation. Use what you have. Use your Imagination. V Value your life. Remember your goal: to get out alive. Remain stubborn. Refuse to give in to problems and obstacles that face you. This will give you the mental and physical strength to endure. A Act like the natives; watch their daily routines. When, where, and how do they get food? Where they get water? L Live by your wits. Learn basic skills.
suspension traverse	used to move personnel and equipment over rivers, ravines, chasms, and up or down a vertical obstacle
TAC	tactical air controller
tamping factor	depends on the location and tamping of the charge
technical climbing	using safe and proper equipment and techniques to climb on a rock formation in parties of two or more
tensionless anchor	used to anchor rope on high-load installations such as bridging
TL	team leader
TLP	troop-leading procedures
TL	team leader
TOC	tactical operations center
triple roll knot	knot used to join branches of detonation cord
TTP	tactics, techniques, and procedures
uli knot	knot used to securely fasten detonation cord to explosive
VIXL	video image crosslink
WARNO	warning order
WBGT	wet bulb globe temperature

WFFs	warfighting functions (fire support, movement and maneuver, protection, command and control, and sustainment)
WIA	wounded in action
XO	executive officer
WSL	weapons squad leader

NOTES

INDEX

Appendix A
RESOURCES

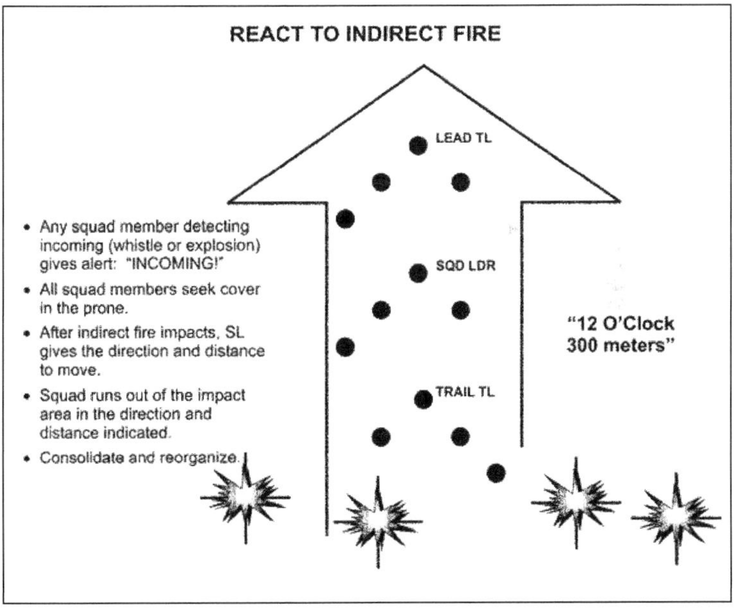

REACT TO CONTACT

- Seek nearest cover.
- Return fire (known or suspected enemy position).
- TM LDRs control fire by using fire commands.
- Report enemy situation (number/position).
- Maintain contact (visual/oral) with team members.
- SQD LDR moves to team in contact and makes an assessment of the situation.
- Factors of his assessment:
  - Can squad move out to engagement area.
  - Can squad gain and maintain suppressive fire.
  - Location of enemy.
  - Size of enemy.
  - Vulnerable flanks.
  - Covered/concealed flanking routes.
- SQD LDR determines COA (break contact, squad attack, etc.)
- Report situation to PL.

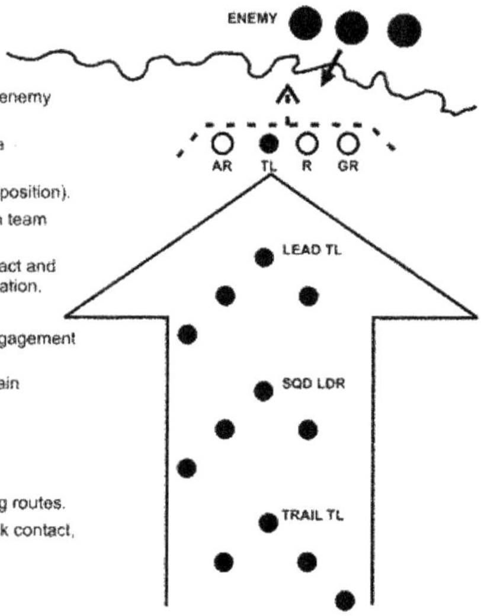

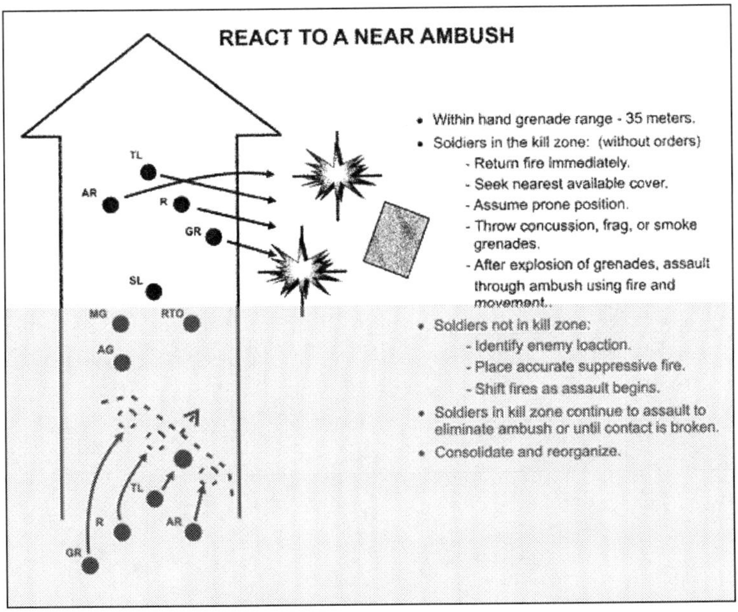
REACT TO A NEAR AMBUSH
TL
AR
R
GR
SL
MG
RTO
AG
TL
R
AR
GR
• Within hand grenade range - 35 meters.
• Soldiers in the kill zone: (without orders)
- Return fire immediately.
- Seek nearest available cover.
- Assume prone position.
- Throw concussion, frag, or smoke grenades.
- After explosion of grenades, assault through ambush using fire and movement..
• Soldiers not in kill zone:
- Identify enemy loaction.
- Place accurate suppressive fire.
- Shift fires as assault begins.
• Soldiers in kill zone continue to assault to eliminate ambush or until contact is broken.
• Consolidate and reorganize.

REACT TO A FAR AMBUSH

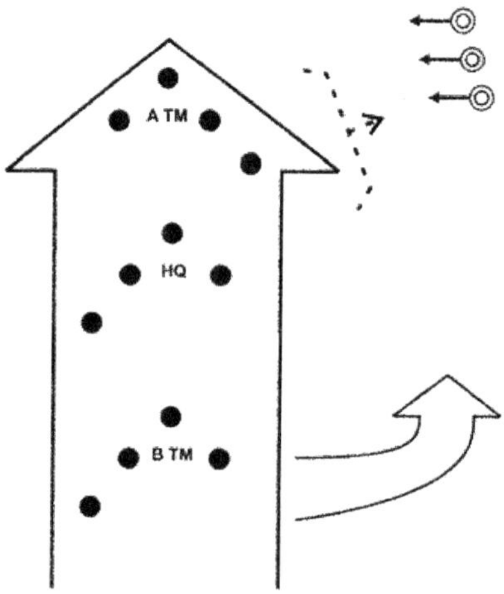

- More than 35 meters.
- TM in the kill zone: (without orders)
  - Return fire immediately.
  - Seek cover and concealment.
  - Suppress enemy (overwatch).
- SL assess situation.
  - Determines COA (flank).
- TM not in contact:
  - Moves along covered and concealed route.
  - Assaults enemy on weak flank.
  - Suppress enemy (overwatch).
- Overwatch TM continues to suppress shifts/cease fire as bounding team enters sector.
- Bounding team continues to assault through enemy.
- SL may request indirect fire.
- Consolidate and reorganize.

BREAK CONTACT

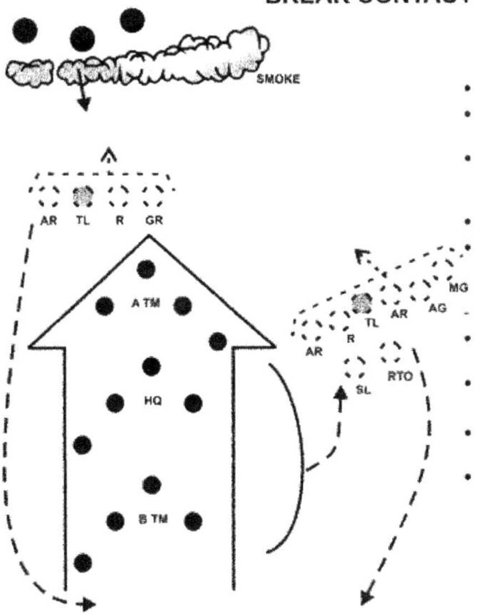

- Squad Leader orders: "Break Contact".
- Squad Leader designates SPT element and maneuver element.
- SL issues distance and direction or a terrain feature for the maneuver element.
- SBF suppresses enemy position.
- Maneuver uses smoke to mask movement.
  - Takes up overwatch position.
  - Begins to suppress enemy
- Squad Leader directs SBF to break contact.
- SBF uses smoke to screen movement.
  - Takes up overwatch position.
- Squad continues to bound away until contact is broken.
- Consolidate / reorganize.

FORMATIONS AND ORDERS OF MOVEMENT

I. Movement Formation: Fire Team Wedge; MG Team attached.

II. Three Movement Techniques utilized:

A. Traveling technique used behind FFL when contact is not likely.

B. Traveling Overwatch forward of the FFL when enemy contact is possible.

C. Bounding Overwatch used forward of the FFL when enemy contact is expected.

III. Distances are based on but not dictated by visibility, terrain, and vegetation.

IV. Actions at Night: Modified Wedge

V. Actions at the Halt: Short and Long Halt (GV/LV)

VI. Leader Location: Fixed/Unfixed

FIRE TEAM WEDGE

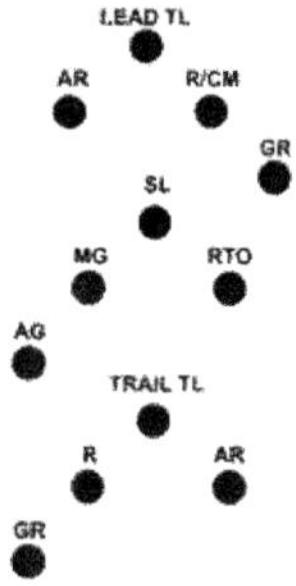

MODIFIED WEDGE

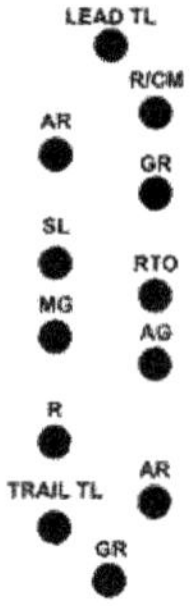

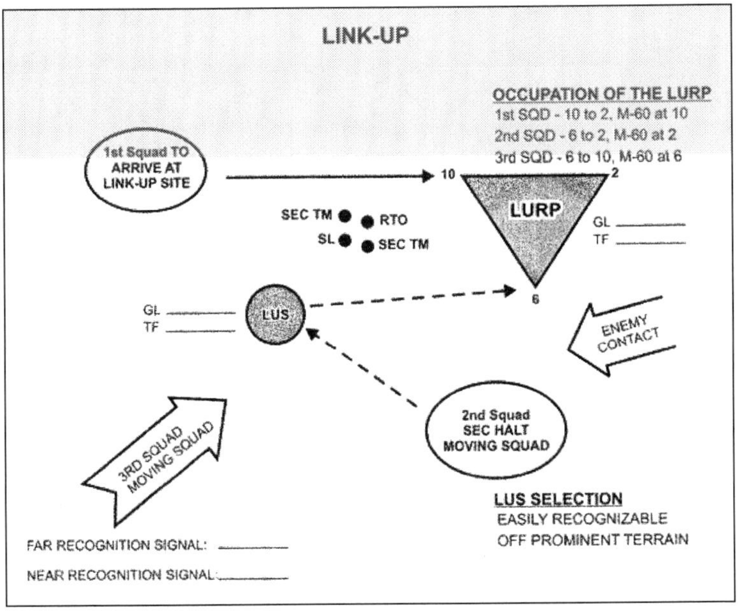
LINK-UP
OCCUPATION OF THE LURP
1st SQD - 10 to 2, M-60 at 10
2nd SQD - 6 to 2, M-60 at 2
3rd SQD - 6 to 10, M-60 at 6
1st Squad TO ARRIVE AT LINK-UP SITE
10
2
6
LURP
GL
TF
SEC TM
RTO
SL
SEC TM
LUS
GL
TF
ENEMY CONTACT
2nd Squad SEC HALT MOVING SQUAD
3RD SQUAD MOVING SQUAD
LUS SELECTION
EASILY RECOGNIZABLE
OFF PROMINENT TERRAIN
FAR RECOGNITION SIGNAL:
NEAR RECOGNITION SIGNAL:

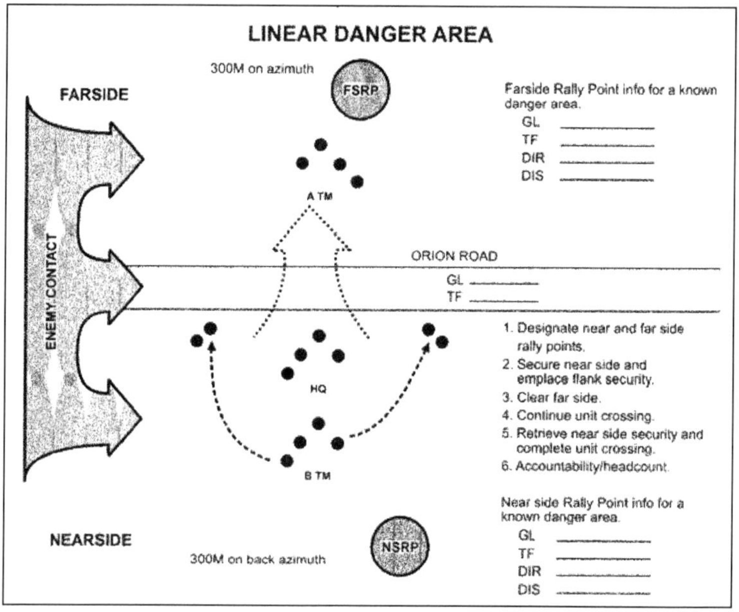
LINEAR DANGER AREA
300M on azimuth
FSRP
FARSIDE
Farside Rally Point info for a known danger area.
GL
TF
DIR
DIS
A TM
ENEMY CONTACT
ORION ROAD
GL
TF
1. Designate near and far side rally points.
2. Secure near side and emplace flank security.
3. Clear far side.
4. Continue unit crossing.
5. Retrieve near side security and complete unit crossing.
6. Accountability/headcount.
HQ
B TM
Near side Rally Point info for a known danger area.
NEARSIDE
300M on back azimuth
NSRP
GL
TF
DIR
DIS

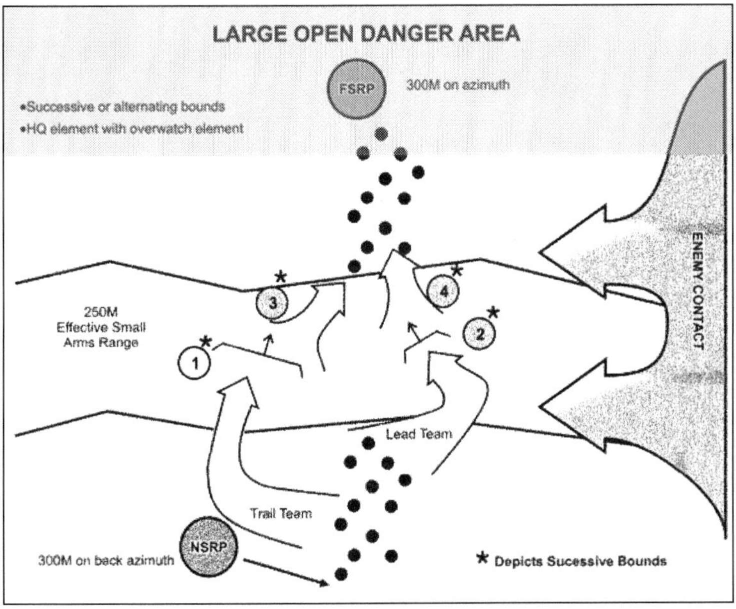
LARGE OPEN DANGER AREA
FSRP
300M on azimuth
•Successive or alternating bounds
•HQ element with overwatch element
250M
Effective Small
Arms Range
ENEMY CONTACT
1
2
3
4
Lead Team
Trail Team
NSRP
300M on back azimuth
* Depicts Sucessive Bounds

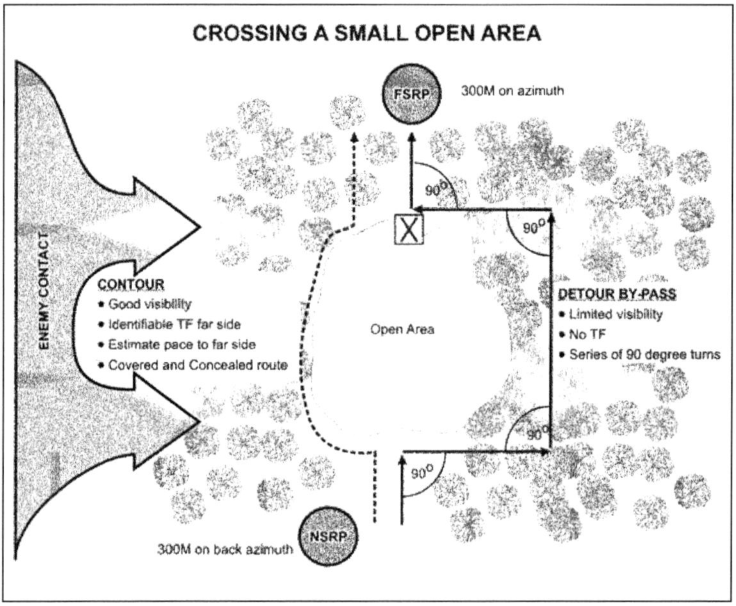
CROSSING A SMALL OPEN AREA
FSRP
300M on azimuth
90°
90°
ENEMY CONTACT
CONTOUR
• Good visibility
• Identifiable TF far side
• Estimate pace to far side
• Covered and Concealed route
Open Area
DETOUR BY-PASS
• Limited visibility
• No TF
• Series of 90 degree turns
90°
90°
NSRP
300M on back azimuth

SQUAD ATTACK

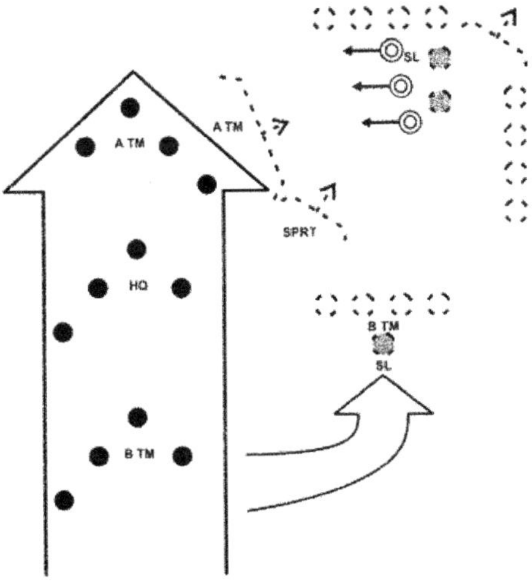

- React to contact.
- TM in the kill zone: (without orders)
  - Return fire.
  - Seek cover and concealment.
  - Suppress enemy calls out 3 D's becomes overwatch.
- SL assess situation.
  - Determines COA (flank/attack).
- TM not in contact: (with SL)
  - Moves along covered and concealed route.
  - Assaults enemy on weak flank.
- Overwatch TM continues to suppress shifts/cease fire as bounding team enters sector.
- Bounding team continues to assault through enemy.
- SL may request indirect fire.
- Consolidate and reorganize.

RAID BOARDS (LEFT)
GENERAL INFO

Raid References:
- RHB Chapter 5
- FM 3-21.8

Purposes:
1. Destroy
2. Liberate
3. Collect Intel

Planning Considerations:
- Mission
- Enemy
- Troops
- Terrain / OAKOC
- Time
- Civilians

STUDENT INPUT WRITTEN HERE

Definition:
A surprise attack on a fixed position or installation ending in a planned withdrawal.

Characteristics:
1. Surprise - Gain
2. Coordinated fires - Maintain
3. Violence of action - Retain
4. Planned withdrawal

INITIATIVE

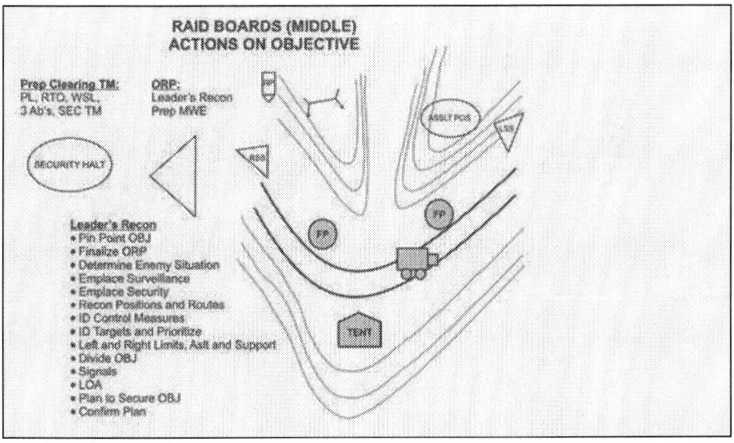

RAID BOARDS (MIDDLE)
ACTIONS ON OBJECTIVE
Prep Clearing TM:
PL, RTO, WSL,
3 Ab's, SEC TM
ORP:
Leader's Recon
Prep MWE
SECURITY HALT
ASSLT POS
FP
FP
TENT
Leader's Recon
• Pin Point OBJ
• Finalize ORP
• Determine Enemy Situation
• Emplace Surveillance
• Emplace Security
• Recon Positions and Routes
• ID Control Measures
• ID Targets and Prioritize
• Left and Right Limits, Aslt and Support
• Divide OBJ
• Signals
• LOA
• Plan to Secure OBJ
• Confirm Plan

RAID BOARDS (MIDDLE)
TASK ORGANIZATION

ELEMENT	WHO	WHY	WHAT
HQ	PL, APL RTO, FO, MEDIC	C2/FACILITATE SIGNALS, PEQ-2, STANO	1x119, ICOM, SKEDCO, PLUGGER
ASSLT 1	SQUAD (+)(-)	DESTROY	ICOM, STANO, M18A1, AT4, SPECIAL TEAM KITS
ASSLT 2	SQUAD (+)(-)	DESTROY	ICOM, STANO, M18A1, AT4, SPECIAL TEAM KITS
SUPPORT	WPN S SQD (+)(-)	SUPPRESS	3xM240B COMPLETE, ICOM, PEQ-2, PVS-4, AT, STANO, SIGNALS, MAP
SECURITY	SQUAD (+)(-)	DELAY	ICOMs, M18A1, AT, STANO, MAP SIGNALS

RAID BOARDS (RIGHT)
SOP'S

CONTINGENCIES

1. COMPROMISE
   a. ORP
   b. LEADERS RECON
   c. OCCUPATION
2. MASS CAL
3. COUNTERATTACK

SOP's

1. EPW SEARCH
2. AID LITTER
3. MED EVAC
4. CCP
5. WITHDRAW PLAN

AMBUSH SOPS (LEFT)

References:
- RHB Chapter 5
- FM 3-21.8

Ambush:
A surprise attack from a covered and concealed position on a moving or temporarily halted target.

Types:
- Point
- Area

Fundamentals:
1. Surprise - Gain
2. Fire superiority - Maintain
3. Violence of action - Retain

Purposes:
1. Disrupt/Destroy
2. Collect Intelligence
3. Block or Deny Access
4. Canalize

Categories:
Deliberate: has specific target at a predetermined time and location.

Hasty: Platoon makes visual contact with the enemy and has time to establish an ambush without being detected.

Planning Considerations:
- Mission - Task/Purpose
- Enemy - MPCOA, MDCOA
- Terrain - Map Recon/Ldrs Recon
- Time - 1/3 2/3 Rule
- Troops - Task Org
- Civilians -

AMBUSH BOARDS (MIDDLE)

HQ	PL, PL RTO, PSG, MEDIC, FO	COMMAND AND CONTROL	PRC-119s, ICOM, NVGs, BINO's, SKEDCO, PEQ-2, SIGNALS, MAP
SECURITY	MINIMUM OF 3-2 MAN TEAMS	EARLY WARNING, SEALS OFF OBJ, LFS, RFS, ORP SECURITY	COMMO, NVGs, CLAYMORES, AT WPNS, BINO's, SIGNALS, MAP
SUPPORT	WSL, 3 GUN TEAMS COMPLETE +/-	SUPPRESSIVE FIRE ON KILL ZONE, SEC ON OBJ DURING RECON/REORG	MG's, COMMO, NVGs, AT WPNS, BINO's, SKEDCO, PEQ-2, SIGNALS., MAP
ASSAULT	2 SQUADS +/-	MAIN EFFORT, BLOCK, DESTROY, CANALIZE, CAPTURE PIR	NVGs, BINO's, CLAYMORE, AT, FLEX CUFFS, POLELESS LITTER, MARKINGS, SIGNALS

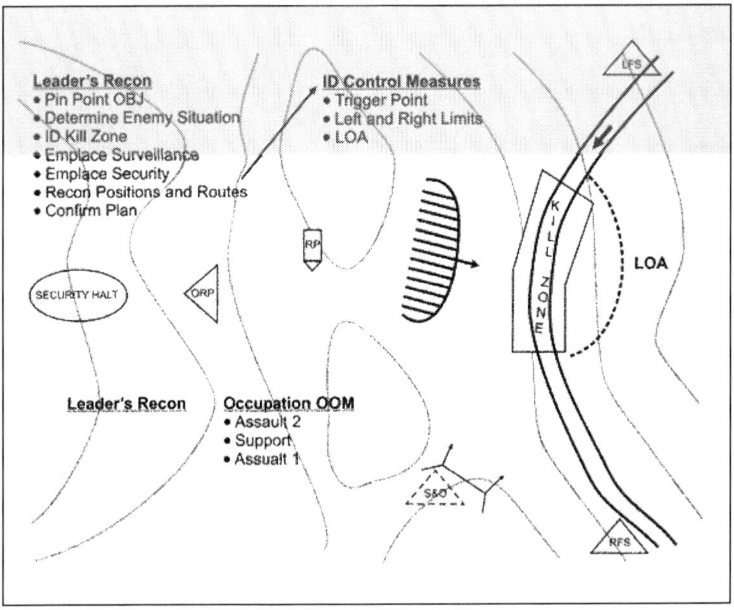
Leader's Recon
• Pin Point OBJ
• Determine Enemy Situation
• ID Kill Zone
• Emplace Surveillance
• Emplace Security
• Recon Positions and Routes
• Confirm Plan
ID Control Measures
• Trigger Point
• Left and Right Limits
• LOA
LFS
RP
KILL ZONE
LOA
SECURITY HALT
ORP
Leader's Recon
Occupation OOM
• Assault 2
• Support
• Assualt 1
S&O
RFS

AMBUSH BOARDS (RIGHT)
SOP'S

CONTINGENCIES

1. COMPROMISE
   a. ORP
   b. LEADERS RECON
   c. OCCUPATION
2. MASS CAL
3. COUNTERATTACK

SOP's

1. EPW SEARCH
2. AID LITTER
3. MASS CAL
4. MED EVAC
5. CCP

NOTES

Appendix B

QUICK REFERENCE CARDS

CASUALTY FEEDER CARD

NAME ____________________ ROS # __________

COMPANY ________ ALLEGERIES______________

LOCATION/EVENT ____________________________

TIME ______________ TIME ______________

TEMP ______________ TEMP ______________

PULSE ______________ PULSE ______________

RESP ______________ RESP ______________

B/P ______________ B/P ______________

MENTAL STATUS ____________________________

A & O x 1 2 3

NPA OPA ET Tube

SAO^2 ______________ GLUC __________________

IV _____ AC FOREARM HAND OTHER

1000-1000-500-500-500- NS RL D$_5$W

NOTES: ______________________________________

____

GTA 07-01-038
INFANTRY LEADERS' REFERENCE CARD
January 1995

HEADQUARTERS, DEPARTMENT OF THE ARMY

References: FM 3-3, FM 6-30, FM 6-121, FM 7-8, FM 7-10, FM 7-90

(Supersedes GTA 7-1-31)

DISTRIBUTION RESTRICTION: Distribution authorized to US Government agencies only to protect technical or operational information from automatic dissemination under the International Exchange Program or by other means. Other requests for this document will be referred to Commander, US Army Training Support Center, ATTN: ATIC-DMF, Ft Eustis, VA 23604-5168.

DESTRUCTION NOTICE: Destroy by any method that will prevent disclosure of contents or reconstruction of the document.

DISTRIBUTION: US Army Training Support Centers (TSCs).

TROOP-LEADING PROCEDURES

1. Receive mission.
2. Issue warning order.
3. Make a tentative plan.
4. Start movement.
5. Reconnoiter.
6. Complete plan.
7. Issue plan.
8. Supervise.

FIRE REQUEST (FM 6-30)

1. Identification.
2. Warning order.
3. Target location.
4. Target description.
5. Method of engagement.
6. Method of control.

ESTIMATE OF SITUATION (FM 7-10)

1. Conduct a detailed mission analysis.
2. Analyze the situation and develop courses of action.
3. Analyze courses of action (wargame).
4. Compare courses of action.
5. Make a decision.

OPERATION ORDER (FM 7-10)

Task Organization
1. Situation
   a. Enemy
   b. Friendly
   c. Attachments/detachments
2. Mission
   - Who, what, where, when, and why.
   - Task and purpose
3. Execution.
   a. Concept of operation.
      (1) Maneuver.
      (2) Fire support.
      (3) Engineer.
   b. Tasks to maneuver units
   c. Tasks to combat support units.
   d. Coordinating instructions.
4. Service support.
   a. General.
   b. Materiel and services.
   c. Casualty evacuation.
   d. Miscellaneous.
5. Command and signal
   a. Command.
   b. Signal.

SPOT REPORT

1. Size.
2. Activity.
3. Location.
4. Unit/uniform.
5. Time.
6. Equipment.

SHELL REPORT (FM 6-121)

1. Observer identification.
2. Location (coded).
3. Azimuth to flash or sound.
4. Time (from and to).
5. Area shelled.
6. Nature of fire.
7. Type rounds received (artillery, mortars, etc.).
8. Damage (coded).

WEAPONS (FM 7-8)

TYPE	MAX EFF RANGE (m)
M16A2	580 (pt) 800 (area) 200 (mov)
M203	150(pt) 350 (area)
M249	600 (pt) 800 (area)
M136 (AT4)	300
M47 (Dragon)	1,000 (sta) 100 (mov)
MK19	1500 (pt) 2,212 (area)
M3 RAAWS	700 (sta) 60 (mov)
M60 MG	1,100 (600 grazing)
.50 Caliber MG	1,800 (1,000 grazing)
TOW	3,000 (plng purposes)
TOW 2	3,750
105-mm	11,500
105-mm Tank	*2 to 2.5 km
120-mm Tank	*2 to 2.5 km
25-mm BFV	2,200
155-mm M109A3	18,100
M198	24,000
8-in Howitzer	22,900

WEAPONS (MORTAR) HE ONLY (FM 7-90)

	MIN	MAX
60-mm	70m	3,500m
81-mm (M252)	80m	5,800m
81-mm (M29A1)	73m	4,790m
4.2-inch	770m	6,840m
120-mm	200m	7,200m

FPFs (FM 7-90)

GUNS	MORT	WIDTH		DEPTH
2	60-mm	60m	x	30m
4	81 (M252)	150m	x	50m
4	81 (M29A1)	140m	x	40m
6	4.2-in	240m	x	40m
6	120-mm	360m	x	60m
Plt	155-mm	200m	x	50m
Btry	155-mm	400m	x	50m

Optimum engagement ranges

MEDEVAC REQUEST

1. Requesting unit identification.
2. Location.
3. Number of patients by type (litter or ambulatory).
4. Type of injuries.
5. Special equipment needed.
6. Tactical situation.

AIRCRAFT REQUEST

1. Identification.
2. Precedence/priority.
3. Target description.
4. Target location.
5. Target time/date.
6. Desired ordnance/results.
7. Final control.

DELIBERATE ATTACK CONSIDERATIONS

1. Reconnoiter, pinpoint objective/enemy positions/obstacles.
2. Determine weak points; designate supporting positions.
3. Assign platoon/squad objectives—identify the decisive point.
4. Determine main attack, supporting attack, reserve.
5. Assign breach, support, assault missions.
6. Designate fire control measures.
7. Coordinate indirect/direct fires and CAS to time of attack.
8. Control measures during attack.
9. Secure (ground and air).
10. Consolidate and reorganize.

NBC 1 OBSERVER'S INITIAL OR FOLLOW-UP REPORT

Instructions

1. Line items D and H are mandatory for NBC 1 reports.
2. Line items A, E, G, I, K, L, M, S, Y, and ZA are optional for NBC 1 reports.
3. Line Items B, C, F, PAR, and PBR are reported if data is available.

Section I. Chemical or Biological Only

A. Strike serial number, if known (assigned by NBCE).
B. Position of observer.
C. Azimuth of attack from observer (state degrees or mils).
D. Date and time attack started (Zulu, local, or letter zone).
E. Time attack ended, if known.
F. Location of attack (UTM or place) (state actual or estimated).
G. Means of delivery, if known.
H. Type of agent and height of burst, if known.
I. Type and number of munitions or aircraft (state which).
K. Description of terrain (bare, scrubby vegetation, wooded, urban, or unknown).
S. Date and time contamination detected (Zulu, local, or letter zone).
Y. Representative downwind direction—4 digits (state degrees or mils), wind speed—3 digits (data kmph or knots).
ZA. Temperature (centigrade)—2 digits, cloud cover—1 digit, significant weather phenomena—1 digit, air stability—1 digit.
ZB. Remarks

Section II. Nuclear Only

A. Strike serial number, if known (assigned by NBCE).
B. Position of observer.
C. Azimuth of attack from observer (state degrees or mils and grid or magnetic).
D. Date and time attack started (Zulu, local, or letter zone).
F. Location of attack (UTM or place) (state actual or estimated).
G. Means of delivery, if known.
H. Type of burst (state air, surface, or unknown).
J. Flash-to-bang time (seconds).
K. Crater diameter (meters), if known.
L. Cloud width at H +5 minutes (degrees or mils).
M. Cloud angle(top or bottom) or cloud height(top or bottom) at H +10 minutes (state degrees, mils, meters, or feet).
PAR. Location of radioactive cloud outline (UTM).
PAB. Downwind direction of radioactive cloud (state degrees or mils).
ZB. Remarks.

DEFENSE PLANNING CONSIDERATIONS

1. Establish security (OP/patrols, PWs, M8).
2. Position key weapons:
   a. Coordinate with units on left and right.
   b. Establish FPF or PDF for machine gun.
   c. Ensure mutual support between machine guns.
   d. Cover armor approaches with antiarmor systems.
   e. Establish fire control measures.
3. Prepare positions:
   a. Check sectors of fire.
   b. Check overhead cover and view positions from enemy vantage.
   c. Position in depth and achieve mutual support between positions.
   d. Select/prepare alternate and supplementary positions.
4. Integrate indirect fires, CAS, and obstacles with direct and indirect fire.
5. Check communications and establish emergency signals.
6. Designate ammunition, supply, PW, and casualty points.

IED / UXO

Procedures when IEDs are found

Security - Maintain 360 degrees. Scan close in and far out, up high and down low.

Always - Scan you immediate surroundings for more IEDs.

Move - Move away. Vary safe distances, but plan for 300m minimum safe distance and adapt to your METT-TC.

Attempt - To confirm suspected IEDs using optics while staying back as far as possible.

Cordon - Off the area. Direct people out of danger area. Do not allow anyone to enter except for EOD. Question, search, and detain suspects as defined by your Existing ROE.

Report - Your situation using the 9 line IED / UXO spot report format.

This could be your hand if you try to dispose of UXOs or IEDs. The enemy has developed anti-handling to catch you when you try defusing. Leave it to experts!

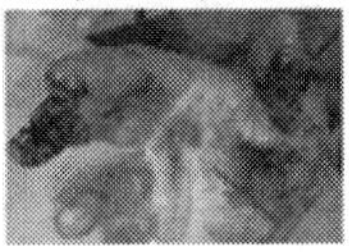

Call EOD - Don't be a Hero!

IED / UXO Report

LINE 1. DATE-TIME-GROUP: When the item was discovered.

LINE 2. REPORT ACTIVITY AND LOCATION: Unit and grid location of the IED/UXO.

LINE 3. CONTACT METHOD: Radio frequency, call sign, POC, and telephone number.

LINE 4. TYPE OF ORDNANCE: Dropped, projected, placed, or thrown. Give the number of items, if more than one.

LINE 5. NBC CONTAMINATIONS: Be as specific as possible.

LINE 6. RESOURCES THREATENED: Equipment, facilities, or other assets that are threatened.

LINE 7. IMPACT ON MISSION: Short description of current tactical situation and how the IED/UXO affects the status of the mission.

LINE 8. PROTECTIVE MEASURES: Any measures taken to protect personnel and equipment.

LINE 9. RECOMMENDED PRIORITY: Immediate, Indirect, Minor, No Threat.

Priority

Immediate: Stops unit's maneuver and mission capability or threatens critical assets vital to the mission.

Indirect: Stops the unit's maneuver and mission capability or threatens critical assets important to the mission.

Minor: Reduces the unit's maneuver and mission capability or threatens non-critical assets of value.

No Threat: Has little or no effect on the unit's capabilities or assets.

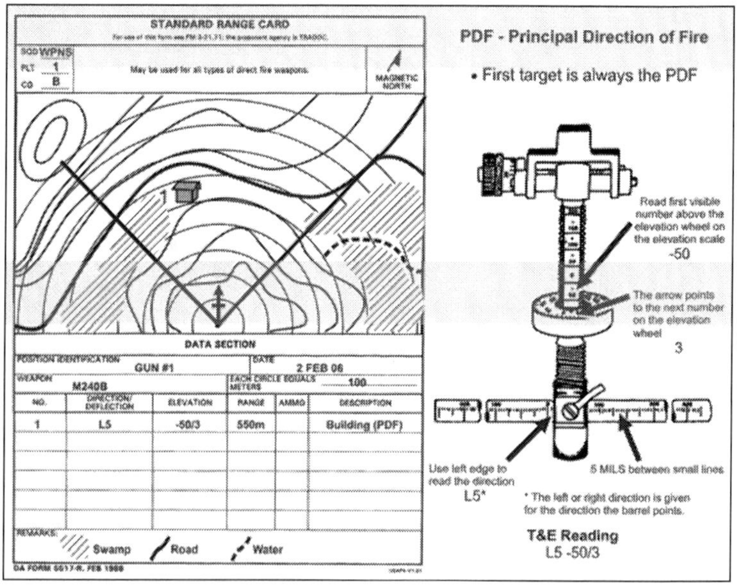
STANDARD RANGE CARD
SQD WPNS
PLT 1
CO B
May be used for all types of direct fire weapons.
MAGNETIC NORTH
1
DATA SECTION
POSITION IDENTIFICATION GUN #1
DATE 2 FEB 06
WEAPON M240B
EACH CIRCLE EQUALS METERS 100
NO. | DIRECTION/DEFLECTION | ELEVATION | RANGE | AMMO | DESCRIPTION
1 | L5 | -50/3 | 550m | | Building (PDF)
REMARKS: Swamp Road Water
DA FORM 5517-R, FEB 1986
PDF - Principal Direction of Fire
• First target is always the PDF
Read first visible number above the elevation wheel on the elevation scale
-50
The arrow points to the next number on the elevation wheel
3
Use left edge to read the direction
L5*
5 MILS between small lines
* The left or right direction is given for the direction the barrel points.
T&E Reading
L5 -50/3

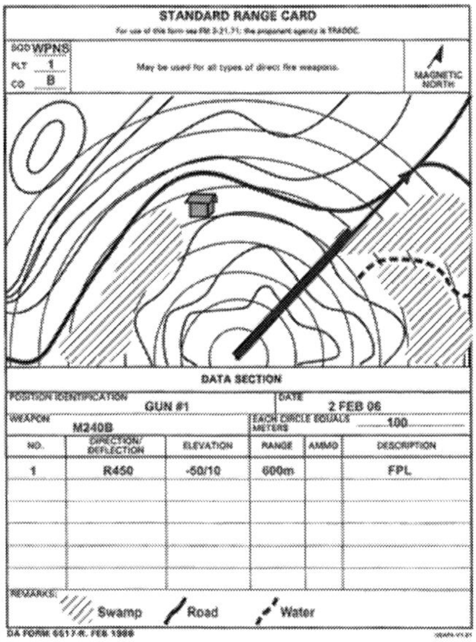

STANDARD RANGE CARD

For use of this form see FM 3-21.71; the proponent agency is TRADOC.

SQD WPNS
PLT 1
CO B

May be used for all types of direct fire weapons.

MAGNETIC NORTH

DATA SECTION

POSITION IDENTIFICATION: GUN #1
DATE: 2 FEB 06
WEAPON: M240B
EACH CIRCLE EQUALS 100 METERS

NO.	DIRECTION/ DEFLECTION	ELEVATION	RANGE	AMMO	DESCRIPTION
1	R450	-50/10	600m		FPL

REMARKS: Swamp Road Water

DA FORM 5517-R, FEB 1986

FPL - Final Protective Line

- FPL is always target #1
- FPL will always be metal-to-metal
- Represented by thick line
    - Break in thick line for dead space out to 600 meters
    - The gap is equal to the width of the dead space

FPL - Final Protective Line

STANDARD RANGE CARD

For use of this form see FM 3-21.71; the proponent agency is TRADOC.

SQD ______

PLT ______

CO ______

May be used for all types of direct fire weapons.

MAGNETIC NORTH

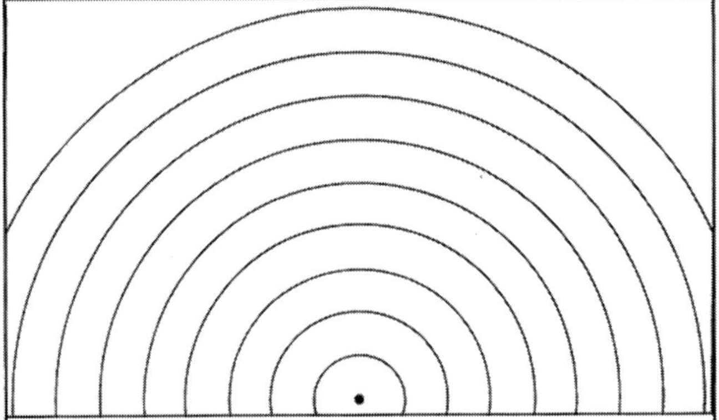

DATA SECTION

POSITION IDENTIFICATION | DATE

WEAPON | EACH CIRCLE EQUALS ______ METERS

NO.	DIRECTION/ DEFLECTION	ELEVATION	RANGE	AMMO	DESCRIPTION

REMARKS:

DA FORM 5517-R, FEB 1986

USAPA V1.01

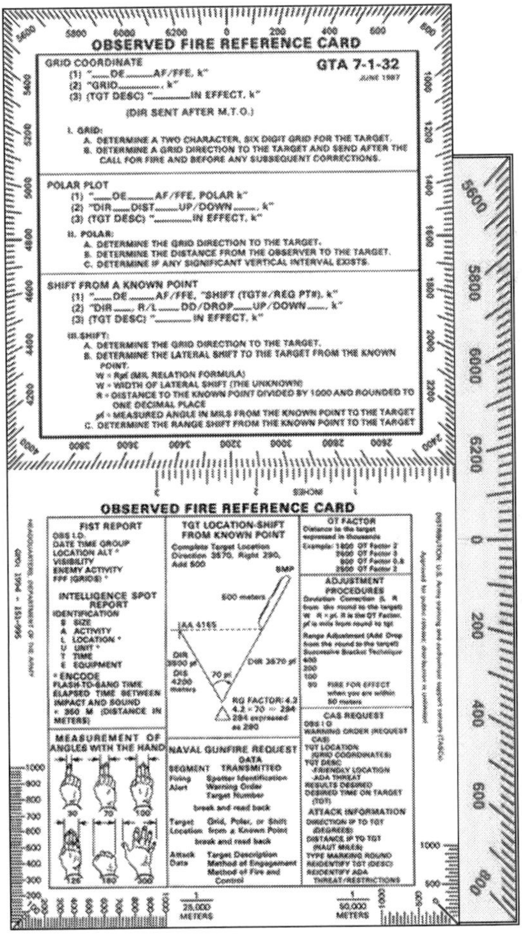
OBSERVED FIRE REFERENCE CARD
GRID COORDINATE
GTA 7-1-32
JUNE 1987
(1) "____DE____AF/FFE, k"
(2) "GRID________, k"
(3) (TGT DESC) "______IN EFFECT, k"
(DIR SENT AFTER M.T.O.)
I. GRID:
A. DETERMINE A TWO CHARACTER, SIX DIGIT GRID FOR THE TARGET.
B. DETERMINE A GRID DIRECTION TO THE TARGET AND SEND AFTER THE CALL FOR FIRE AND BEFORE ANY SUBSEQUENT CORRECTIONS.
POLAR PLOT
(1) "____DE____AF/FFE, POLAR k"
(2) "DIR____DIST____UP/DOWN____, k"
(3) (TGT DESC) "______IN EFFECT, k"
II. POLAR:
A. DETERMINE THE GRID DIRECTION TO THE TARGET.
B. DETERMINE THE DISTANCE FROM THE OBSERVER TO THE TARGET.
C. DETERMINE IF ANY SIGNIFICANT VERTICAL INTERVAL EXISTS.
SHIFT FROM A KNOWN POINT
(1) "____DE____AF/FFE, "SHIFT (TGT#/REG PT#), k"
(2) "DIR____, R/L____DD/DROP____UP/DOWN____, k"
(3) (TGT DESC) "______IN EFFECT, k"
III. SHIFT:
A. DETERMINE THE GRID DIRECTION TO THE TARGET.
B. DETERMINE THE LATERAL SHIFT TO THE TARGET FROM THE KNOWN POINT.
W = Rø (MIL RELATION FORMULA)
W = WIDTH OF LATERAL SHIFT (THE UNKNOWN)
R = DISTANCE TO THE KNOWN POINT DIVIDED BY 1000 AND ROUNDED TO ONE DECIMAL PLACE
ø = MEASURED ANGLE IN MILS FROM THE KNOWN POINT TO THE TARGET
C. DETERMINE THE RANGE SHIFT FROM THE KNOWN POINT TO THE TARGET
INCHES
OBSERVED FIRE REFERENCE CARD
FIST REPORT
OBS I.D.
DATE TIME GROUP
LOCATION ALT *
VISIBILITY
ENEMY ACTIVITY
FPF (GRIDS) *
INTELLIGENCE SPOT REPORT
IDENTIFICATION
S SIZE
A ACTIVITY
L LOCATION *
U UNIT *
T TIME
E EQUIPMENT
* ENCODE
FLASH-TO-BANG TIME
ELAPSED TIME BETWEEN IMPACT AND SOUND
× 350 M (DISTANCE IN METERS)
TGT LOCATION-SHIFT FROM KNOWN POINT
Complete Target Location
Direction 3870, Right 290, Add 500
SMP
500 meters
AA 4165
DIR 3800 ø
DIS 4200 meters
70 ø
DIR 3870 ø
RQ FACTOR: 4.2
4.2 × 70 = 294
294 expressed as 290
OT FACTOR
Distance to the target expressed in thousands
Example: 1800 OT Factor 2
2500 OT Factor 3
500 OT Factor 0.5
2500 OT Factor 2
ADJUSTMENT PROCEDURES
Deviation Correction (L, R from the round to the target)
W = R × ø, R is the OT Factor, ø is mils from round to tgt
Range Adjustment (Add, Drop from the round to the target)
Successive Bracket Technique
400
200
100
50 FIRE FOR EFFECT when you are within 50 meters
CAS REQUEST
OBS I.D
WARNING ORDER (REQUEST CAS)
TGT LOCATION (GRID COORDINATES)
TGT DESC
-FRIENDLY LOCATION
-ADA THREAT
RESULTS DESIRED
DESIRED TIME ON TARGET (TOT)
ATTACK INFORMATION
DIRECTION IP TO TGT (DEGREES)
DISTANCE IP TO TGT (NAUT MILES)
TYPE MARKING ROUND
REIDENTIFY TGT (DESC)
REIDENTIFY ADA THREAT/RESTRICTIONS
MEASUREMENT OF ANGLES WITH THE HAND
30
70
100
125
180
300
NAVAL GUNFIRE REQUEST
SEGMENT DATA TRANSMITTED
Firing Alert Spotter Identification
Warning Order
Target Number
break and read back
Target Location Grid, Polar, or Shift from a Known Point
break and read back
Attack Data Target Description
Method of Engagement
Method of Fire and Control
1/25,000 METERS
1/50,000 METERS
HEADQUARTERS, DEPARTMENT OF THE ARMY

9 Line MEDEVAC

LINE ITEM	EXPLANATION
1. Location of Pickup Site.	Encrypt grid coordinates. When using *DRYAD Numeral Cipher, the same SET line* will be used to encrypt grid zone letters and coordinates. To preclude misunderstanding, a statement is made that grid zone letters are included in the message (unless unit SOP specifies its use at all times).
2. Radio Frequency, Call Sign, Suffix.	Encrypt the frequency of the radio at the pickup site, *not* a relay frequency. The call sign (and suffix if used) of person to be contacted at the pickup site may be transmitted in the clear.
3. No. of Patients by Precedence.	Report only applicable info & encrypt brevity codes. A = Urgent, B = Urgent-Surg, C = Priority, D = Routine, E = Convenience. (If 2 or more categories reported in same request, insert the word "*break*" between each category).
4. Spec Equipment.	Encrypt applicable brevity codes. A = None, B = Hoist, C = Extraction equipment, D = Ventilator.
5. No. of Patients by Type.	Report only applicable information and encrypt brevity code. If requesting MEDEVAC for both types, insert the word "*break*" between the litter entry and ambulatory entry: L + # of Pnt - Litter; A + # of Pnt - Ambul (sitting).
6. Security Pickup Site (Wartime).	N = No enemy troops in area, P = Possibly enemy troops in area (approach with caution), E = Enemy troops in area (approach with caution), X = Enemy troops in area (armed escort required).
6. Number and type of Wound, Injury, Illness (Peacetime).	Specific information regarding patient wounds by type (gunshot or shrapnel). Report serious bleeding, along with patient blood type, if known.
7. Method of Marking Pickup Site.	Encrypt the brevity codes. *A* = Panels, *B* = Pyrotechnic signal, *C* = Smoke Signal, *D* = None, *E* = Other.
8. Patient Nationality and Status.	Number of patients in each category need not be transmitted. Encrypt only applicable brevity codes. A = US military, B = US civilian, C = Non-US mil, D = Non-US civilian, E = EPW.
9. NBC Contamination (Wartime).	Include this line only when applicable. Encrypt the applicable brevity codes. N = nuclear, B = biological, C = chemical.
9. Terrain Description (Peacetime).	Include details of terrain features in and around proposed landing site. If possible, describe the relationship of site to a prominent terrain feature (lake, mountain, tower).

Reference: FM 8-10-6, *Medical Evacuation in a Theater of Operations, pages 7-7 through 7-9.*

TABLE OF CONTENTS

PART I. SCOUTING

Part I. Scouting

Chapter 1. Fundamentals of Scouting

Scouting involves observing terrain and/or the enemy, and accurately reporting those observations. Scouting requires proficiency in the use of weapons, cover and concealment, route selection, and skill in unobserved day or night movement.

1001. PURPOSE

When an infantry unit is not actively fighting the enemy, it should be actively searching for the enemy, which is the primary purpose of scouting. The unit attempts to keep the enemy off balance while making preparations for further attacks. Physically locating and keeping the enemy off balance are normally accomplished by small units ranging from a two-person scouting party to a squad-size patrol.

Infantrymen are sent out as scouts or as members of a patrol because the commander needs information about the enemy, terrain, and the location of friendly troops. The lives of the entire unit may depend upon the success or failure of a scout or patrol and the accuracy and timeliness of the report. The success of the scout or patrol will depend upon their training, preparation by the commander, and understanding their mission and the commander's requirements.

To wage combat successfully, a commander must have accurate, detailed, and timely information about the enemy, the terrain, and adjacent friendly units. Well-trained scouts and capably led patrols are among the most effective means the commander has for acquiring the information necessary to plan tactical actions and make decisions in execution.

1002. REQUIRED SCOUTING SKILLS

To be effective, a scout must be able to—

- Recognize terrain features.
- Read a map and determine direction.
- Practice and implement the principles of cover and concealment.
- Fully utilize movement and route selection.
- Know the enemy (estimate enemy unit composition and strength).
- Observe and report information accurately.
- Select routes and move through numerous types of terrain.

CHAPTER 2. TERRAIN, MAPS, AND DIRECTION

A scout must understand map symbols, identify elevations from contour interval lines, scale distance on a map, relate natural and man-made features shown on the map to the actual features on the ground, plot a course from one point to another, and locate his current position. To relate a map to the actual terrain and its features, a scout must be able to orient it to the ground using a compass, two points, a watch, and the sun or the stars.

2001. TERRAIN FEATURES

Since the infantry works and fights on the ground, terrain ashore information that scouts gather and report on is of great importance. Hills, valleys, woods, and streams are the forms and growths commonly referred to as natural land features. Artificial or man-made features include houses, bridges, and railroads. Figure 2-1 shows some important terrain features.

2002. THE LENSATIC COMPASS

The best method of finding direction, during both day and night, is with a compass. The lensatic and M2 are the two types of compasses issued to infantry units. Both work on the same principle.

The standard compass for general use in the Marine Corps is the pivot-mounted lensatic compass, so called because azimuths are read through a magnifying lens in the eyepiece. Figure 2-2, on page 2-2, shows the lensatic compass and its nomenclature. The meter graphic scale on the side of the compass is 1:50,000, which is the most used scale in military mapping. The graphic scale is useful in the field as a straightedge, an aid in orienting the map, and a means of reading map grid coordinates. The plastic dial is graduated in both degrees and mils. Numbers on the dial are printed in black. There are luminous markings on the bezel, floating dial, and on both ends of the sighting wire, plus a 3-degree bezel serration and clicking device that permit reading azimuths at night. The compass is carried in a nylon case that may be attached to the cartridge belt.

Magnetic compasses are affected by the presence of iron or magnetic fields. Consequently, the scout should not be within the influence of local magnetic

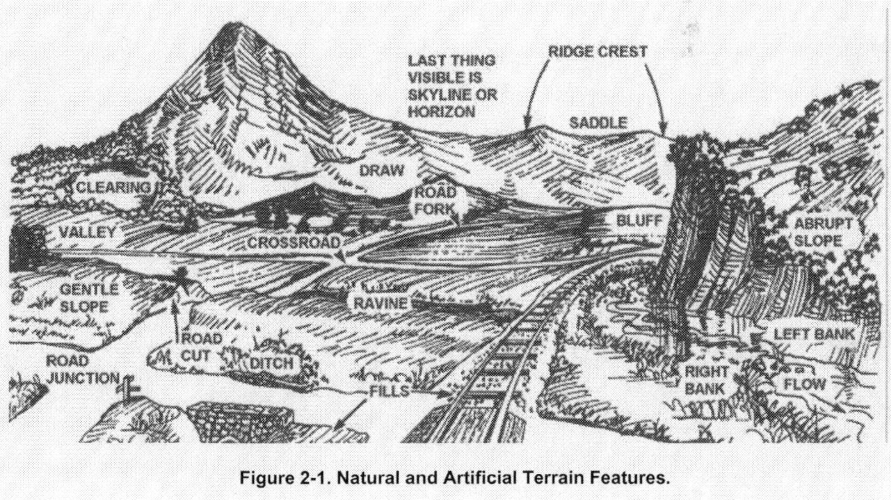

Figure 2-1. Natural and Artificial Terrain Features.

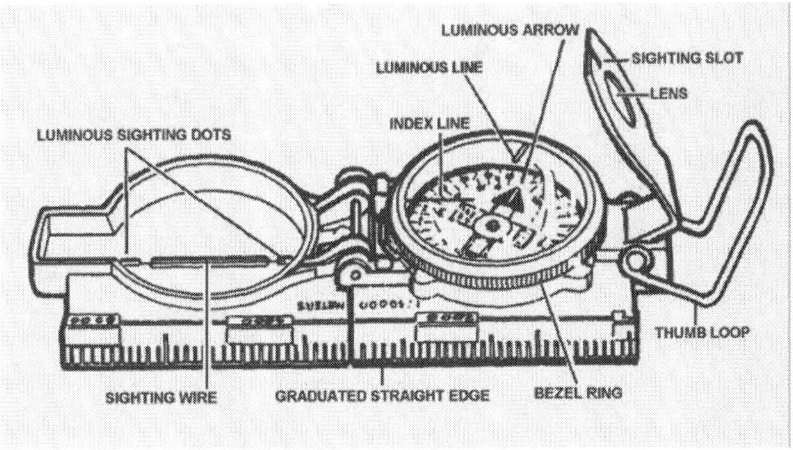

Figure 2-2. Lensatic Compass.

attraction while using a compass to determine direction. The rifle, pistol, and other metal objects must be laid aside when reading the compass. The minimum distances the scout should be from visible masses of iron and electrical fields of magnetism for the compass to provide accurate readings are provided below.

Magnetic Fields	Minimum Distances (meters)
High tension powerlines	60
Field guns	20
Vehicles (wheeled or tracked)	20
Telephone and telegraph wires	10
Barbed wire	10
Machine gun	3
Rifle, pistol	1

Reading an Azimuth

To read an azimuth to any point, the cover of the compass is raised to an angle of 90 degrees in relation to the index face, and the eyepiece is lifted to a 45-degree angle in relation to the bezel, or so the numbers on the dial can be seen. The thumb of either hand is placed in the thumb loop, the index finger extended along the side of the compass case, and the remainder of the hand closed. The closed hand and wrist are grasped with the other hand. The elbows are drawn in close to the body, forming a firm foundation for the compass. The eye is placed to the lens on the eyepiece. The compass is pointed at the object or point to which the azimuth is to be read. A sighting is taken through the sighting slot in the eyepiece, and the point is lined up with the sighting wire in the cover. The compass is held until the dial steadies; then the reading is taken through the lens of the eyepiece. This reading is the magnetic azimuth of the line from the observer to the point.

Reading a Back Azimuth

A back azimuth is the direction opposite the line of sight. If the azimuth is less than 180 degrees, the back azimuth is obtained by adding 180 degrees. If the azimuth is greater than 180 degrees, the back azimuth is obtained by subtracting 180 degrees. Back azimuths are used to determine a return route or to resection to determine a current position.

Circumventing Obstacles

When a scout is traveling on an azimuth and comes upon an obstacle—such as a contaminated area, minefield or swamp—the following steps (sometimes referred to as the *90-degree offset* method) is employed to go around or circumvent the obstacle and resume movement along the original azimuth (see fig. 2-3). The steps are as follows:

1. Move up to the obstacle and make a full 90-degree turn to the right (or left).

2. Walk beyond the obstacle, keeping track of the distance in paces or meters.

3. Stand at the end of the obstacle, face in the original direction of march, and follow that azimuth until the obstacle has been passed.

4. Make a 90-degree turn to the left (or right) and move the distance previously measured to return to the original line of march.

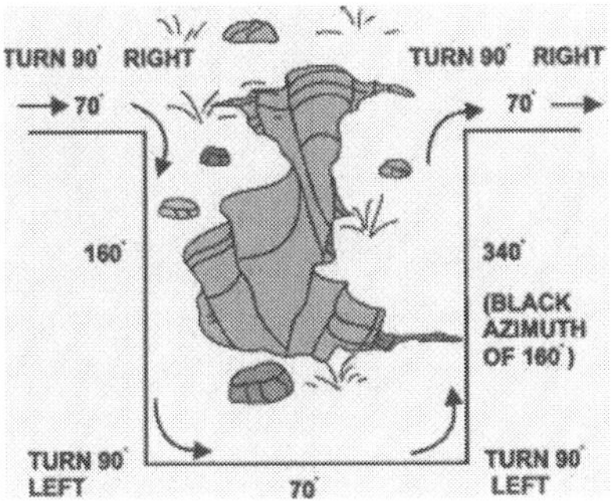

Figure 2-3. Circumventing an Obstacle.

Following an Azimuth During the Day

The eye is placed to the lens in the eyepiece and the compass moved until the desired azimuth reading is visible beneath the fixed index. Without moving the compass, the vision is shifted from the lens through the sighting slot in the eyepiece, and a sighting is taken out beyond the sighting wire in the cover. A prominent terrain feature on this line of sight is selected, the compass closed, and the landmark approached. When the scout reaches the landmark, the procedure is repeated.

Following an Azimuth at Night

It is necessary to prepare and set the compass before departing on a night movement because at night only the luminous parts of the compass can be seen. To prepare the compass for night use, the luminous parts must be fully charged by sunlight or artificial light, such as a flashlight. To set a compass—

1. Move the compass so that the desired azimuth on the dial is directly under the index line on the lower glass.

2. Rotate the upper movable glass so that the luminous line is directly above the north arrow of the dial.

3. Set the compass for marching at night on the specified azimuth.

—OR—

1. Face the general direction of movement.

2. Line up the north arrow and the luminous line on the bezel with the luminous sighting dots.

3. Hold the compass still with one hand and grip the knurled bezel ring with the other hand.

4. Turn the bezel ring the prescribed number of clicks in the proper direction, remembering that each click equals 3 degrees. For example, to set an azimuth of 21 degrees, the bezel ring would be turned seven clicks to the left.

5. Turn the whole compass until the north needle lines up with the luminous line. The compass is then set on the desired azimuth. The azimuth is the line formed by the two luminous sighting dots on the inside of the cover.

To march on a preset azimuth during night movement, open the compass and move it so the north arrow is directly below the luminous line (see fig. 2-4). Move

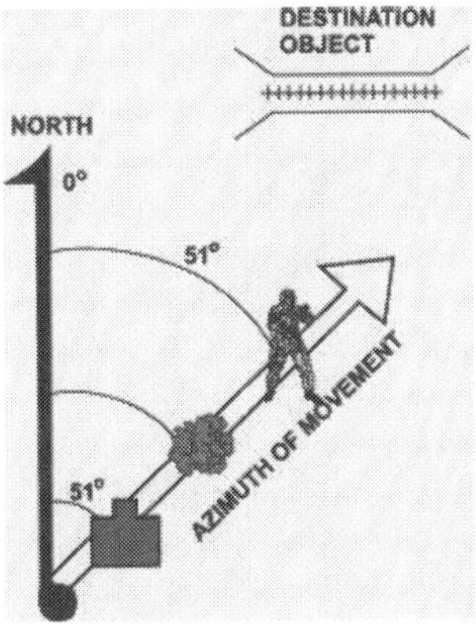

Figure 2-4. Following a Night Azimuth.

in the direction of the line formed by the two luminous sighting dots. It is necessary to refer to the compass more frequently at night than during the day. If stars are visible, find a prominent star along the azimuth of movement to use as a reference point. When the view of the sky is restricted by overcast conditions or vegetation, send a scout forward along the azimuth of movement to the limit of visibility. This scout is guided along the azimuth of movement by a stationary navigator. When the scout reaches the limit of visibility, the navigator moves to the scout's location. This process is repeated until the destination is reached.

A more rapid method for reaching the scout's destination is to equip the navigator with a compass. The navigator can set the compass as explained earlier and the scout proceeds providing security 180 degrees to the front on the specified azimuth, receiving right and left corrections from the navigator while both are on the move. The point scout must stay within visual range of the navigator. If available, a strip of white or luminous tape on the back of the point scout's helmet will assist.

Intersection

Intersection is the location of an unknown point by successively occupying at least two, preferably three known positions and sightings on the unknown point. It is used to locate features not depicted on the map or not readily identifiable. To determine an intersection, perform the following steps (see fig. 2-5):

1. Orient the map using the compass.

2. Locate and mark your position on the map.

3. Measure the magnetic azimuth to the unknown position; then convert to grid azimuth.

4. Draw a line on the map from your position on this grid azimuth.

5. Move to a second known position from which unknown point is visible. Locate this position on the map and again orient the map using the compass. The second unknown position should be a minimum of 30 degrees offset from the first position.

6. Repeat steps 4 and 5.

Figure 2-5. Intersection.

To check accuracy, move to a third position and repeat steps 1 thorough 4. Where the lines cross is the location of the unknown position. Using three lines, a triangle is sometimes formed—called the *triangle of error*—instead of an intersection. If the triangle is large, recheck your work to find the error. Do not assume that the position is at the center of the triangle.

Resection

Resection is the location of the user's unknown position by sighting on two or three known features that are identifiable on the map. To determine a resection, perform the following steps (see fig. 2-6):

1. Orient the map using the compass.

2. Locate two or three known positions on the ground and mark them on the map.

3. Measure the magnetic azimuth to a known position then convert to grid azimuth.

4. Change the grid azimuth to a back azimuth and draw a line on the map from the known position back toward the unknown position.

5. Repeat step 3 and step 4 to determine a second known position.

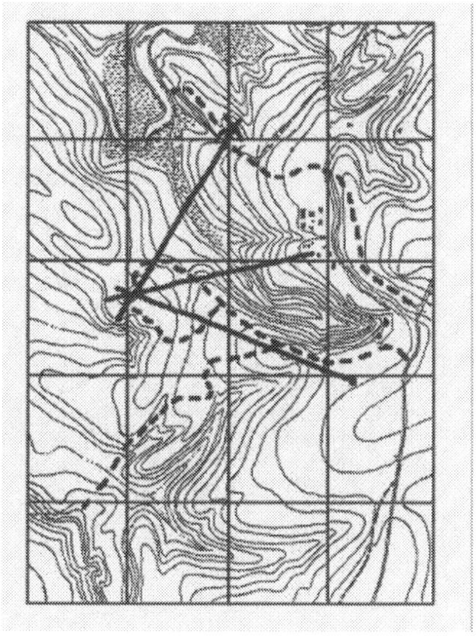

Figure 2-6. Resection.

To check accuracy, repeat the steps above for a third known position. The intersection of the lines is your location. Using three lines, a triangle of error may be formed. If the triangle is large, recheck.

2003. STARS, SUN, AND OTHER FEATURES

In rare cases when a scout is without a compass, the following examples are alternate means to determine direction. When using constellations to determine direction, identify your location's Temperate Zone. A Temperate Zone is the area between the tropics and the polar circles.

At night, the stars provide an excellent means of maintaining a line of march. In the North Temperate Zone (north of the equator), the Big Dipper constellation is one key to determining direction of true north. It is made up of seven fairly bright stars in the shape of a dipper with a long curved handle (see fig. 2-7). The two stars that form the side of the cup farthest from the handle, used as pointers, are situated in the direction of a bright star that is about five times the distance between the two stars of the dipper cup. This bright star is the North Star and is directly over the North Pole. The pointers always designate the North Star, which is the direction of true north.

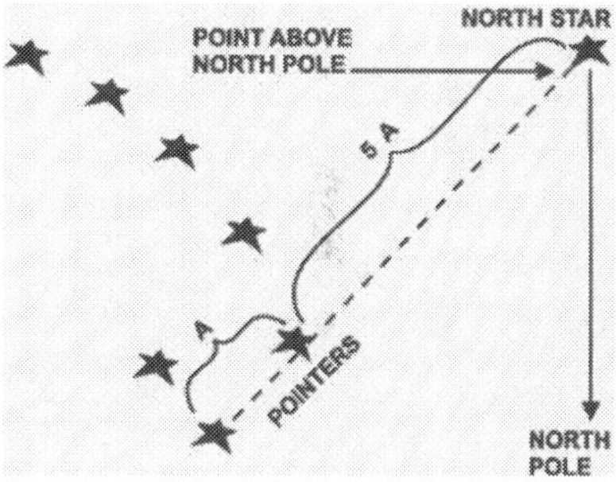

Figure 2-7. Locating the North Pole.

In the Southern Hemisphere, true south is determined in relation to the Southern Cross, a constellation composed of five stars. Two bright pointer stars in the vicinity of the Southern Cross serve as locators to help locate true south (see fig. 2-8). The outer four stars are

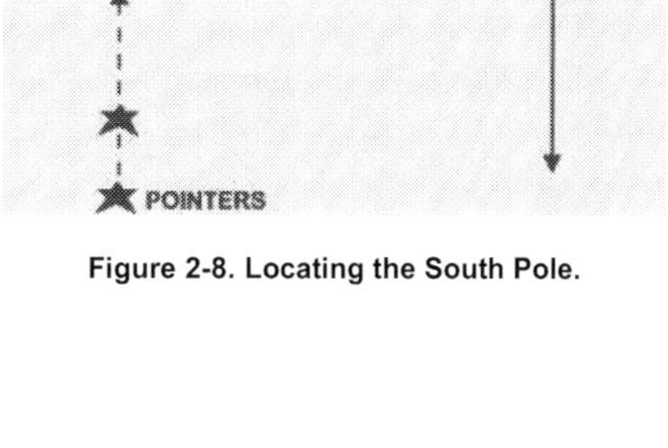

Figure 2-8. Locating the South Pole.

fairly bright and form a cross. This cross is imagined as the frame of a kite. A straight tail, four and one half times as long as the length of the kite itself, is put on the kite using finger widths for a measuring stick. The end of this tail will be close to a position directly over the South Pole. Usually, it will not be possible to see a star in the immediate vicinity, because there is no bright star visible directly above the South Pole.

During daylight hours, a watch and the sun can be used to determine direction within 8 degrees. In the North Temperate Zone, the watch is held horizontally, face up, and the hour hand pointed at the sun (see fig. 2-9). The north-south line and the direction of south can be found midway between the hour hand and the number 12, if the watch is set on standard time. If in daylight savings time, the direction of south is found midway between the hour hand and the number 1.

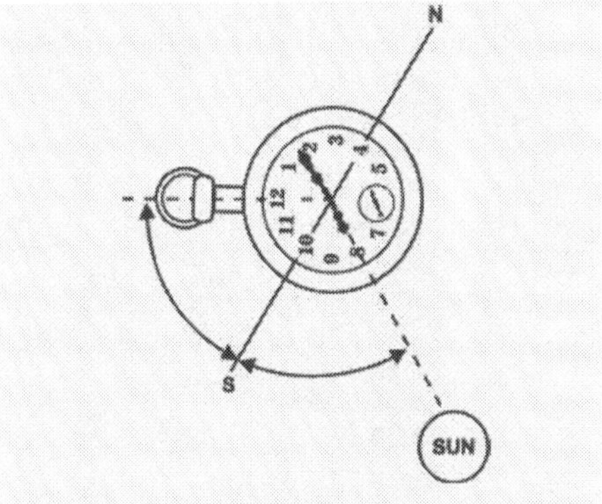

Figure 2-9. Determining Direction by Watch and Sun (North Temperate Zone).

In the South Temperate Zone, if the watch is set on standard time, the number 12 on the watch is pointed at the sun; if the watch is set on daylight savings time, the number 1 is pointed at the sun. North is midway between 12 (or 1) and the hour hand (see fig. 2-10).

When laying in a north-south line, if any doubt exists as to which end of the line is north, remember that the sun is in the east before noon and in the west in the afternoon.

In addition to the sun and stars, other methods a scout without a compass can use to determine direction include determining prevailing wind direction and

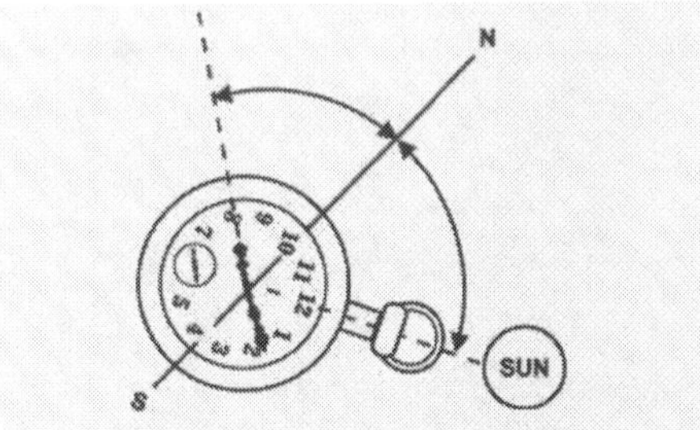

Figure 2-10. Determining Direction by Watch and Sun (South Temperate Zone).

using a mountain for orientation. By previous study of maps and photographs, a scout can keep informed of location and direction by using a distinctive edge of woods, a deep ravine or the direction of a stream's flow. A scout should constantly evaluate and memorize both the immediate terrain and general area for prominent features and landmarks.

2004. RANGE DETERMINATION

Range determination is the method of finding the distance between an observer and an enemy target or an object. By accurate range determination, the members of a given unit can set their sights correctly and place effective fire on enemy targets. The degree of accuracy is dependent on several factors, such as terrain relief, time available, and experience of the observer.

Mental Estimation

A mental distance estimate is made using a known unit of measure. Distance is estimated to the nearest 100 meters by determining the number of known units of measure between the observer's position and a target. For example, a football field, which is 100 yards, can be used as a known unit of measure for determining the distance between an observer's position and a target. For longer distances, progressive estimation may be necessary. To do this, the observer determines the number of units of measure to an intermediate point and doubles the value. The observer should consider the effects in table 2-1 in estimating distances.

Table 2-1. Effects to Consider in Mental Estimation of Distances.

Objects Appear Nearer	Objects Appear More Distant
In bright light.	In poor light or in fog.
In clear air at high altitude.	Only a small part of the object can be seen.
The background is in contrast with the color of the object.	The background is similar in color to that of the object.
The observer is looking down from a height.	The observer is looking over a depression, most of which is visible.
The observer is looking over a depression, most of which is hidden.	The observer is kneeling or sitting, especially on a hot day, when the ground is moist.
The observer is looking down a straight feature such as a road.	
The observer is looking over water, snow, or a uniform surface such as a cultivated field or desert.	

Estimating in Good Visibility

When visibility is good, distances can be estimated by using the appearance of tree trunks, branches, and foliage (as seen by the naked eye) in comparison with map data. Table 2-2 is a guide for wooded terrain. Table 2-3 is a guide for urban environments.

Estimating From a Terrain Study

The Marine should always use terrain/map analysis to assist in estimating distances. When the Marine is looking in a specific direction, the estimation of distance can be enhanced by studying the terrain and comparing it with the map. Particular emphasis should be given to color contrasts of terrain features seen along the observer-target line (OTL). For example, the distance across successive ridge lines or depressions in the distance may be identifiable by only slight changes of color to the eye. Different colors of grass might reveal a hidden terrain feature such as a stream.

Table 2-2. Estimating Distance in Wooded Terrain.

Distance in Meters	Tree Description
1,000	Trunk and main branches are visible. Foliage appears in cluster-like shape. Daylight may be seen through the foliage.
2,000	Trunk visible, main branches distinguishable, foliage appears as smooth surface. Outline of foliage of separate trees distinguishable.
3,000	Lower half of trunk visible. Branches blend with foliage. Foliage blends with adjoining trees.
4,000	Trunk and branches blend with foliage and appears as a continuous cluster, smooth in appearance. Movement of foliage due to wind cannot be deleted.
5,000 and beyond	Whole area covered by trees and appears smooth and dark.

Table 2-3. Estimating Distance in Urban Terrain.

Distance in Meters	Object Identified by the Unaided Eye
1,000	Lone tree trunk
1,500	Individuals and horsemen
3,000	Chimneys on rooftops
4,000	Windows in houses
4,000–5,000	Individual houses in populated area
8,000–9,000	Villages and individual houses
15,000–18,000	Large houses, towers, and steeples

Chapter 3. Enemy Activity

A commander often acts on information furnished by scouts. Therefore, scouts must aim at absolute accuracy in reporting enemy activity. This chapter discusses estimating enemy strengths, interpreting signs and tracks, and knowing the enemy.

3001. ESTIMATING ENEMY STRENGTHS

If troops cannot be counted, their strength may be estimated by: noting the length of time it takes various types of moving columns to pass given point, the area required of a unit in camp or bivouac, or the front on which they are deployed. When the ground is dry, infantry on the march raise a low, thick cloud of dust, and motor vehicles or mechanized units raise a thick, rapidly moving cloud. Additionally, through practice, a scout may gain information as to the strength and composition of enemy forces by listening to noises and observing lights, fires, and smoke. A scout gains valuable experience in estimating enemy strengths by observing friendly forces in camp, on the march, and deployed. The knowledge scouts gain during field exercises of the appearance and tactical dispositions of squads, platoons, companies, and larger units will be of great assistance in estimating the strength and composition of enemy units observed under various conditions.

3002. INTERPRETING SIGNS AND TRACKS

In addition to estimates made through direct observation, a scout may often be able to estimate size, composition, direction, rate of movement, condition, discipline, state of training, and morale of enemy forces through signs and tracks left behind.

Signs

The examination of vacated enemy positions provides valuable information. The size of a bivouac or defense area ordinarily indicates the number of enemy occupants. Clothing, ration containers, dumps, etc., further indicate the quantity of the departed enemy force. The condition of the bivouac area and amount of material abandoned give an indication of the enemy morale, training, and discipline. A well-policed area indicates good discipline. Rubbish, ration and smoking residue, and nonessential personal items of equipment adrift indicate a lower state of morale, training, and discipline. Stores and material left behind in good condition may indicate a hasty movement or withdrawal. Burned or destroyed materials indicate a deliberate, orderly withdrawal or movement. Letters, insignia, and other articles may reveal the identity of the enemy unit.

In the case of a moving enemy, the distance between periodic halts indicates the rate of march if enemy habits relative to marches and halts are known. Condition of the halt areas indicates the state of morale, training, and discipline.

The physical condition of enemy dead and wounded and their personal equipment and weapons are reported. The general condition and state of maintenance of destroyed or abandoned vehicles should also be reported.

Tracks

A track is a mark left on the ground by the passage of a person or object. Examination of tracks reveals information about the enemy.

Troops

A few tracks overlapping each other on both sides of a road or trail may indicate a patrol in staggered formation. A large number of tracks indicates troops in column formation. A large column will wear a dry road smooth and flat. In damp terrain, a freshly made track will have sharp edges; ordinarily, signs of moisture will disappear in about 15 minutes. A runner's toes are dug into the ground; a walker's footprint is fairly even.

Vehicles

The type of track indicates whether the vehicle is wheeled or tracked. A scout acquires the necessary experience to make the proper determination by observing vehicle tracks during training.

The direction of travel can be determined by the way tracks pass across ruts, by impressions on the edges of holes in the ground, how water is splashed from puddles, or by the way grass, twigs, and branches are bent; for example—

- A vehicle (wheeled or tracked) entering a rut pushes dirt into the rut and leaves an indentation on the exit side of the rut.
- A wheel going over holes in the ground leaves a deeper impression on the edge toward the direction of travel.
- The side of a puddle with the greater splash indicates the direction of travel of the vehicle.
- When traveling cross-country, the direction in which grass is bent and/or twigs, branches, and bushes are broken indicates direction of travel.

A general rate of speed can be estimated by the amount of water or mud splattered. A fast-moving vehicle will throw larger amounts of water or mud a greater distance to the front and sides than a slow-moving vehicle, and it leaves a deeper impression on the exit edges of holes. The faster the travel, the deeper the impression.

3003. KNOWING THE ENEMY

A scout should learn as much as possible about enemy psychology, habits, organization, and tactics. The more knowledge gained about the enemy, particularly the enemy's normal security measures, the better the scout's chances are to observe and obtain accurate information with minimum risk to the mission's success. Scouts gain much of this knowledge through experience, but they also gain a great deal of their preliminary information, particularly that pertaining to enemy organization and tactics, during training and may be updated by unit commanders and intelligence officers.

Chapter 4. Daylight Scouting

A scout must be able to operate in all types of terrain and under all conditions of visibility. He must be thoroughly familiar with the principles for using cover and concealment, camouflage, individual movement, and route selection, both to and from the objective.

4001. COVER AND CONCEALMENT

Cover is protection from the fire of hostile weapons. Concealment is protection from observation or surveillance from hostile air and ground observation, but not from hostile fire. Both cover and concealment are divided into two main categories: natural and artificial. Natural cover includes small hills, ditches, rocks or vegetation. Fighting holes, bunkers, and brick walls are examples of artificial cover. Some features, such as buildings, provide both cover and concealment. In deciding whether to seek cover or concealment, a scout must make the best choice to complete the mission (see fig. 4-1).

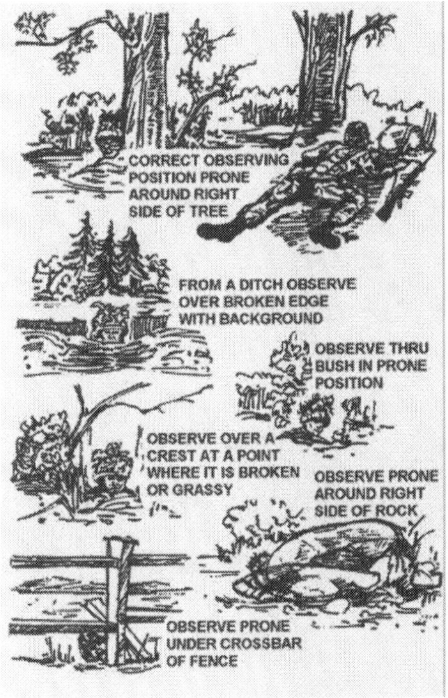

Figure 4-1. Correct Use of Cover.

Concealment Principles

Concealment principles are as follows:

- Remain motionless while observing. Anything in motion attracts the eye.
- Use all available concealment.
- Observe from the prone position (it offers a low silhouette and makes detection by the enemy difficult.
- Expose nothing that reflects light.
- Blend with the background because contrasting colors are noticeable.
- Remain in the shade because moving shadows attract attention.
- Distort or change the regular outline of objects. Most military objects have distinctive shapes that make obvious shadows and silhouettes.
- Avoid the skyline. Figures on the skyline can be seen from great distances and are easily identified by their outlines.

Concealment Techniques

Concealment techniques are as follows:

- When observing, the scout looks around an object's side (unless it is transparent) and prepares to fire, if necessary, around the side of or, if possible, through an object.
- Looking or firing over an object can make the scout an easily visible target for the enemy. If the scout must fire over the top of concealment or cover, the outline of the head or helmet should be broken or distorted.
- Upon the approach of an airplane, the scout takes a prone position, turns face-down, and remains motionless. If surprised by an airplane, the scout remains in place and does not look up.

- The scout covers exposed body parts such as the face, back of the neck, and hands with grease paint, mud or other materials to reduce sun reflection.
- Camouflage for equipment can be improvised from garnishing or sandbags to prevent sun reflection.
- In snowy terrain, white overgarments are worn.
- The helmet cover outline should be distorted.

4002. CAMOUFLAGE

Camouflage is the use of concealment and disguise to minimize the possibility of detection and/or identification of troops, material, equipment, and installations. The purpose of camouflage is to provide concealment of military objects from enemy observation. Camouflage is also used to conceal an object by making it look like something else. A scout's mission usually requires individual and equipment camouflage. If natural camouflage is not adequate, the position is camouflaged. In using camouflage, remember that objects are identified by their form (outline), shadow, texture, and color. The principal purpose of camouflage in the field is to prevent direct observation and recognition.

Individual Camouflage

Successful individual camouflage involves the ability to recognize and take advantage of all forms of natural and artificial concealment available (vegetation, soil, debris, etc.) and knowledge of the proper use of artificial camouflage materials.

Aids to Individual Camouflage

A scout must recognize the terrain's dominant color and pattern and must change the appearance of clothing and equipment accordingly in order to blend and not contrast with the terrain (see fig. 4-2).

The helmet is camouflaged by breaking up its shape, smooth surface, and shadow. Use of a helmet cover works best. In the absence of a helmet cover, mud can be irregularly blotched on the helmet to disguise its form and dull the surface. A helmet cover may be improvised from irregularly colored cloth or burlap to blend with the background. Foliage can be draped to prevent the visor of the helmet from casting a dark shadow across the face. Foliage should not stick up like plumes because any head movement will give away the position.

A small, thin bush in the shadow of a large bush makes a good observation point. Lone trees, rocks, fence corners, and outstanding landmarks are easily picked up by the enemy as obvious observation posts.

If camouflage clothing is not available, other available clothing can be attached in irregular splotches of appropriate colors.

Exposed skin reflects light and attracts the enemy's attention. Even very dark skin will reflect light because of its natural oil. The buddy system is recom-mended when applying camouflage. Standard Marine Corps issue camouflage face paint sticks are two toned:

- Loam and light green for light-skinned troops, in all but snow regions.
- Sand and light green for dark-skinned troops.
- Loam and white for troops in snow-covered terrain.

Shiny areas (forehead, cheekbones, nose, and chin) are painted with a dark color. Shadow areas (around the eyes, under the nose, and under the chin) are painted

Figure 4-2. Avoid Contrasting Backgrounds.

with a light color. Skin that is exposed on the back of the neck and hands is painted with a two-color combination in an irregular pattern (see fig. 4-3).

When standard issue face paint sticks are not available, burnt cork, charcoal or lamp black can be used to tone down exposed areas of skin.

Mud is used only in an emergency because it changes color as it dries and may peel off, leaving the skin exposed. Since mud may contain harmful bacteria, mud should be washed off as soon as possible.

Any equipment that reflects light should be covered with a nonreflective material that aids in the concealment of the weapon (for example, black electrical tape or mud). The straight line of the rifle or other infantry weapons may be very conspicuous to an enemy observer. The barrel and hand guard should be wrapped with strips of contrasting colored cloth or tape to break the regular outline. Mud or dirt dulls the reflecting surface of the stock, barrel, and bayonet where coloring has been worn. Lamp black may also be used on metal parts. The function of the weapon must not impaired.

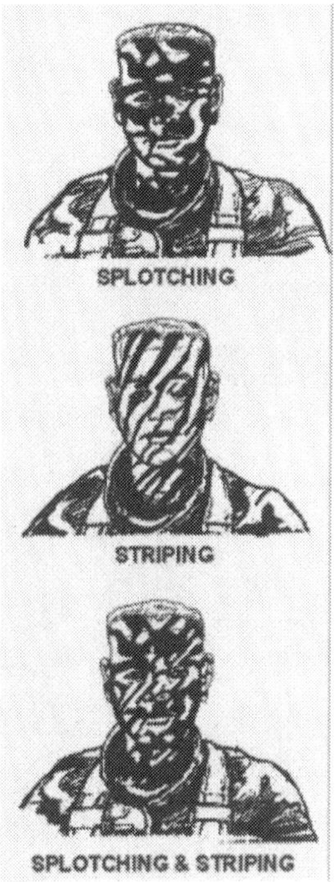

Figure 4-3. Face Camouflage.

If time, material, and surroundings permit, a ghillie suit should be constructed. (Refer to MCWP 3-15.3, *Scout Sniping*.)

Aids to Camouflage a Position

To successfully camouflage a position, the scout must remember to—

- Camouflage the position as soon as it is occupied.
- Avoid using too much material for camouflage. Even though natural materials are used, too much may make the object and its shadow stand out from its surroundings, thus attracting the attention of a hostile observer.
- Inspect completed camouflage work from the enemy's point of view to check effectiveness.

Continuous Camouflage

Camouflage around and on the scout's position must be maintained in a fresh condition as wilted and dead foliage can give the position away. If the mission dictates that the position should be occupied for longer periods, wilted foliage should be replaced during periods of reduced visibility.

4003. INDIVIDUAL MOVEMENT

Principles

The principles of individual movement are as follows:

- Scouts move from one concealed position to another. When not changing positions, they remain motionless.
- The scout's head is lifted slowly but steadily, without abrupt movements, to search for a new position.
- Scouts select the next stopping place before moving and ensures it is not contained by the enemy.
- Scouts change position on the run: springs up, runs with the body bent low, zigzags, quickly drops to the ground slightly to the right or left of the objective, then rolls or crawls to the desired position.

(Remember the phrase, "I'm up—he sees me—I'm down.")

Rushing

When starting from the prone position—

- Raise the head slowly and steadily and select a new position.
- Lower the head slowly, draw arms inward, cock right leg forward, and prepare to rush.
- Use one movement to raise the body by straightening both arms.
- Spring to your feet, stepping off with the left foot.
- Bend forward as low as possible when running. Never advance directly to the next position; always zigzag.

When hitting the deck—

- Stop.
- Plant both feet in place.
- Drop quickly to the knees and slide the hand to the heel of the rifle.
- Fall forward, breaking your fall with the butt of the rifle. (To confuse the enemy, roll over after hitting the deck and roll into firing position with feet, knees, and stomach flat on the ground.)
- Keep head down if you do not intend to fire.

When rolling over—

- Hit the deck and assume the prone position.
- Bring the rifle in close to the body, placing the rifle butt in the crotch.
- Roll over swiftly to confuse any enemy observers as to final intended location. Never reappear at the same place you went down.

Low Crawl

The low crawl is used when—

- Cover and concealment are scarce.
- The enemy has good observation over the area in which the scout is moving.
- Speed is not essential.

To perform the low crawl, keep the body as flat as possible against the ground. Grasp the rifle sling at the upper sling swivel. Let the balance of the rifle rest on the forearm and let the butt of the rifle drag on the ground. Keep the muzzle off the ground.

To start forward, push arms forward and pull right leg forward. To move forward, pull with arms and push with right leg. Change the pushing leg frequently to avoid fatigue.

High Crawl

The high crawl is used when—

- Cover and/or concealment are available.
- Poor visibility reduces enemy observation
- Greater speed of movement is required.

To perform the high crawl, keep body off the ground. Rest weight on forearms and lower legs. Cradle rifle in arms, keeping the muzzle off the ground. Keep knees well behind the buttocks to stay low.

Move forward, alternately advancing right forearm and left knee; then left forearm and right knee.

Movement Aids

Aids to movement include—

- Carrying only necessities. Additional weight causes premature fatigue and impedes free movement.
- Not disturbing birds or animals whose flight would betray your presence. If birds or animals are alerted, remain motionless under cover for a few minutes, as the enemy's attention may also be attracted.
- Moving during an incident that diverts attention, such as an airplane flight, a distant disturbance or sudden bursts of fire.
- Fog, smoke, or even light haze offer concealment for movement; however, the enemy may have thermoimagery and night vision devices. Therefore, darkness and smoke cannot be used as easily.
- Following a stream or road by staying as far away from them as possible while still keeping them in sight. Keep close to the dune line when moving along a beach.
- When moving through tall grass or similar growth, move when the wind blows, changing direction frequently. A straight route will be more readily noticed.
- Whenever possible, avoid areas of soft ground so as not to leave tracks.

- When crossing a road or water obstacle, choose crossing sites where the enemy's observation is restricted (an area in shadows or near a bend) and cross rapidly.

4004. ROUTE SELECTION

Prior to Movement

A scout and the immediate commander conduct a map reconnaissance before starting on a mission. This assists them in selecting the route according to available cover and concealment and any indicated enemy activity.

Prior to and during the course of the mission, move to an observation point to visually reconnoiter the terrain for movement and select the tentative route. It may be necessary to make wide detours around open spaces or those containing enemy patrols or other enemy activity.

Carefully study the country to be traversed and pay close attention to the general features, streams, ridges to be crossed, and their relation to the general direction to be taken (see fig. 4-4).

Make notes of terrain features and landmarks along the proposed route and rely on notes for guidance (see fig. 4-5 on page 4-6). Additionally, determine the compass direction and readings for each change of direction at the start. Finally, learn the location of unit boundaries and observation/listening posts as well as general location of other friendly or scouting parties. Be sure to avoid man-made and natural obstacles as they will slow progress and overall success of the mission. If possible, use the local populous as a source of intelligence. When returning to friendly lines, avoid using the same route.

En Route

En route, the actual advance will be a series of movements from one observation point to the next. The distance and route will depend on cover and terrain. Assess the cover, terrain, and any enemy or civilian activity to determine whether or not to modify the approach or return routes. Unless the mission requires it, avoid danger areas (for example, houses, villages, potential assembly or bivouac areas, roads, and streams) that may give away your position by being observed by the enemy. When required to reconnoiter danger areas, choose a covered approach and return, and make entry or passage as quietly and quickly as possible. If part of a larger effort, the approach and return should be covered by observation and fires of the other members of the scouting party or patrol.

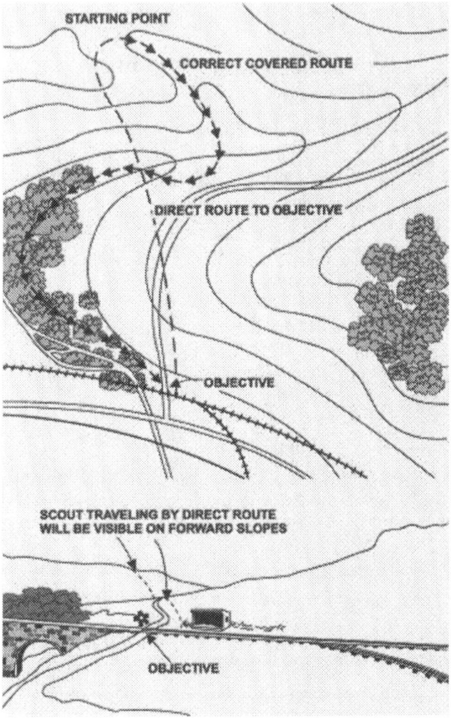

Figure 4-4. Choosing a Concealed Route of Advance from a Map.

Stream Crossings

When the crossing does not appear to be held by the enemy, advance upon it rapidly. If there are two or more scouts, one crosses while the other(s) provide protection. Note the length, width, depth, and approaches to a crossing. Observe the condition of the road or trail that crosses the stream, and report on the suitability of the crossing for use by tracked and wheeled vehicles. If the crossing is under observation by enemy, seek another crossing site or dash across to avoid detection.

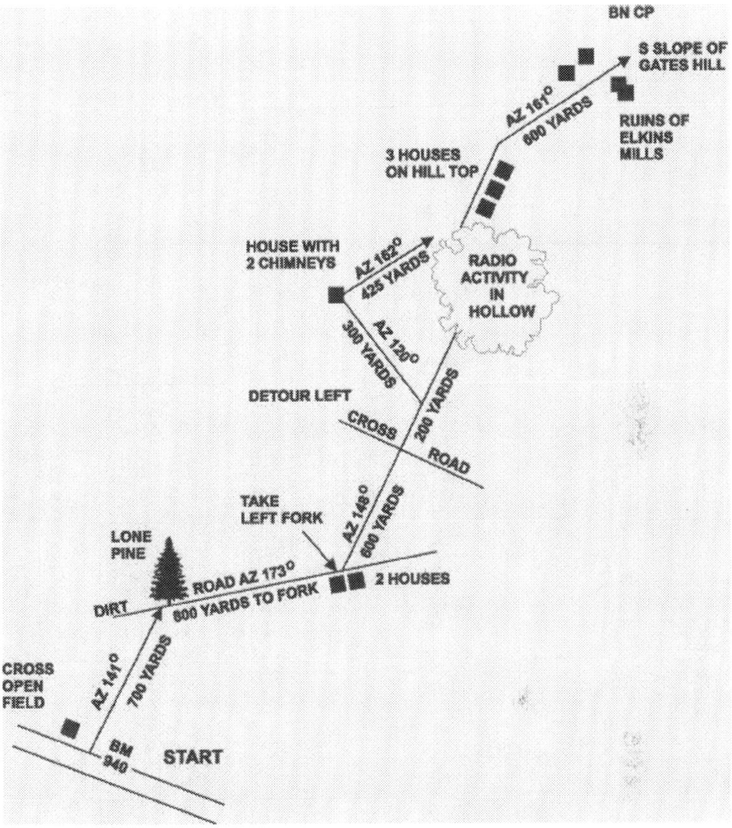

Figure 4-5. Proposed Route Sketch.

CHAPTER 5. NIGHT SCOUTING

Night scouting presents many of the same problems encountered in day operations-such as cover, concealment, movement, and camouflage-as well as additional considerations. Knowledge of human eye construction and operation will enable maximum advantage under night conditions or poor visibility.

5001. NIGHT VISION

Certain parts of the eye correspond to parts of a simple camera (see fig. 5-1). The lens focuses light entering the eye similar to a camera lens. The iris (colored part of eye) corresponds to the diaphragm of a camera, opening and closing to regulate the amount of light entering the eye through the pupil. The retina corresponds to camera film. Light rays strike the retina, form an image, and cause an impression to be transmitted to the brain through the optic nerve. In a camera, the image is formed and fixed on film.

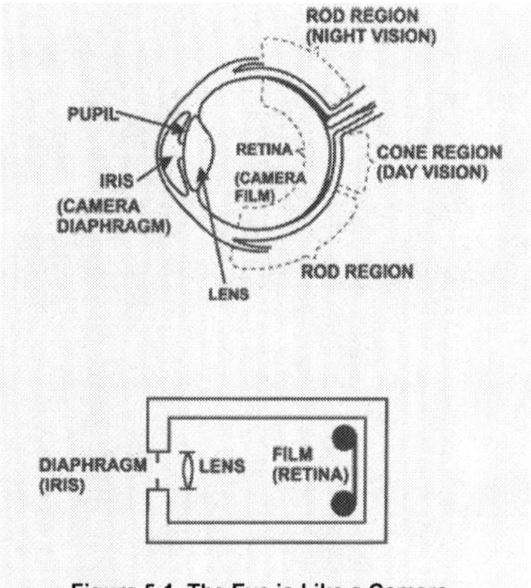

Figure 5-1. The Eye is Like a Camera.

The retina is composed of cone cells and rod cells, so-called because of their shapes. Cone cells distinguish color, shape, and sharp contrast. Because they are activated by light conditions, they are blind during periods of low illumination. Rod cells produce a chemical substance called visual purple that makes them active in darkness, low illumination or night conditions. Rod vision distinguishes black, white, shades of gray, and general outlines.

Principles

To effectively "see" at night, the principles of night vision dark adaptation, off-center vision, and scanning are applied.

Dark Adaptation

Allowing the eyes to become accustomed to low levels of illumination is called dark adaptation. It takes the rod cells about 30 minutes to produce enough visual purple to activate them and enable the eye to distinguish objects in dim light. This may also be accomplished by staying in a red-lighted area, or by wearing red goggles for 20 minutes, followed by 10 minutes in darkness (which allows the pupils to open wide). This method saves valuable time by allowing Marines to be in a lighted area to receive orders, check equipment, or perform some other function before moving into darkness.

Off-Center Vision

The technique of focusing on an object without looking directly at it is called off-center vision. When looking directly at an object, the image is formed on the cone region, which is not sensitive at night (see fig. 5-2 on page 5-2). When looking slightly to the left, right, above or below an object, the image is formed on the area of the retina containing rod cells, which are sensitive in darkness. The most sensitive area varies in individuals, but is usually found by looking 6 to 10 degrees away from an object; in effect, out of the corner of the eye (see fig. 5-3 on page 5-2).

Scanning

Off-center vision used to observe an area or an object is called scanning. When using rod vision, the visual purple in the rod cells bleaches or blacks out in 4 to 10 seconds and the object observed disappears. As the visual purple in the rod cells in one area bleaches out, the eyes must slightly shift to use fresh rod cells. Eyes

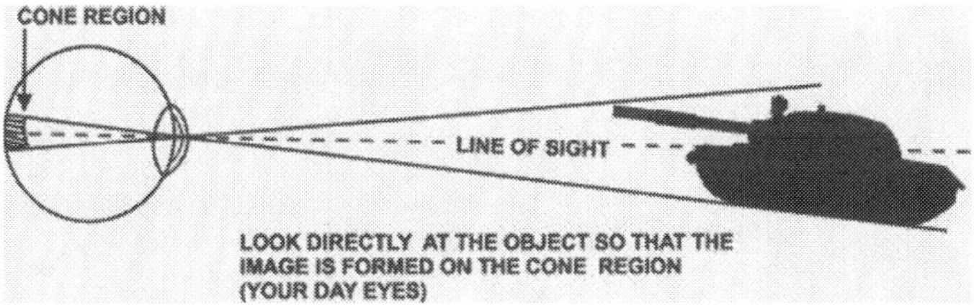

Figure 5-2. Day Vision.

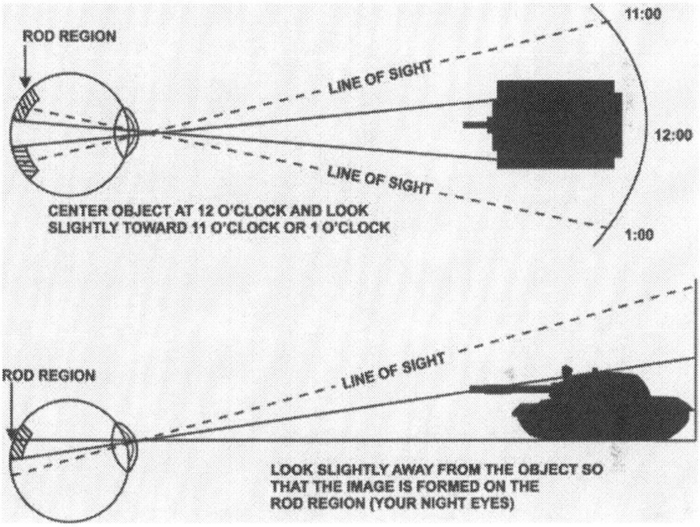

Figure 5-3. Night Vision.

should be moved in short, abrupt, irregular movements over and around the target (see fig. 5-4).

Preserving Night Vision

Night vision is quickly destroyed if bright light is allowed to enter the eye. When entering a lighted area or when observing in a temporarily lighted area (illumination, flares), one eye should be closed and covered to preserve its night vision. When the light goes off, fades or the lighted area is exited, the night vision retained by the protected eye enables it to see until the other eye adapts to the darkness. Red light helps preserve night vision, but like white light, it can be observed at long distances.

Factors that decrease night visual acuity include fatigue, lack of oxygen, long exposure to sunlight, alcohol, nicotine (within the past 48 hours), and age. When night vision has been attained, straining will not improve effectiveness; however, practice identifying objects at night will improve perception.

5002. APPEARANCE OF OBJECTS

Darkness not only makes objects difficult to see but also changes their appearance, distorts size, and blots out details. A tree visible against the night sky appears smaller than in the daytime because the twigs at the end of branches cannot be seen. A scout must train to identify objects by block outlines at night and cannot rely on details visible in daylight. Binoculars enlarge objects or parts of objects otherwise too small to be seen and help identify objects already spotted. Night observation devices increase night visibility and should be used whenever possible.

5003. SOUNDS

At night, sounds become very important. By listening, a scout gains information about the enemy and by exercising care, keeps information from the enemy. A scout stops frequently to listen. Scouts must listen for long periods in perfect silence. Hearing is amplified with the mouth open; removing the helmet will reduce sound distortion. Sounds are transmitted a greater distance in wet weather and at night than in dry weather and in the daytime. By holding the ear close to the ground sounds of people walking and vehicles moving can be heard. Sound travels approximately 370 meters a second. When a flash from a fired weapon is observed, the range to the weapons can be easily estimated by counting the time interval between the flash and hearing the report. For example, counting to three (one thousand one, one thousand two, one thousand three), indicates the distance is 1,110 meters. The cadence is determined by actual practice at known ranges.

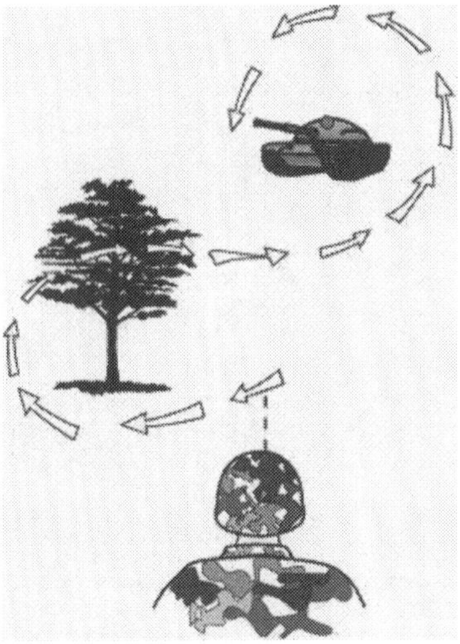

Figure 5-4. Scanning

5004. SMELLS AND TOUCH

A scout's sense of smell can warn of enemy fires, cooking, motor parks, gasoline and diesel engines, and bodies of water. A scout must feel and recognize objects in the dark, and adjust and operate equipment quietly by sense of touch.

5005. CLOTHING AND WEAPONS

All loose clothing must be secured (string or tape can be used) to prevent snagging on barbed wire, brambles, and brush. Helmet covers are worn to muffle sounds made by low branches.

The belt buckle should be turned around to the side in order to move in a prone position without scraping the buckle against stones or hard surfaces. Identification tags can be taped together to prevent rattling. Hands, face, and neck can be blackened so skin does not reflect light or appear as white spots in the darkness. (Refer to para. 4002.)

Scouts are normally armed with rifles. Rifle slings should be taped to prevent rattling. All weapons parts should be checked for glare elimination measures.

5006. CONCEALMENT

Although total darkness provides concealment, scouts must observe the same principles of concealment during moonlight conditions as in the daytime. Scouts should assume enemy employment of night observation devices and observe the principles of night movement such that presence will not be disclosed by noise when close to the enemy.

5007. AIDS TO NIGHT SCOUTING

Aids to night scouting include the following:

- Carry out scouting missions close to or within hostile positions on dark or rainy nights.
- Stifle a threatening sneeze by pressing fingers upward against the nostrils.
- Stifle a threatening cough by applying slight pressure with the finger on the Adam's apple.
- Stop a ringing sound that interferes with hearing by yawning.
- Speak softly rather than whispering if voice communication is necessary.
- Move boldly and rapidly when taking advantage of any sound-such as shelling, rustling wind or distant firing-to push forward if firing is taking place.
- Avoid shell craters and depressions in damp weather conditions if the enemy has employed chemical munitions.
- Move the eyes constantly; concentrating on one object too long will strain them.
- Take notice of the enemy's use of flares. When the enemy employs flares, few enemy patrols are apt to be out; when flares are not employed, the enemy's patrols are likely to be numerous.
- Drop to a prone position upon hearing a flare being fired and before it illuminates. Remain motionless while it is burning. If open or moving when a flare bursts in the air, freeze or drop quickly in the split second after the flare illuminates while the enemy is blinded. You are an easy target for the enemy if the flare bursts in the air or on the ground behind you. Never look at a flare. If you activate a trip flare, drop to the ground and crawl away from the illuminated area.
- Consider all patrols or individuals encountered as hostile until proven friendly. If encountering someone, crouch low, silhouetting the approaching person against the sky. At the same time, make yourself an indistinct target in case the person encountered is an enemy.
- Return fire only to avoid capture if fired on when close to enemy positions.

5008. AIDS TO NIGHT MOVEMENT

- Aids to night movement include the following:
- Move silently.
- Advance in stealthy legs. Each leg should follow some terrain feature that serves as a guide. When there are no terrain features to serve as guides, move in a straight or nearly straight line from one defined point to another, or maintain direction by using a compass.
- Avoid running, except in an emergency.
- Take advantage of sounds that may distract the enemy.
- Fall silently without making an outcry.

Walking

When walking at night—

- Place the heel down first. Balance the weight of the body on the rear foot until a secure spot is found.
- Lift the forward foot high to clear any stiff grass, brush, or other obstruction.
- Continue to balance body weight on the rear foot, lower the forward foot gently, toe first, to explore the ground for objects that might make noise. Step over fallen logs and branches, not on them.
- Lower the heel of the forward foot slowly; gradually transferring body weight to that foot.

Creeping

The low crawl and high crawl are not suitable at night when very near the enemy because an easily heard shuffling noise results. Creeping is the recommended method of movement:

- Creep at night on the hands and knees.
- Use your hands to feel for twigs, leaves or other substances that might make a noise. Clear a spot to place your knee. Keeping your hand at that spot,

bring your knee forward until it meets your hand. Then place your knee on the ground and repeat the action with the other hand and knee.

- Lay the rifle on the ground at your side and clear an area for it. Lift the rifle up and move it forward. Movement is slow and tedious, since it must be done silently.

Hitting the Deck at Night (Right-Handed Shooter)

To safely hit the deck at night from the standing position—

- Advance your left leg, place the butt of the rifle in your right armpit with the hand remaining on the pistol grip, and grasp it with the right hand at the balance.
- Quietly drop down on the right knee and left hand.

- Move the left leg carefully to the rear, and then move the right leg to the rear.

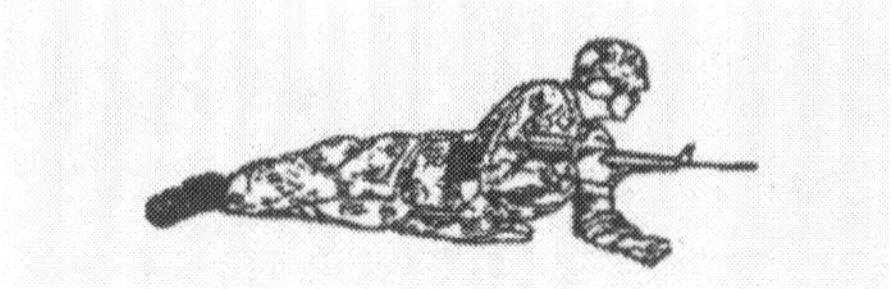

- Lie flat on the ground, or take up a firing position if necessary.

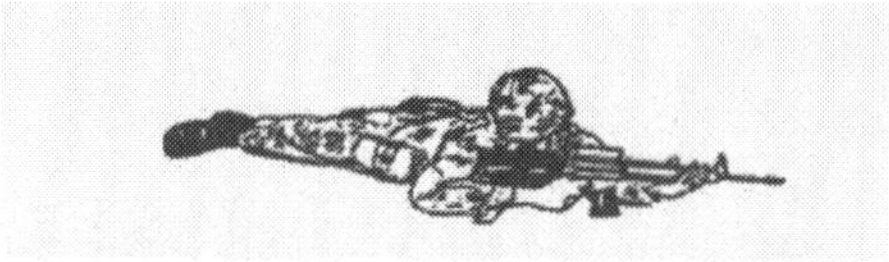

Wire Obstacles

A mission often requires a scout to pass through and work behind enemy positions. To accomplish this, the scout must be able to quietly pass through enemy wire obstacles and cross trenches. Cutting a gap in wire is time-consuming. If possible, walk over the low bands of enemy wire and crawl under the high bands (see fig. 5-5). Avoid movement along wire barriers, as enemy

Figure 5-5. Crossing Wire Silently at Night.

covering fires are generally planned parallel to them to take advantage of canalization and enfilade fire.

To step over low wire at night, crouch low to view the strands against the sky. Grasp the top strand with one hand; with the other hand, reach forward and feel for a clear spot for foot placement without stepping on other strands or any object apt to make a noise. Raise the body up, still grasping the top strand of wire. To avoid catching the foot in another strand, lift the foot up and over, passing it close to the hand grasping the wire.

If a high wire obstacle is encountered at night and wire cutters are not available, pass under the wire with your back on the ground. Grasp the lowest strands in your hands and hold them clear of the body while you slide under them.

When cutting wire and working solo, cut a wire near a post (see fig. 5-6), then dispose of all but one loose end. Grasp the wire close to a post and cut between your hand and the post, muffling the sound and keeping the loose wire in your grasp. When cutting wire in tandem, one firmly holds the wire with the hands positioned close to the cutters, in order to muffle the sound and prevent the loose ends from flying back, while the other one cuts. In both instances, the loose ends of the wire are bent back to form a passage.

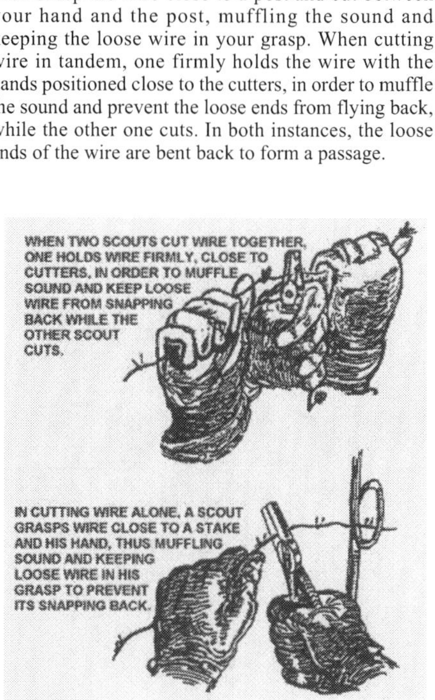

Figure 5-6. Cutting Wire Silently at Night.

Wrap a sandbag around the wire cutters and wire to deaden the sound.

Do not cut a complete gap in the wire; cut only the bottom wire(s). Leave the top wire(s) intact to lessen the chance of discovery by the enemy.

Crossing Trenches

Before approaching a trench, wait outside the trench for awhile and listen. Do not enter or cross a trench near its junction with a communication trench. Crawl silently up to the edge of the trench and look into it. Remove all loose dirt and rocks from the edge. If it is a narrow trench, spring up and jump across, sinking quietly to the ground on the other side and remaining there a moment to listen before proceeding. If the trench is wide, climb silently and slowly down into it and out the other side, using the revetment for support (see fig. 5-7). Do not enter enemy trenches unless it is absolutely necessary in order to accomplish the mission. Ordinarily, work is better accomplished from outside the trench. Sentries usually pay more attention to sounds in front of them; therefore, if it is necessary to enter a trench, cross it first at the place where enemy observation is restricted, then approach from the rear.

5009. LOCATING AND PLOTTING THE ENEMY AT NIGHT

For night work, a scout must understand the use of a lensatic compass. (Refer to para. 2002.) Using the lensatic compass, the scout can guide the platoon into position, locate adjoining elements of the command, keep direction when on patrol, determine the location of gaps in the enemy wire and the position of enemy out guards.

Locating Gaps in Enemy Wire

When searching for gaps in enemy wire, at least two lensatic compasses are needed: one to register the gap in the wire and the other for navigation. When a gap in enemy wire is located, lie outside the gap, keeping a distance of 10 meters from the barbed wire. Sight with the lensatic compass on a prominent point on the skyline in line with the gap. In selecting the prominent point in the skyline, pick one that appears on the map (i.e., hill mass, house, road junction). If the only prominent point available is one not identifiable on a map (i.e., a tree, destroyed vehicle, enemy position),

Figure 5-7. Crossing Trenches Silently at Night.

use it. The exact location of this point can be fixed the next day by visual reconnaissance of the area from an observation point. When the needle rests—

- Clamp it in place by lowering the eyepiece to the closed position.
- Rotate the movable bezel ring until the luminous line is directly over the north end of the needle. The azimuth of the gap from the prominent point is now registered.

The compass is carried back without further adjustment. The azimuth setting can be recorded later on a map (see fig. 5-8).

Locating Enemy Out Guards

At night, locate enemy guards by their sounds and failure to maintain light discipline. As sounds of the enemy are heard, and/or observations of the enemy made, shoot an azimuth with one compass. When the needle comes to rest, clamp the compass by lowering the eyepiece to the closed position.

Figure 5-8. Locating a Gap in Enemy Wire at Night.

Note the time and nature of each sound, the estimated distance, and which compass was used to fix the location. Plot this data on a map then wait until the debrief to turn in the notes and compasses. See figure 5-9.

Figure 5-9. Locating Enemy Outguards at Night.

50010. ROUTES OF MOVEMENT

Prior to beginning a night missions, a scout studies the ground in detail from an observation point, air photos, and a map during daylight. The route of advance should be below the skyline. Avoid becoming a silhouette (see fig. 5-10).

Unless the moon is bright, avoid, if possible, passing through woods, ditches, ravines, and brush, because noises of movement may lead to discovery. If the enemy is known to have night observation device capability, avoiding these kinds of terrain may not be possible. To avoid enemy ambushes, return by a different route from the advance and change routes on successive nights.

Figure 5-10. Difference Between Correct Routes by Day or Night.

CHAPTER 6. OBSERVING AND REPORTING

When sent out on a mission, a scout's duties are to observe and report. the first five chapters covered scout protection measures and movement in enemy territory. This chapter discusses performing the assigned mission and reporting the mission after its completion.

6001. OBSERVATION POSTS

Positions

The following guidelines apply to observation posts. See figure 6-1.

- When selecting observation posts, scouts should chose the least prominent position.
- Scouts may occupy one or more observation posts.
- An observation post should not be manned for more than 24 hours.
- A selected observation post should be observed for 10 to 15 minutes to ensure it is not occupied.
- Scouts move to the chosen observation post by a concealed route.
- If the post is located on a hill, crawl to a position where the skyline is broken.
- If a tree is used, the position should have a background so as not to be silhouetted against the sky while climbing or observing.
- When leaving the observation post, a different route from that of the approach should be used.
- If a radio is used, its antenna should be located to provide clear communication to the controlling commander but masked from enemy observation and direction-finding equipment. Upon departure, scouts should remove the antennae from the observation post so as not to give away the position.

Observing

Using all senses available, be particularly alert for movement, objects, sounds, and smells inappropriate to the surroundings. While observing, avoid all unnecessary movement. If observing from a building, keep back from doors and windows.

In daylight, look first at the ground nearest you. Begin observing close to your post and search a narrow strip 50 meters or less deep, going from right to left parallel to your front; then search from left to right a second and similar strip farther away but overlapping the first. Continue to observe until the entire field of view has been searched (see fig. 6-2 on page 6-2).

At night, use a night observation device. If one is not available, search the horizon with short, jerky movements, and short pauses. Look a little to one side of an object and then to the other. Lower the head close to the ground to view the object more clearly. Use low-powered field glasses to increase sight range.

Figure 6-1. Method of Approaching an Observation Post.

Figure 6-2. Method of Searching Ground.

6002. REPORTING

It is imperative that the scout accurately and completely report who, where, when, and what was observed to the proper authority upon mission completion. In most cases, a scout will not be equipped with a radio.

Verbal Reports

Verbal reports should be made when writing is impractical, when the information is not complicated or when the enemy is likely to intercept a messenger carrying a written message.

Written Messages

Written messages, preferred to verbal reports, are recorded in message book blanks issued for that purpose (see fig. 6-3) and delivered to a higher authority as soon as possible. The NATO spot report (SPOTREP) should be reviewed as the written message is a NATO format.

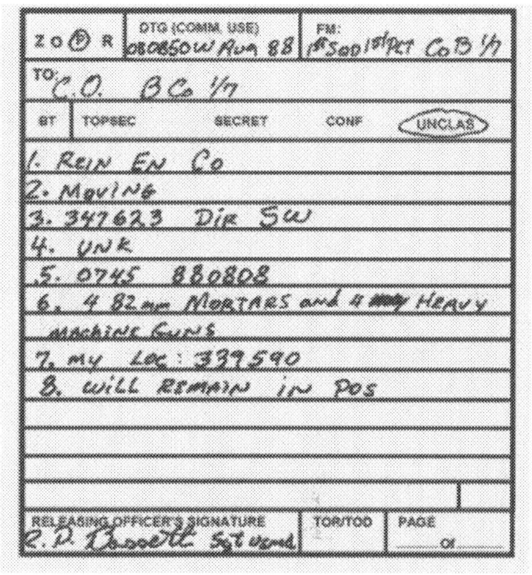

Z O P R	DTG (COMM. USE) 080850W Aug 88	FM: 1st Sqd 1st Plt Co B 1/7
TO: C.O. B Co 1/7		
BT	TOPSEC SECRET CONF UNCLAS	
1. REIN EN CO		
2. MOVING		
3. 347623 DIR SW		
4. UNK		
5. 0745 880808		
6. 4 82mm MORTARS and 4 HEAVY MACHINE GUNS		
7. MY LOC: 339590		
8. WILL REMAIN IN POS		
RELEASING OFFICER'S SIGNATURE R.P. Bassett Sgt USMC	TOR/TOD	PAGE ___ OF ___

Figure 6-3. Sample Field Message.

The message body is brief, accurate, and clear; facts and opinions are distinguished. If secondhand information is reported, its source is included. Reports include all information of value, first about the enemy, and then about the originator. Information about the enemy should cover—

- Size and/or strength.
- Actions or activity.
- Location and direction of movement.
- Unit identification. (The designation of the enemy unit may be derived from unit markings, uniforms worn or through prisoner interrogation.)
- Time of observation.
- Equipment and weapons.

Messages are printed in block letters. Individual items of information are numbered and separated into paragraphs. If doubt exists as to message receipt by the commander, a summary of its contents is included in the next message. Information about the originator or writer should cover—

- Location at the time of enemy observation (reference to an important terrain feature, by map coordinates, by the back azimuth from each of two definitely located points, or the back azimuth and distance from one known point).
- Intentions. (Remain in position? Continue on the mission? Take other action?)

The message is carefully reread and if possible read by another person to ensure understandability. If a messenger is used, the messenger must read and understand the message in order to answer any questions the commander might have.

Sketch

Information difficult to describe may be given accurately on a simple sketch. The sketch may give all the necessary information or it may be used to supplement a written message. A military sketch is generally one of two types: simple or panoramic. The simple sketch is easily made and read.

Figure 6-4 shows a simple sketch that has been included in the message itself. The sketch may be on a separate sheet of paper, but all of the necessary information must be contained in either the sketch, the message or both.

The panoramic sketch is a picture of the terrain's elevation in perspective, as seen from one point of observation. Although a panoramic sketch is not difficult to create, skill and training are necessary to enhance usefulness. Figure 6-5 on page 6-4 illustrates how to make a panoramic sketch.

Overlay

The same information sent back to higher headquarters on the sketch may be sent on an overlay, if the sender and the person to whom the message is to be sent have copies of the same map. Figure 6-6 on page 6-6 illustrates a simple overlay. The overlay is drawn on transparent paper as follows:

- Orient the map and place it on a hard, flat surface.
- Place the transparent paper over the part of the map of the object or information to be transmitted and hold the paper in this position.
- Orient the overlay to the map by tracing in the intersecting grid lines at two opposite corners of the overlay. Write the correct number designation on the overlay. The cross made by the intersection is called a *tick mark* and enables the receiver to locate the exact area on the map covered by the overlay.
- Sketch the object seen or the information to be transmitted on the tracing paper (the sheet on top of the map) in the exact location it would appear on the map (the sheet underneath the tracing paper).

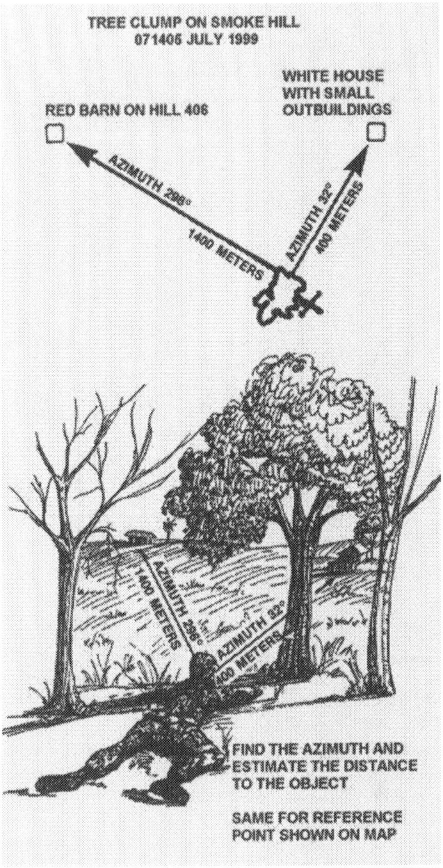

Figure 6-4. Making a Simple Sketch.

Explanatory notes are annotated in the overlay's margin, arrows point to the objectives.

- Indicate with an "X" and an appropriate explanation the position from which the observer saw the object or obtained information.
- Include title and scale of the map from which the overlay was made, date and hour the information was obtained, and signature of the observer on the overlay in the lower right-hand corner.

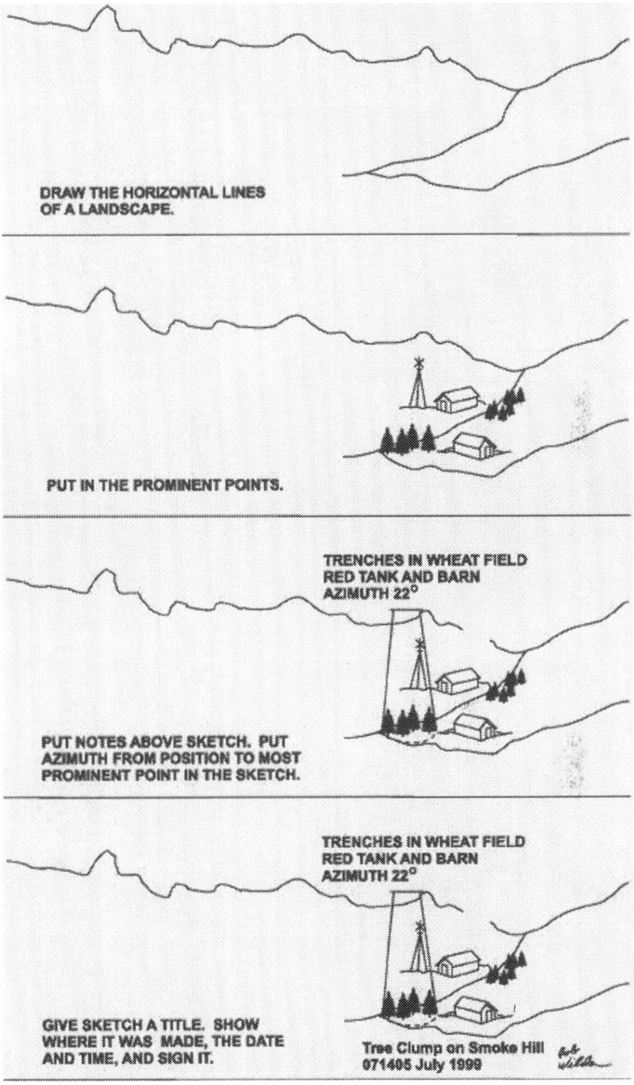

Figure 6-5. Panoramic Sketch.

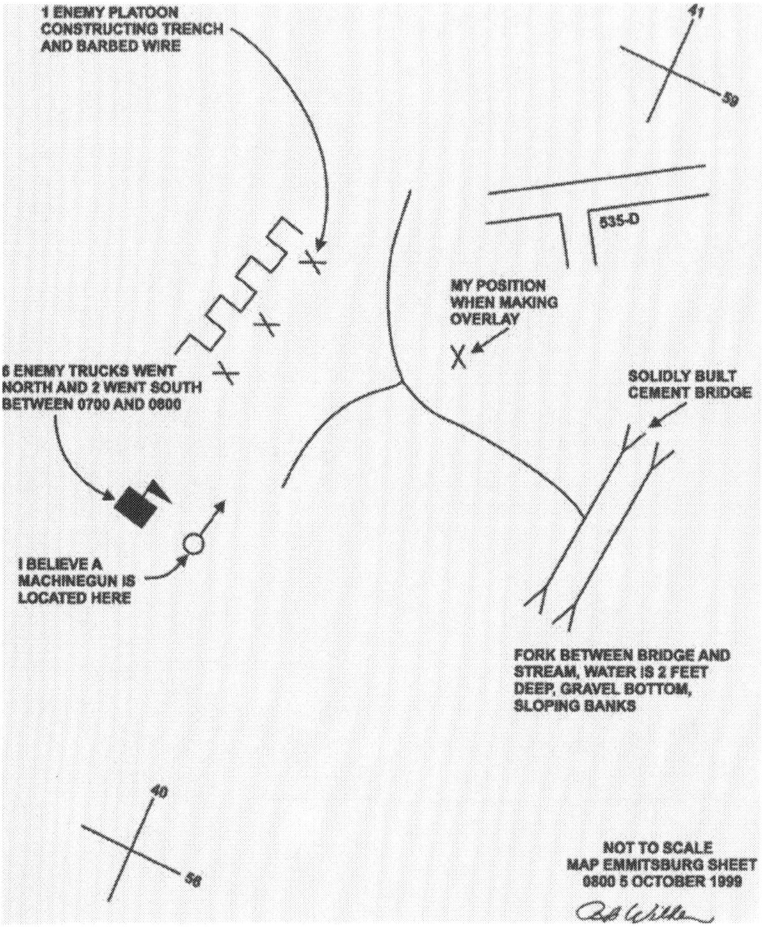

Figure 6-6. Simple Overlay.

CHAPTER 7. SCOUTING FIRE TEAMS

Upon completion of individual training, a Marine adapts this training to the functioning of the unit. The basic tactical unit for scouting is the fire team. Within the fire team, scouts normally work in pairs to furnish security and gather information. Information gathered by the individual scouts is passed to the fire team leader.

7001. POSITIONING

When a rifle platoon in the approach march is not preceded by adjacent units (left, right, front, and rear), it employs its own scouting elements. The scouting element for a platoon is usually one fire team; however, an entire squad may be used.

A fire team used as a scouting element is called a scouting fire team and is controlled by the platoon commander, assisted by the squad leader. A squad leader whose squad is providing the scouting fire teams normally marches near the platoon commander to assist in the control of the scouting fire teams.

A scouting fire team moves aggressively to cover the front of the advancing platoon and to locate the enemy's position(s). It generally moves in a wedge or skirmisher's formation. Normally, a scouting fire team is deployed on a frontage of 50 to 75 meters (10 to 17 meters between each individual scout). The entire squad may be employed to cover a wider frontage. The platoon commander coordinates the movement of the scouting fire team(s) so as to protect the main body of the platoon from enemy fire from points within 400 to 600 meters away, or in close terrain from points within the limits of enemy observation.

Scouting fire team(s) should have enough firepower to overcome resistance from small enemy advanced posts and patrols; the intent is to make enemy riflemen and machine gunners open fire and disclose their positions. Without scouts in advance, the platoon may move into areas where enemy fire may prevent further advance or maneuver and inflict heavy casualties (see fig. 7-1).

Scouting fire teams are covered by the platoon or, when the platoon is masked, the fire team leader ensures individual scouts are maneuvered and coordinated so that the fire team covers its own advance. The fire team leader constantly watches for signals from the platoon commander and remains in visual contact at all times.

The distance between the scouting fire team and the platoon is terrain-dependent. The scouting fire team should not be beyond visibility of the platoon. In open terrain, the platoon commander usually directs the scouting fire team to move by bounds along a succession of locations designated by the platoon commander as intermediate objectives.

Individual scouts should advance as stealthily as possible, while remaining consistent with their mission of reconnaissance to the front, taking advantage of cover without delaying the advance. An occasional glimpse of scouts constantly advancing over a wide front can make the enemy uneasy. It is this activity, and not the target the scouts offer, that may cause the enemy to open fire and disclose its location.

When fired upon, scouts must drop to cover and return fire only when necessary to complete the mission. If

Figure 7-1. Position of Scouts Preceding an Attacking Platoon.

necessary, one scout reports back to the fire team leader who informs the platoon commander.

7002. LOCATING ENEMY POSITIONS

The key terrain of defensive positions are those points that afford extended observation over the ground where the attack must advance. The enemy will place machine guns and infantry to defend critical points.

Members of a scouting fire team preceding an attacking platoon identify the probable enemy infantry and machine gun positions (see fig. 7-2). They use concealment and cover to conduct their advance in order to discover the exact location of enemy positions.

7003. ACTION WITH AN ATTACKING PLATOON

The scouting fire team reconnoiters to the front of the advancing platoon. As soon as the scouting fire team leader indicates the area is secure, the platoon advances and the scouting fire team moves forward. Squads within the platoon advance by bounds; at least one squad is positioned to support the other(s) by fire. Successive positions along the line of advance are selected and designated by the platoon commander as intermediate objectives, and reconnoitered by the scouting fire team before occupation. By conducting proper reconnaissance, surprise by the enemy or movement in the wrong direction may be prevented.

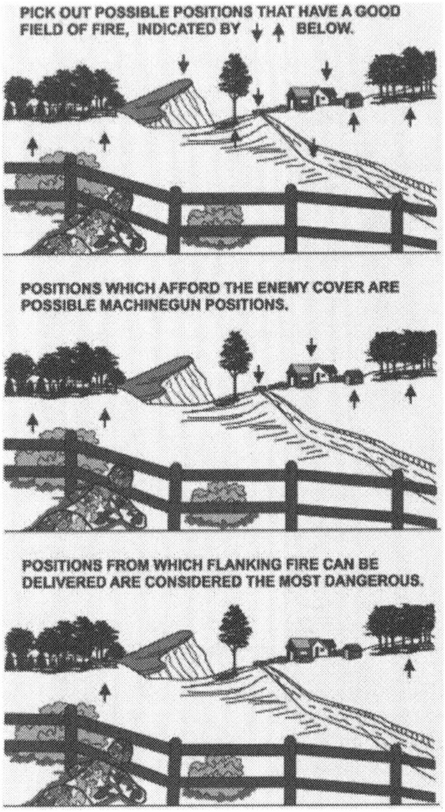

Figure 7-2. Assessing Probable Enemy Machine Gun Positions.

Movement

The distance between the scouting fire team and the front of the main body of the platoon is dependent on the mission, enemy, terrain and weather, troops and support available, and time available (METT-T). In close terrain, such as dense woods, the scouting fire team's movements closely resemble those used for night operations. In approaching houses, woods, and villages, one scout of each pair covers the other while the latter reconnoiters (see fig. 7-3).

A scouting fire team moves forward aggressively to cover the front of the advancing platoon, usually adopting either skirmisher's or wedge formation in order to be prepared to go into action immediately and to cover a wide frontage of 50 to 75 meters.

As a scouting fire team advances in open terrain, it is supported when possible by elements of the platoon; in close terrain, by mutual support within each fire team. Mutual support within the fire team is accomplished by the fire team leader and the automatic rifleman forming a team that supports by fire the advance of the rifleman and assistant automatic rifleman until they reach a location designated by the fire team leader. The rifleman and assistant automatic rifleman then support by fire the movement of the team leader and automatic rifleman. These successive points to which the sub elements of the fire team move are normally designated as fire team intermediate objectives by the fire team leader. The fire team leader sets as many fire team intermediate objectives as necessary to maintain mutual support within the team. This process is repeated until the team can be covered by other elements of the rifle platoon.

A scouting fire team takes advantage of available cover and concealment without delaying its advance. The orders of the platoon commander govern the distance at which it precedes the platoon. The terrain and the probable position of the enemy affect the scouting fire team's distance in front of the platoon. It may be as much as 400 to 600 meters in advance of the platoon. In open terrain, the platoon commander usually directs that the scouting fire team move by bounds to a succession of intermediate objectives. In close terrain or conditions of limited visibility, the scouting fire team is normally ordered to precede the platoon at the limit of visibility, maintaining visual contact with the platoon commander.

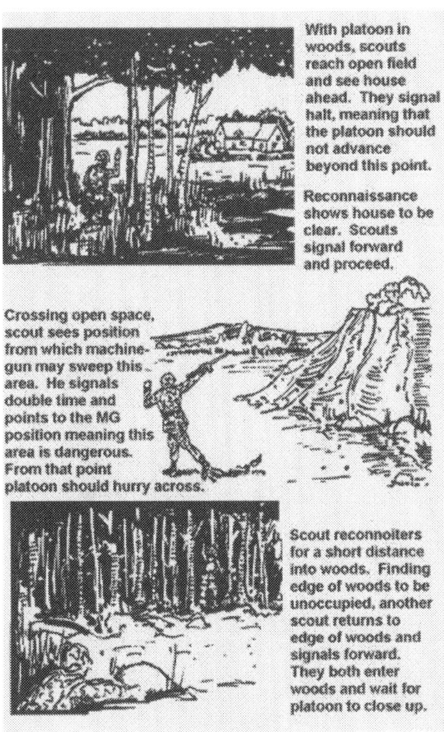

Figure 7-3. Conduct of Scouts Preceding a Platoon.

Action in Woods

When a scouting fire team is directed to advance over open ground to the edge of a woodline, two members of the team, preferably the rifleman and assistant automatic rifleman, reconnoiter inside the woodline while the remainder of the fire team covers them. It is not recommended they separate until finished with their reconnaissance of the far side of the danger area. Both members staying together can cover the same area using a zigzag reconnaissance and they are better equipped to overcome any opposition.

In heavy underbrush and/or poor visibility, the rifleman and assistant automatic rifleman proceed into the woodline together for 50 to 60 meters. The two then separate, searching out either flank to the first high ground or limits of observation, probably 50 to 100 meters (see fig. 7-4 on page 7-4). After the initial search and out posting the limit of advance, the remaining scout signals the fire team forward.

In light underbrush and/or good visibility, the assistant automatic rifleman remains at the edge of the woodline while the rifleman searches the woodline. (see fig. 7-5 on page 7-5). The rifleman searches the woodline in a zigzag pattern, reports back to the assistant automatic rifleman, then moves to an outpost position at the limit of advance. In turn, the fire team leader signals the platoon commander that it is safe for the platoon to move forward.

The fire team leader then moves the remainder of the fire team into the woods, joining up with the forward scout manning the outpost. The scouting fire team occupies and holds a line 50 to 75 meters within the woods and observes toward the direction of movement until the platoon closes up. The scouting fire team leader awaits further word from the platoon commander before moving the team further into the woods.

When directed, the scouting fire team leader moves the team forward until they reach the far edge of the woods. The team is held at the edge of the woods and the fire team leader notifies the platoon commander of the situation. The platoon commander moves the platoon to a position where it can cover the scouting

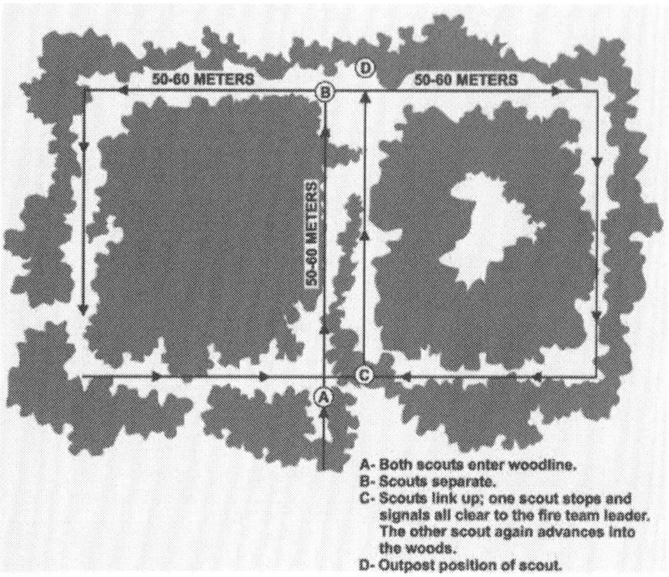

Figure 7-4. Searching Edge of a Woodline (Dense Underbrush/Poor Visibility).

fire team as it exits the woods and directs the team leader to move out and continue the scouting mission.

A scouting fire team passing through woods ahead of its unit maintains a distance allowing visual and oral communications. If an obstacle is encountered, reconnaissance to its front and flanks must be carried out. When advancing along a road or path, scouts precede the platoon to provide necessary protection and to prevent surprise fire on the platoon. When crossing a road or path, they reconnoiter well to the flanks before signaling "all clear" to the platoon.

The scouting fire team will not exit the woods until the arrival of the platoon commander, who will then be given an opportunity to alter the disposition or direction of march. The point where the platoon exits the woods is considered a danger area where the platoon is vulnerable to enemy fire. The scouting fire team is sent ahead to reconnoiter the danger area, as well as the next area to be occupied by the platoon. They signal back whether conditions require a halt, an advance or a quick rush across the open area. The scouting fire team leader must be continually on the lookout for signals from the rear.

Action Under Fire

When a scouting fire team is fired upon, they immediately take cover, locate targets, and return fire. The scouting fire team leader then determines—

- Location of enemy (range and reference points).
- Extent of position (location of flanks).
- Types of positions (obstacles, bunkders, fighting holes, etc.).
- Number of enemy.
- Enemy weapons (machine guns, mortars, tanks, etc.).

The platoon commander assesses the situation as quickly as possible based on the limited information obtained. Usually the platoon commander brings up the remaining squads, sets up a base of fire, and assaults the enemy position. Should the enemy position prove too strong for the platoon, the platoon remains engaged with the enemy as a base of fire until the remainder of the company is committed to clear the enemy resistance.

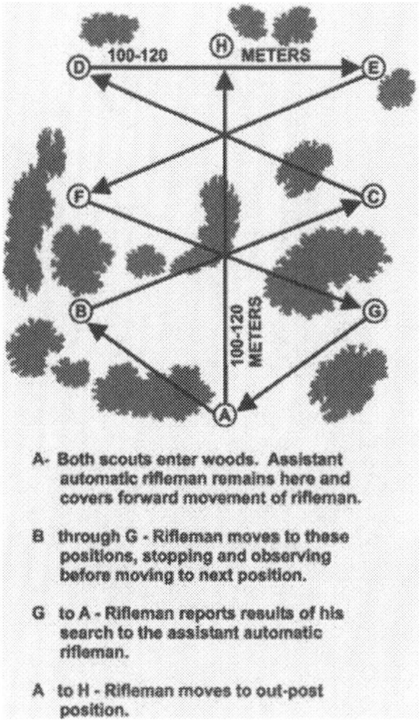

Figure 7-5. Searching Edge of a Woodline (Light Underbrush/Good Visibility).

7004. ACTION WITH AN ENVELOPING UNIT

When a platoon is given the mission to envelop an enemy position, a scouting fire team is employed for protection and reconnaissance in the same way as when the platoon is advancing in the approach march.

Part 2. Infantry Patrolling

Chapter 8. Fundamentals of Infantry Patrolling

This chapter begins Part 2 and provides basic information about infantry patrols; specifically, their purpose, types, and missions. Infantry patrol training is also addressed and keys to successful patrolling are presented. Subsequent chapters of this part cover patrol organization, preparation, movement, and reconnaissance actions.

8001. DEFINITIONS

A patrol is a detachment of ground, sea or air forces sent out for the purpose of gathering information or carrying out a destructive, harassing, mopping-up or security mission (Joint Publication [JP] 1-02). The mission to conduct a patrol may be given to a fire team, squad, platoon or company.

8002. RELATION OF PATROLLING TO SCOUTING

Each patrol member must be knowledgeable in the principles of scouting and maintain membership of a larger team. To develop the teamwork skills required among the members of a patrol, additional training beyond the basic principles is necessary to become a well-trained scout. A patrol member must respond quickly to the decisions and orders of the patrol leader. There must be complete confidence among all members of the patrol and the confidence that they, as a team, will be successful in their mission.

8003. PURPOSE

A commander must have current information about the enemy and the terrain in order to employ the unit effectively. Patrols are an important means of gaining this information and are used to destroy enemy installations, capture enemy personnel, perform security missions or prevent the enemy from gaining information. Modern warfare places a high premium on effective patrolling because units have larger areas of operations and can be threatened from all directions. As distances between units increase, more patrolling becomes necessary to prevent infiltration by guerrillas or small enemy units, as well as to maintain contact with friendly adjacent units. Active patrolling by numerous small groups is needed to locate the enemy and gather information on the enemy's disposition, strength, morale, and weapons, as well as gather and confirm information about the terrain.

8004. TYPES OF PATROLS

Classification as to Mission

Reconnaissance Patrol

Reconnaissance patrols gather information about the enemy, terrain or resources. Relying on stealth rather than combat strength, they gather this information and fight only when necessary to complete the mission or to defend themselves. The distance covered by reconnaissance patrols varies based on the terrain and mission. The squad is ideally suited for reconnaissance patrol missions because of its relative small size and its experience of working together.

Combat Patrol

A combat patrol is a fighting patrol assigned missions that require engagement with the enemy in combat. Larger and more heavily armed than reconnaissance patrols, combat patrols have a mission to capture enemy documents, provide security, and capture or destroy enemy equipment and installations. Such action is ordinarily followed by a return to friendly positions. Regardless of the mission, the patrol reports any information concerning the enemy and terrain acquired during the accomplishment of the assigned mission. There are four types of combat patrols: raid, contact, ambush, and security (normally conducted by a Marine rifle platoon). A rifle platoon reinforced with crew-served weapons is normally considered the minimum size for contact, economy of force or ambush patrols. In some situations, such as the capture of a small enemy outpost, a rifle platoon could conduct a raid. However, a raid is a complex mission and, due to the organization of a raid force (command, reconnaissance, assault, support, security, and reserve elements), a rifle company is normally the smallest force assigned to a raid.

Classification as to Means of Movement

Foot Patrols

Movement by foot is the most common means; however, there are inherent disadvantages. Foot patrols travel slowly and carry limited quantity and types of equipment and supplies. Range and area coverage is relatively restricted. Foot patrols also have apparent advantages in that they have fewer restrictions as to terrain that can be covered; are more difficult for the enemy to detect; provide thorough coverage within limits of range; and are generally not inhibited by weather.

Motorized Patrols

Where terrain and road networks permit, a motorized patrol overcomes the inherent disadvantages of the foot patrol. Mechanized forces require patrolling units that can keep pace with them. However, motorized patrols are restricted to certain types of terrain, and tend to bypass areas that may be advantageous to and occupied by enemy infantry.

Waterborne Patrols

Waterborne patrols move over seas, lakes, rivers and streams, canals, and other inland waterways. The water is either used as a medium of entry to an objective area or is the actual patrol route. Waterborne patrols are limited by the location of water routes in the terrain and tend to bypass areas that may be advantageous to and occupied by the enemy.

Helicopterborne Patrols

Where terrain is extremely difficult or the enemy situation precludes the use of vehicle or motorized patrols, helicopterborne patrols are a method or means to conduct a patrol.

8005. TRAINING

Training is essential to successful patrolling. Premature and unordered actions by members of the patrol destroy coordination and control. Leaders are trained to issue their orders calmly to inspire confidence and discipline, and to avoid misunderstanding. Patrol members must work together and fight as a team. Training should develop the following skills:

- Expertise in handling individual and special weapons, and familiarity with enemy weapons that may be captured.
- Recognize camouflaged personnel, equipment, and defensive positions; ability to pick up fleeting targets fire the rifle from any firing position.
- Understand fire discipline and, after weapon firing, immediately change position (see fig. 8-1).
- Quick and accurate observation skills, and the ability to recall and transmit clearly and briefly, both orally and in writing.

Figure 8-1. Changing Position After Firing.

- Recognize and quickly respond to improvised signals, visually or by sound.
- Ability to swim with weapon and equipment.
- Use issued or improvised camouflage suits and garnish helmet in order to blend with the surroundings. Smudge face, hands, and any bright surfaces of weapons and equipment with some substance, such as mud or charcoal, to prevent the reflection of light.
- Silence self, equipment, and weapon.
- Use antimalarial and water purification tablets.
- Acclimation to temperature extremes.
- Develop a sense of direction and learn how to follow a course by compass, stars, sun, flow of streams, prominent terrain features, and by observing other natural phenomenon. Learn to determine the distance traveled from a known point and to keep a record of azimuths and the distance traveled on each azimuth (dead reckoning).
- Call for and adjust indirect fire assets.
- Familiarization with all communications assets and the use of field expedient antennae.

8006. KEYS TO SUCCESSFUL PATROLLING

Regardless of the category or means of conducting a patrol, the keys to successful patrolling are—

- Detailed planning. Every portion of the patrol must be planned, all possible contingencies considered.
- Productive, realistic rehearsals. Each phase of the patrol is rehearsed, beginning with actions in the objective area. Similar terrain and environmental conditions are used when conducting rehearsals.
- Thorough reconnaissance. Ideally, the patrol leader will physically conduct a reconnaissance of the route and objective. Photographs and/or maps will be used to supplement the reconnaissance.
- Positive control. The patrol leader must maintain positive control, this includes supervision during patrol preparations.
- All-around security. Security must be maintained at all times, particularly near the end of the patrol where there is a natural tendency to relax.

CHAPTER 9. PATROL ORGANIZATION

Organizing a patrol is a two-step process: the general organization of the entire patrol and the task organization of various patrol elements. Normally, the nature of patrolling does not permit long preparation periods and rehearsals to specifically build a unit for each mission. Accordingly, the patrol leader must combine unit integrity considerations with proven concepts of patrol organization.

9001. GENERAL ORGANIZATION

The patrol leader establishes a patrol headquarters and elements to accomplish the mission.

The headquarters is composed of the patrol leader and the personnel who provide support for the entire patrol, such as a forward observer, corpsman, and radio operator.

The major subdivisions of reconnaissance and combat patrols are elements. The existing infantry structure (squads and fire teams) is reinforced as required. For example, a reinforced platoon tasked to conduct a combat patrol that will raid an enemy outpost could be organized as follows:

- Platoon headquarters (command element):
  - Patrol leader (platoon leader).
  - Assistant patrol leader (platoon sergeant).
  - Navigator.
  - Radio operator (company tactical net).
  - Radio operator (patrol tactical net).
  - Corpsman.
- The first squad (security element) provides security en route to the objective area (the point, flank security, and rear security) and at the objective area (the flanks and the objective rally point).
- The second squad (support element) supports by fire for the attack, covering fire for the withdrawal, and supporting fires to cover the crossing of danger areas.
- The third squad (assault element) provides the assault force to attack and seize the objective; searchers to clear the objective; pacers, compass man, navigator, and the assistant patrol leader en route and back from the objective area.

Any attachments a patrol may have (i.e., demolition team, scout snipers, and machine gun squad) will be added to the element that supports its function. For example, the demolition personnel should go with the unit conducting the attack, and scout snipers and machine gunners should stay with the support squad.

9002. TASK ORGANIZATION

The preceding paragraphs described the elements necessary for a patrol to accomplish its mission. These elements reflect the internal functions or tasks required for the patrol to succeed. Depending on the METT-T, there are various methods of grouping these elements together. Task-organization is the further subdivision of patrol elements into teams that are required to perform essential tasks. In creating teams, unit integrity of infantry units should be maintained.

The patrol is organized so each individual, team, and element is assigned a specific task, but capable and prepared to perform other tasks. This may not be possible for certain specialist tasks requiring a trained technician.

CHAPTER 10. PATROL PREPARATION

For a patrol to succeed, all members must be well trained, briefed, and rehearsed. The patrol leader must have a complete understanding of the mission and a thorough understanding of the enemy and friendly situations. The patrol leader should make a complete reconnaissance of the terrain to be covered (either visual or map), and must issue an order to the patrol, supervise preparations, and conduct rehearsals.

10001. MISSION

The mission assigned to a patrol must be clear and oriented toward one objective with a specific task and purpose. More than one primary objective or indefinite missions invites confusion, casualties, and failure.

10002. FACTORS INFLUENCING PATROL SIZE

The circumstances under which patrolling by infantry units is conducted make it necessary for combat patrols to be able to fight offensively, security patrols to defend themselves, and reconnaissance patrols to move quickly and only fight if necessary to break contact or defend themselves. The size of a patrol depends on METT-T.

Generally, a patrol should contain the least number of members needed to accomplish the mission. Combat missions ordinarily require larger patrols than reconnaissance missions.

10003. COMMANDER'S DUTIES

Determining Patrol Requirements

The need for conducting patrols derives from the commander's stated mission (issued by higher level commander) and other specified and implied tasks (secondary or supporting efforts necessary to accomplish the stated mission). This requires analyzing the unit's mission and determining the necessary reconnaissance and/or combat tasks that must be performed. By considering the mission, along with time available, the commander develops the overall concept of operations to include the patrol plan and the specific tasks for each patrol.

Assigning Units

When assigning patrol missions, maintaining the integrity of the existing unit while considering the skills and experience of the unit and its leader are critical factors to the infantry commander. To provide operational depth and equitable apportionment of hazardous assignments, the commander ensures that each of the subordinate leaders and units develops the skills and experience necessary to conduct successful patrols. Assignment of patrol units must consider the commander's concept of operations as a whole and the plans for subsequent employment of assigned forces after completion of the assigned patrolling mission.

Providing Adequate Time

The commander must allow the patrol sufficient preparation time by completing the mission analysis, estimate, and preparation of orders. The commander should use warning orders to alert subordinates to possible requirements and afford them the opportunity for concurrent planning.

Providing the Patrol Leader Information

The following information should be provided:

- A simple, straightforward explanation of the mission, particularly for night patrols.
- General routes (defined by checkpoints) or exact routes (defined by avenues of approach or other terrain features) to follow.
- Enemy composition, disposition, and strength.
- Location and activities of friendly troops.
- Outposts or other security elements through which the patrol is to pass.
- Terrain conditions.
- Missions and routes of other patrols.

- Time patrol is to depart and return.
- Method of reporting information while on patrol (radio, messenger), place where messengers are to be sent, and place where the patrol leader is to report upon completion of the patrol.
- The challenge and password to be used for exiting and reentering friendly lines.
- Special instructions such as locations to be avoided and essential elements of information higher headquarters is seeking.
- Report barrier/obstacle plan of friendly units if known; support available from friendly unit (e.g., medical evacuation [MEDEVAC], react force).
- Fire support available.

Providing Special Skills and Equipment

There will be situations when the unit assigned to conduct a patrol does not have the necessary technical skills or equipment organic to it to successfully accomplish the patrol mission. In these cases, additions (attachments) are made to the unit, such as—

- Machine gun and/or shoulder-launched, multipurpose assault weapon (SMAW) team and/or squad.
- Forward observers (mortar/artillery).
- Radio operator(s).
- Combat engineers.
- Tracked vehicle crewman to assess the ability to traverse the terrain.
- Corpsman.
- Other personnel (snipers, translators) as required.

Providing Miscellaneous Support

The commander must ensure that the patrol leader is provided with the food, water, ammunition, radios and batteries, maps, special clothing, and any other items required by the unit (includin attachments) needed for the mission. Post-patrol support such as debriefings by intelligence personnel must also be planned.

Reviewing the Patrol Leader's Plan and Preparations

Once the patrol leader receives the mission, conducts visual and/or map reconnaissance, and develops the plan, the commander confirms the patrol leader's understanding of the mission and plan for accomplishing it. This discussion between the patrol leader and commander ensures that the patrol leader understands the commander's intent and is properly prepared to provide the patrol leader an opportunity to ask for clarification or additional support, if required.

Debriefing the Patrol

Upon return of the patrol, the commander receives the patrol report at a debriefing attended by the patrol leader and all patrol members. The debriefing should be conducted as soon as possible following the patrol's return, while information is still fresh in the minds of the patrol members. A patrol report, based on the information collected during the debrief, is generated and forwarded to the next higher com-mander.

10004. PATROL LEADER DUTIES

The patrol leader organizes and prepares the patrol by using the six troop-leading steps to make the best use of resources available. These steps are to—

1. Begin planning.
2. Arrange for reconnaissance and coordination.
3. Make reconnaissance.
4. Complete the plan.
5. Issue the order.
6. Supervise.

Begin Planning

The patrol leader begins by evaluating all factors affecting the mission. He looks for possible courses of action that lead to a decision, and then transforms this decision into an order. The first step includes making an initial assessment and decision on using available time, issuing a warning order and initial preparatory tasks, and initiating his estimate.

The initial planning effort assesses the time, assistance, and information available, and plans the proper use of each. Time allowances include reconnaissance; completion of the estimate and order; troop preparation; and such briefings, rehearsals, and inspections as required before beginning the patrol.

The patrol leader reviews the mission and the attachments and/or support available and decides what preparatory efforts must begin immediately. Proper use of subordinates to manage these initial tasks during this period reduces preparation time and frees the patrol leader for proper planning and reconnaissance. A warning order is issued using a modified five-paragraph order format (situation, mission, execution, administration and logistics, and command and signal [SMEAC]) as a checklist. (Refer to app. A for specific information that a patrol warning order should contain within each checklist item.

Once the initial preparations are set in motion, the patrol leader begins estimating by analyzing the mission; considering the friendly, enemy, and operating environments; considering each course of action available against what the enemy might do; comparing the courses of action in terms of mission accomplishment, capabilities, and probable casualties; and choosing one that becomes the basis for concept and order. While the patrol leader should use and organize notes, estimates must be done quickly and accurately, particularly for immediate situations. To organize thinking, the patrol leader uses METT-T, which consists of:

- Mission—the mission assigned to the patrol and how it relates to the mission of the commander who is sending the patrol.
- Enemy—what is known or suspected of enemy presence and capabilities, habits and characteristics, and fighting techniques.
- Terrain and weather—including ground, vegetation, drainage, weather, and visibility.
- Troops and support available—friendly situation and support available.
- Time available—the constraints and impact of time on preparation and mission accomplishment.

The estimate begins with mission analysis, which is the most important part of the entire planning process. Here, the patrol leader considers the specified tasks of the mission assigned and identifies other significant actions (specified and implied tasks) that must be undertaken to accomplish the stated mission. These tasks are arranged in sequence of accomplishment. The tasks and sequence create the framework for developing courses of action for the patrol concept of operations.

Arrange for Reconnaissance and Coordination

The patrol leader arranges a personal reconnaissance to observe as far forward as possible and also coordinates with the appropriate commanders for the patrol's "passage of lines" (see paragraph 11001 for complete definition) and supporting fires. The patrol leader also coordinates with other patrol leaders who may be operating in the same or adjacent areas and requests that the commander assigning him the patrol mission coordinate the patrol action with adjacent commanders, local security, and night defensive fires, as appropriate. The patrol leader may delegate any or all of these arrangements to the assistant patrol leader if the patrol leader requires the time for planning.

Make Reconnaissance and Complete the Estimate

The patrol leader uses personal reconnaissance to answer questions that arise from the map reconnaissance and METT-T evaluation. Specific points include passage points, lanes through obstacles, locations of friendly listening posts and observation posts, possible approach and return routes, enemy positions (if any), and intermediate observation points on the way to the objective.

In selecting approach and return routes, the patrol leader chooses routes that best use concealment and avoid opposition and obstacles. To lessen the chances of ambush by the enemy, the return trip is planned along a different route. In addition to personal reconnaissance and review of the map and aerial photographs, the advice of other patrol leaders who already are familiar with the terrain and the objective area should be considered.

After compiling information about the situation and possible time constraint, the patrol leader completes an estimate. The first step is developing courses of action, each of which will provide for movement to the objective area, mission accomplishment, and the return, based on the tasks and their sequencing identified in mission analysis. While the eventual concept of operations is presented in order of occurrence, the patrol leader must develop the courses of action by either backward or forward planning. In situations where the objective is well defined and there is sufficient information to plan the action for mission accomplishment (reconnaissance or combat), the patrol leader begins the scheme for accomplishing the mission at the objective and then, planning backwards, considers the options for getting there and back.

The following sections on movement to and return from the objective area, reconnaissance missions, and combat missions discuss methods and options available to the patrol leader in developing the courses of action. The principal variables between courses of action will be who, where, and how in the following:

- Patrol task organization.
- Routes to the objective area.
- Observation point(s) (reconnaissance patrol), ambush site, form of maneuver, type of ambush (combat patrol), fire support plan.
- Return routes.

The patrol leader then mentally considers the progress of each course of action (a map or simple sketch is a useful aid) against expected and unexpected enemy action. By comparing the options against each other and prospective enemy opposition, the patrol leader chooses the course of action that has the best chance of success. Includes in this mental preview process is the time to determine the patrol plan for unexpected contingencies (enemy attack/counterattack, casualty handling). These contingency actions, together with the selected course of action, become the patrol concept of operations.

Once the patrol leader determines the scheme of maneuver, the fire support required to accomplish the mission is addressed including the fire power organic to the patrol and what additional indirect fire support will have to be provided by other units. When planning for indirect fire support, the patrol leader considers the following questions:

- Will artillery, mortar, or close air support be required at the objective area (combat patrols)?
- What artillery and mortar targets exist along the routes to and from the objective area that can be employed by the patrol if it encounters the enemy during movement (reconnaissance patrols and combat patrols)?
- What additional fire support will be required to cover the patrol's movement from the objective area back to the friendly area once the enemy is aware of the patrol's actions at the objective area (combat patrols)?

The effect that casualties have upon the patrol depends upon many factors. Generally, more casualties can be expected in a combat patrol than in a reconnaissance patrol. A patrol may continue on to the objective carrying its casualties, send them back with a detail of Marines, abort the mission and return the entire patrol with the casualties, or call their parent unit for assistance.

Some factors that determine what action the patrol leader takes are: patrol's mission; unit's standing operating procedure for handling wounded; number of casualties and nature of their injuries; availability of aid, helicopters or other means of casualty evacuation. Helicopter evacuation should only be used for the most serious casualties. For infantry units conducting patrols in proximity to the enemy, helicopter evacuation of casualties may compromise the patrol's mission and force the patrol to return to friendly positions before the mission is completed.

The patrol leader determines the requirement for nuclear, biological, and chemical (NBC) defense equipment. Gas masks should always be carried due to the availability of riot control agents (RCAs) to the enemy. If chemical or biological agents have been employed in the area that the patrol must pass through, protective garments will have to be worn by patrol members for part of or the entire patrol. Wearing extra clothing and carrying extra equipment affects the speed of the patrol's movement. A contingency plan for post patrol decontamination must be developed.

Complete the Plan

At this point, the patrol leader has completed the basic thinking necessary for accomplishing the assigned mission. The patrol leader prepares the patrol order to spell out the details, assign tasks to subordinates, and explain the entire endeavor for ease of understanding by the other members of the patrol.

Prepare the Order

The patrol leader's order contains more detailed information than discussed in the warning order. Orders follow the prescribed five-paragraph order format but contain greater detail. The patrol order is a modified 5-paragraph order; the major modifications are to paragraphs 3a and 3c. The format for the order is contained in appendix B.

Issue the Order

The patrol leader asks for a status report on the initial preparatory tasks assigned to subordinate leaders and specialists when the warning order was issued. When the patrol leader has completed planning and initial preparations have progressed to the point where the patrol order may be issued, the members of the patrol are assembled. Roll call is taken to ensure all patrol members are present, then the prepared order is issued.

This will be the only opportunity for the patrol leader to issue detailed instructions. The mission, in particular, must be unmistakably clear so that once the patrol is committed, all subordinate leaders can act with unity of purpose.

Whenever possible, the patrol leader should have a Marine, such as the navigator, build a terrain model using dirt, sand, twigs, etc., explaining the concept of operations for movement to the objective area, actions at the objective area, and the return. Terrain models provide patrol members with a clear and simple layout of the area of operations and key terrain.

Supervise

Inspections and rehearsals, vital to proper preparation, are conducted even though the patrol leader and patrol members are well experienced in patrolling. Inspections determine the patrol's state of physical and mental readiness.

The patrol leader inspects before rehearsals to ensure completeness and correctness of uniform and equipment. The following areas are checked:

- Camouflage.
- Identification tags, Geneva Convention cards.
- Prescribed equipment, weapons, and ammunition are available and serviceable.
- Tape and other items are used to "silence" equipment (prevent noise produced during movement).
- Items that could provide information to the enemy (e.g., letters and papers) remain behind.
- Unnecessary equipment and excess weight remain behind.

The patrol leader questions each patrol member to ensure the following is known:

- The mission, planned routes (primary and alternate), and the fire support plan of the patrol.
- The individual's role: what to do and when to do it.
- What others are to do and how their actions impact.
- Challenges and passwords, codes, reporting times, radio call signs, frequencies, and any other pertinent details.

There is usually a period of time between final rehearsal and departure. The patrol leader reinspects just before departure to ensure all equipment is still in working order and the unit is ready to embark on the mission.

Rehearsals ensure the operational proficiency of the patrol. Plans are checked and needed changes are made. The patrol leader verifies the suitability of equipment. It is through rehearsals that patrol members become thoroughly familiar with the actions to take during the patrol.

If the patrol will operate at night, both day and night rehearsals are conducted. Terrain similar to that over which the patrol will operate is used. All actions are rehearsed. If time is limited, the most critical phases are rehearsed. Action at the objective is the most critical phase of the patrol and is always rehearsed.

An effective method is to talk the patrol through each phase, describing the actions and having each member perform individual duties. When satisfied, the patrol leader walks the patrol through all phases of the patrol using only the signals and commands to be used during the actual conduct of the patrol. Rehearsals continue until the patrol is thoroughly familiar with the plan. The rehearsal is also used to test the soundness of the patrol order and patrol organization.

After the rehearsal, the patrol leader makes final adjustments to the plan and patrol organization based on what was learned during the rehearsal and from other sources, such as the S-2 and adjacent patrols. When this is completed, the patrol leader issues final instructions to subordinate leaders noting any changes made in the patrol organization or plan. While the subordinate leaders are briefing the remainder of the patrol members, the patrol leader reports to the commander stating that the patrol is ready to begin the mission. The patrol leader also coordinates the location and time that the patrol can test fire all weapons prior to departure.

CHAPTER 11. MOVEMENT TO AND RETURN FROM THE OBJECTIVE AREA

This chapter provides guidance to patrol leaders for movement to and return from the objective area. Action in the objective area depends on whether the patrol is assigned a reconnaissance mission, combat mission or security mission. Chapters 12 and 13 provide detailed guidance on these types of missions.

11001. PASSAGE OF LINES

A passage of lines is an operation in which a force moves forward or rearward through another force's combat positions with the intention of moving into or out of contact with the enemy. (JP 1-02)

During the initial preparation for the patrol, the patrol leader selects a patrol assembly area and reconnoiters the area of passage designated by the commander. In coordination with the unit commander responsible for the area of passage, the patrol leader identifies gaps or lanes in minefields and wire obstacles and locates local security elements through which the patrol will pass. The patrol leader also checks the route from the patrol assembly area to the passage point or contact point where the patrol will depart friendly lines. If possible, both the route to the passage point and the route through the frontlines should be concealed from the enemy's view.

The patrol leader also reconnoiters the area for return passage of lines and coordinates with the unit commander responsible for the area of passage for passage points and lanes as necessary. The patrol leader observes these points from the direction that the patrol will use upon return to friendly lines, if possible, to aid in recognition upon return. The patrol leader provides the forward unit with information about the size of the patrol, general route, and expected time of return. The manner of challenge and recognition of the returning patrol should be coordinated in detail.

Upon return to friendly forward local security squads and/or frontlines, the patrol leader leaves the patrol in a covered position and moves forward with a radio operator and at least one Marine for security to make contact with the friendly unit in the manner agreed. After contact is made and recognized, the patrol leader rejoins the patrol and takes them to the passage point, personally checking in each member.

11002. ORGANIZATION FOR MOVEMENT

The patrol's task organization establishes the elements and teams needed to accomplish the mission in the objective area and to and from the objective. The patrol leader determines the formation(s) in which the patrol moves to the objective area.

Formations

The proper use of patrol formations is critical to the patrol's success. The squad and fire team formations described in FMFM 6-5 (proposed MCWP 3-11.2), *Marine Rifle Squad*, also apply to infantry patrolling formations. Because the movement of the patrol must be concealed from the enemy, the patrol normally moves through terrain that provides concealment. Control of the patrol in this type of terrain is difficult; thus the column formation, which is easily controlled, is normally used. However, as various types of terrain are encountered, the patrol leader uses the same considerations in determining the appropriate formation used in other infantry operations.

The standard squad and fire team formations are adaptable to a patrol. The patrol may change formations en route to match the situation and terrain. The patrol leader may have to sacrifice some control for better dispersion or give up some speed for greater stealth and more security. Other considerations include—

- Visibility, weather, terrain, and vegetation will influence dispersion and control of individuals and units. These factors may also affect the enemy; if visibility is good for the patrol, it is also good for the enemy. Two pieces of luminous tape worn on the back of the collar will aid in control and movement on dark nights. The collar is turned down when near the enemy. The tape can also be worn on the back of the cap, but should be covered or removed when near the enemy.

- Preserving the integrity of fire units (fire teams and, if attached, machine-gun teams) is of primary importance. If team members are detached from a machine-gun team, the unit integrity is lost and effectiveness is reduced. The patrol leader must also position fire units so as not to mask their fires.
- All-around defense of the patrol must not be sacrificed under any circumstances. The conventional squad and platoon formations provide adequate firepower in any direction required. When attachments are made to the patrol, the attachments are positioned within the formation to enhance the firepower of the patrol. If a fire unit, such as a machine-gun team or squad, is attached, it is incorporated into the all-around defense of the patrol by modifying the conventional formation (see fig. 11-1). It is permissible to employ machine-gun teams individually during movement. However, assault rocket squads and teams should be employed primarily as rifle fire units for movement. Firing the SMAW or AT-4 from within a patrol formation can present a serious back-blast danger to patrol members and should only be fired on the patrol leader's direction.
- Time allotted for mission accomplishment is also a major consideration. In selecting the formations, the patrol leader must consider the speed of the movement required to meet the time constraints (if any) imposed on the patrol. If required to meet a time schedule, a formation that permits rapid movement should be used. Speed, however, must never be permitted to force the patrol leader to make rash tactical decisions.

Exercise of Control

The patrol leader is positioned for best control over the patrol. The assistant patrol leader moves at or near the rear of the patrol. Other subordinate leaders move with their elements. All patrol members assist by staying alert and passing on signals and orders. A signal to halt may be given by any patrol member, but the signal to resume movement is given only by the patrol leader.

Arm-and-hand signals are the primary means of communication within a patrol and should be used exclusively when near the enemy. All members must know the standard infantry signals (refer to FMFM 6-5), as well as any special signals required, and be alert to receive and pass them to other members.

The patrol leader should speak just loudly enough to be heard. At night, or when close to the enemy, the patrol leader halts the patrol and has subordinate leaders come forward. They speak in a low voice and then pass the information to their subordinates by moving from member to member.

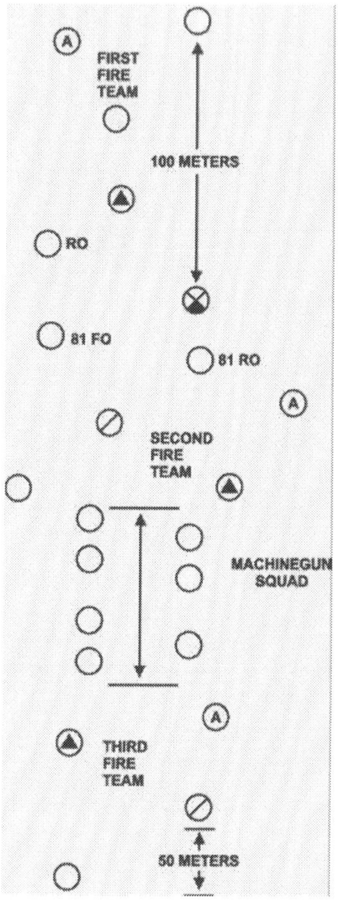

Figure 11-1. Squad Patrol with Attached Mortar Forward Observer Team and Machine Gun Squad.

Radios provide a means of positive control within a large patrol; however, radios should be used only when arm-and-hand signals or face-to-face contact between the patrol leader and subordinate leaders is impractical. When close to the enemy, words are spoken into the microphone with hands cupped over it in a low voice.

Other sound signals may be used if the patrol leader is sure they serve the purpose intended. Planned sound signals are rehearsed before starting on the patrol. Sound signals used must be simple, natural sounds that are few in number and easily understood. Bird and animal calls are seldom satisfactory.

Infrared equipment, such as the sniper scope and infrared filters for the flashlight, may be used as means of sending and receiving signals and maintaining control at night.

Luminous tape may be used to assist in control at night. Small strips on the back of the cap or collar of patrol members aid in keeping visual contact with the front member. However, the luminous tape must be covered when near the enemy.

An important aspect of control is accounting of personnel, especially after crossing danger areas, enemy contacts, halts, and exiting and re-entering friendly lines. The patrol leader may arrange for the last member to send up the count automatically after crossing danger areas, enemy contact, and halts. In large patrols or when moving in a formation other than a column, subordinate leaders check members and report the count to the patrol leader by the quickest method appropriate to the circumstances.

Navigation

One or more members are assigned as navigators for the patrol. Their function is to assist the patrol leader in maintaining direction by the use of the compass.

The patrol leader assigns at least two members as pacers to keep track of the distance from point to point. The average of their count is used to approximate the distance traveled. Pacers are separated so they do not influence each other's count. Pacers must know how to convert their own paces to meters.

The route is divided into increments, with each increment starting at a recognizable point on the ground. The pacers begin their counts from zero at the beginning of each leg. This makes the pace count easier to keep and checks for accuracy.

The pace count is sent forward when the patrol leader turns to the member behind and in a low voice says, "send up the pace" or uses the arm-and-hand signal of tapping his boot. This signal is passed to both pacers, who in turn send up the pace count in meters; for example, "two-hundred" or "one-seven-five."

Patrol members must understand that the counts of both pacers are sent forward. The patrol leader must know the counts of both men in order to check them.

Security

The patrol is dispersed consistent with control, visibility, cover, and concealment. Scouts are employed to the front, flanks, and rear of the patrol to provide security. For the patrol members traveling in the main body of the patrol, areas of responsibility are assigned to the front, flanks, and rear. Scouts are the eyes and ears of the patrol leader. They move when and where directed by the patrol leader and maintain contact with the patrol leader at all times, except when momentarily obscured by vegetation or other terrain features.

Front

Small patrols (squad size) may employ from one scout up to a fire team as the point, depending on the enemy situation, terrain, and route being followed by the patrol. Normally, squad-size patrols will use two scouts as the point. The point is responsible for investigating the route of advance immediately to the front of the patrol.

The point moves as far ahead of the patrol as visibility and terrain permit. When visibility is good, the point may precede the main body by as much as 100 meters. The point travels right and left ahead of the patrol, searching the area over which the patrol will pass.

The point maintains direction by knowledge of the general route to be followed and visual contact with the patrol leader. The patrol leader or the navigator ensures that the point is proceeding correctly.

The point, which stays far enough ahead of the patrol to provide security, is not a trail breaker for the patrol. If the point loses contact with the patrol, the point waits for the main body to catch up or moves rearward if contact is not quickly regained.

One of the navigators may be positioned with the point. One or more members works as the point while the other is the navigator.

Flank

Flank security for a patrol of squad size or less may be provided by using one or two members on either flank. If two scouts are assigned to a flank, one is positioned to observe the patrol leader and the other works farther out from the patrol. The scout who must observe the patrol leader remains within a maximum distance of 100 meters. The scout farther out remains in sight of the inside scout but normally does not move more than 20 or 25 meters away and remains prepared to relieve flank security regularly. Moving through dense woods or jungle may render the use of flank security impractical because of reduced visibility. In such cases, it moves with the patrol itself, but maintains observation to its assigned flanks.

Rear

A small patrol normally has only one rifleman assigned as rear security. An interval between the member assigned as rear security and the last Marine of the patrol is maintained at the limit of visibility, up to 50 meters. This member maintains rear security for the patrol by constantly observing to the rear.

Halting

Speed of movement is slower at night than in day patrols and reduces the danger of a Marine becoming separated from the patrol. The patrol occasionally halts to observe and listen for enemy activity; this is called a security halt. Upon signal, when reaching a danger area and periodically throughout movement en route, every member freezes in place, remains quiet, observes, and listens. It may be necessary to call a security halt just after departing friendly areas and just before entering friendly areas.

The patrol may halt briefly to send a message, eat, rest, check direction, or make a reconnaissance. The area selected should provide adequate concealment and cover, as well as favor the defense. All-round security is established and the patrol leader ensures all members move out when the patrol resumes movement. (For extended halts, see paragraph 11005.)

Infiltration

The disposition of enemy forces may sometimes prevent a patrol from entering the enemy occupied area as a unit; however, pairs of scouts or fire teams may slip through without being discovered. (Refer to FMFM 6-5 for infiltration techniques and procedures).

11003. CONTROL MEASURES FOR MOVEMENT

Checkpoints

A checkpoint is a predetermined point on the surface of the Earth used as a means of controlling movement, a registration target for fire adjustment or reference for location (JP 1-02). Checkpoints are a means of control between the parent unit and the patrol. These locations are decided upon and coordinated before the patrol leaves, so that both the patrol members and parent unit will know the patrol's location when it reports in. The parent unit can follow the progress of the patrol without transmitting coordinates over the radio that the enemy could monitor.

Rally Points

A rally point is an easily identifiable point on the ground where units can reassemble and reorganize if they become dispersed (MCRP 5-12C). It should provide cover and concealment, be defensible for at least a short time, and be easily recognized and known to all patrol members. All rally points are considered and identified as tentative rally points until they are occupied, found to be suitable, and designated as rally points by the patrol leader.

Initial Rally Point

The initial rally point is within friendly lines where the patrol can rally if it becomes separated before departing the friendly area or before reaching the first en route rally point. It may be the assembly area where the patrol waits while the patrol leader contacts the last friendly position through which the patrol will pass. The location of the initial rally point must be coordinated with the forward unit commander.

En route Rally Points

En route rally points are between the initial rally point and the objective rally point; and from the objective rally point back to the point where the patrol reenters friendly lines. They are determined as the patrol passes through a likely area that is suitable for a rally point.

Objective Rally Point

The objective rally point is located nearest the objective where the patrol makes final preparations prior to approaching the objective. It also serves as a location where the patrol reassembles after completing actions on the objective. The objective rally point must be suitable to accommodate those activities accomplished prior to actions on the objective. This position must provide the patrol concealment from enemy observation and, if possible, cover from enemy fires. It may be located short of, to a flank or beyond the objective. It should be out of sight, sound, and small-arms range of the objective area. The patrol leader's reconnaissance of the objective is made from this position; it is the release point from which patrol elements and teams move to the objective to accomplish the mission.

Rally Point Selection

The patrol leader selects likely locations for tentative rally points during reconnaissance or map study. A tentative rally point must be confirmed and announced after examination proves suitability. A tentative initial rally point and a tentative objective rally point are always selected and identified in the patrol order. If necessary, the patrol leader selects additional rally points en route as suitable locations are reached. When the patrol reaches a danger area that cannot be bypassed, such as an open meadow or stream, the patrol leader selects a rally point on both the near and far side. If good locations are not available, the patrol leader designates the rally points in relation to the danger area. The patrol leader will say, for example, "50 meters this side of the trail" or "50 meters beyond the stream."

Rally Point Use

The initial rally point and en route rally points are designated to enable the patrol to reassemble if it is unavoidably separated or dispersed. Identifying features are pointed out. The patrol leader ensures that the information is passed to all patrol members. When crossing a danger area, a near side rally point and a far side rendezvous point are designated. These rally points should only be used when all other methods of retaining control of the patrol have failed. The success of the patrol may be jeopardized if it is dispersed and cannot rally expeditiously.

If the patrol has left the friendly area and becomes dispersed, patrol members return to the last designated rally point (the initial or an en route rally point) unless the patrol leader gives other instructions.

As previously noted, the patrol leader selects two rally points at the near and far sides of danger areas that cannot be bypassed. If the patrol becomes separated or dispersed at a danger area, and there has been no enemy contact, the patrol should reassemble at the rally point on the far side of it. If, however, the patrol is separated or dispersed at a danger area as a result of enemy contact, members who have already crossed the danger area assemble and reorganize at the rally point on the far side; members who have not crossed assemble and reorganize at the rally point on the near side. In the absence of the patrol leader and unless directed otherwise, the senior member at the rally point on the near side takes charge, attempts to move the rallied patrol members to the rally point on the far side, and rejoins the remainder of the patrol.

Rally Point Actions

The patrol leader plans the actions to be taken at rally points and instructs the patrol accordingly in the patrol order. Planned actions at the initial rally point and en route rally points must provide for the continuation of the patrol as long as there is a reasonable chance to accomplish the mission. Plans for actions at rally points should provide for—

- Recognition signals for assembly at rally points.
- Minimum number of members and maximum amount of waiting time required before the senior member at the rally point moves the rallied patrol members onward toward the objective or returns to friendly lines.
- Instructions for patrol members who find themselves alone at a rally point.

11004. PRECAUTIONS AT DANGER AREAS

A danger area is any place where the patrol is vulnerable to enemy observation or fire (open areas, roads, trails, and obstacles such as barbed wire, minefields, rivers and streams, and lakes). Any known or suspected enemy position the patrol must pass is also a danger area. The patrol leader plans for crossing each danger area and includes these plans in the order.

The patrol reconnoiters the near side of a danger area first, then the patrol leader sends scouts to reconnoiter the far side. Once the scouts report that the far side is clear of the enemy, the remainder of the patrol crosses the danger area. As each individual or group crosses the danger area, they are covered by those remaining and by those who have successfully crossed. Enemy obstacles are avoided since they are usually covered by fire.

In crossing a river, the near bank is reconnoitered first; then the patrol is positioned to cover the far bank. Scouts are sent across to the far bank. After the far bank has been reconnoitered and the scouts report that it is clear of the enemy, the patrol crosses as rapidly as possible. This may be done individually or in pairs. If crossing the river requires swimming, the patrol uses improvised rafts to float equipment, weapons, and ammunition across. (Refer to MCRP 3-02C, *Water Survival Handbook.)*

A road or trail is crossed at or near a bend or where the road is narrow. Observation is restricted and, if the enemy is present, the patrol is exposed as short a time as possible. The near side is reconnoitered first, then scouts are sent across to reconnoiter the far side. This includes reconnoitering the tentative rally point on the far side. Once the scouts report "all clear," the remainder of the patrol crosses rapidly and quietly.

If the patrol must pass close to an enemy position, it takes advantage of battlefield noises to cover the sounds of movement. If supporting fires are available, the patrol leader can call for them to divert the enemy's attention as the patrol passes.

11005. HIDE

When a patrol is required to halt for an extended period in an area not protected by friendly troops, the patrol moves into a location which, by the nature of the surrounding terrain, provides passive security from enemy detection. Such an assembly area is termed a *hide*. To establishment a hide—

- Cease all movement during daylight hours to avoid detection.
- Hide the patrol for an extended period while the patrol leader conducts a detailed reconnaissance of the objective area.
- Rest and reorganize after extended movement.
- Reorganize after a patrol has infiltrated the enemy area in small groups.

The patrol leader's plan must include tentative hide locations when the patrol's mission dictates an extended halt within enemy areas. These tentative locations must be confirmed by actual ground reconnaissance prior to occupation by the patrol. The plan for a hide includes both passive and active security measures.

Passive security measures are—

- Avoid built-up areas.
- Select an area remote from all human habitation.
- Avoid known or suspected enemy positions.
- Avoid ridge lines, topographic crests, valleys, lakes, and streams.
- Avoid roads and trails.
- Avoid open woods and clearings.
- Select areas offering dense vegetation, preferably bushes and trees that spread out close to the ground.

Active security measures—

- Establish security covering all likely avenues of approach into the site.
- Establish communications (wire, radio, signal, runner) with posted security to provide early warning of enemy approach.
- Select an alternate area for occupation if the original hide is compromised or found unsuitable.
- Plan for withdrawal in the event of discovery.
- Establish an alert plan with a certain percent of the personnel awake at all times.
- Organize the elements of the patrol so necessary activities can take place with a minimum amount of movement.

The size of the area physically occupied by a patrol in a hide and the number of security posts required are governed by the terrain, quantity and quality of cover and concealment, and size of the patrol.

If the situation permits, a hide can also be used as the final preparation position and/or objective rally point.

11006. IMMEDIATE ACTIONS UPON ENEMY CONTACT

A patrol may make contact with the enemy at any time. Contact may be through observation, a meeting engagement or ambush. Contact may be visual, in which the patrol sights the enemy but is not itself detected. When this is the case, the patrol leader can decide whether to make or avoid physical contact, basing his decision on the patrol's assigned mission and capability to successfully engage the enemy unit.

When a patrol's assigned mission prohibits physical contact (except that necessary to accomplish the mission), its actions are defensive in nature. Physical contact, if unavoidable, is broken as quickly as possible and the patrol, if still capable, continues its mission.

When a patrol's assigned mission permits or requires it to seek or exploit opportunities for contact (as in the case of a combat patrol), its actions are offensive in nature, immediate, and positive.

In patrolling, contacts (visual or physical) are often unexpected at very close ranges, and short in duration. Effective enemy fire often provides leaders little or no time to fully evaluate situations and issue orders. In these situations, immediate action provides a means for swiftly initiating positive offensive or defensive action, as appropriate.

Two types of physical contact with the enemy are meeting engagement and ambush. Meeting engagement is a combat action that occurs when a moving force, incompletely deployed for battle, engages an enemy at an unexpected time and place. It is an accidental meeting where neither the enemy nor the patrol expect contact and are not specifically prepared to deal with it. An ambush is a surprise attack from a concealed position.

Immediate Actions

Immediate actions are designed to provide swift and positive small unit reaction to visual or physical contact with the enemy. They are simple courses of action in which all Marines are well trained. Minimal signals or commands are required and they are developed as needed for the combat situation. The signals can, in many cases, be initiated by any member of the unit. It is not feasible to attempt to design an immediate action drill to cover every possible situation. It is better to know the immediate action drill for each of a limited number of situations that may occur during a patrol. Arm-and-hand signals associated with immediate actions—such as FREEZE, ENEMY IN SIGHT, and HASTY AMBUSH RIGHT or LEFT—are contained in FMFM 6-5 (proposed MCWP 3-11.2).

Immediate Halt

When the patrol detects the enemy but is not itself detected, the situation requires the immediate, in-place halt of the patrol. The first member visually detecting the enemy gives the silent signal for FREEZE. Every member halts in place, weapon at the ready, and remains absolutely motionless and quiet until further signals or orders are given.

Air Observation and/or Attack

These actions are designed to reduce the danger of detection from aircraft and casualties from air attack.

When an enemy or unidentified aircraft that may detect the patrol is heard or observed, the appropriate immediate action drill is FREEZE. The first member hearing or sighting an aircraft that may be a threat signals FREEZE. Every member halts in place until the patrol leader identifies the aircraft and gives further signals or orders. Members of the patrol must not look up at the aircraft as sunlight can reflect off their faces even when camouflaged.

When an aircraft detects a patrol and makes a low level attack, the immediate action drill air attack is used. The first member sighting an attacking aircraft shouts, "AIRCRAFT," followed by the direction of the incoming attack: FRONT, LEFT, REAR or RIGHT. The patrol moves quickly into line formation, well spread out, at right angles to the aircraft's direction of travel. As each member comes on line, the member hits the ground, using available cover, then positions the body perpendicular to the aircraft's direction of travel, to present the shallowest target possible (see fig. 11-2 on page 11-8). Between attacks (if the aircraft returns or if more than one aircraft attacks), patrol members seek better cover. Attacking aircraft are fired upon only on command of the patrol leader.

Meeting Engagement

Hasty Ambush. This immediate action is used to avoid contact and to prepare to initiate an unplanned ambush on the enemy. It may often be a subsequent action after the command freeze. When the signal HASTY AMBUSH is given (by the point member, patrol leader or another authorized patrol member), the entire patrol moves quickly to the right or left of the line of movement, as indicated by the signal, and takes up the best available concealed firing positions (see fig. 11-3 on page 11-9). The patrol leader initiates the ambush by opening fire and shouting, "FIRE"; thus ensuring initiation of the ambush if the weapon misfires. If the patrol is detected before this, the first member aware of detection initiates the ambush by firing and shouting. The patrol leader may decide not to initiate the ambush in order to avoid contact unless the patrol is detected. When used as an offensive

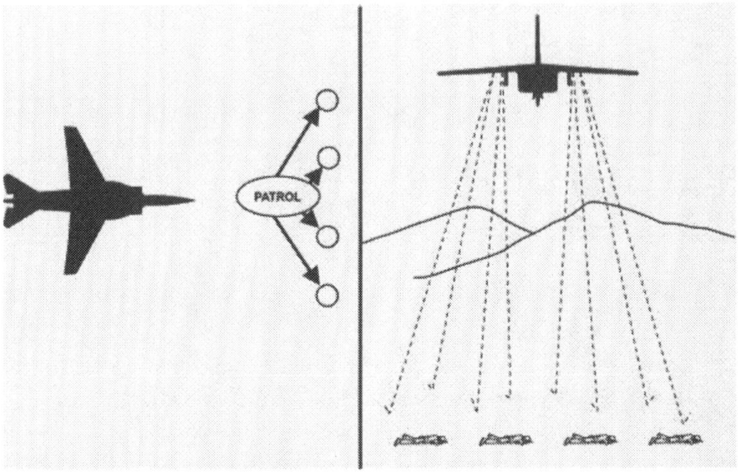

Figure 11-2. Immediate Action, Air Attack.

measure, the enemy is allowed to advance until he is in the most vulnerable position before the ambush is initiated. An alternate means for initiating the ambush is to designate an individual (for example, point or last member) to open fire when a certain portion of the enemy unit reaches or passes that member.

Immediate Assault. This immediate action drill is used defensively to make and quickly break undesired but unavoidable contact (including ambush) and offensively to decisively engage the enemy (including ambush). When used in a meeting engagement, members nearest the enemy open fire and shout, "CONTACT," followed by the direction of the incoming attack: FRONT, LEFT, REAR or RIGHT. The patrol moves swiftly into line formation and assaults (see fig. 11-4).

Defensive Measures

When used defensively, the assault is stopped if the enemy withdraws and contact is broken quickly. If the enemy stands fast, the assault is carried through the enemy positions and movement is continued until contact is broken.

Offensive Measures

When used offensively, the enemy is decisively engaged. Escapees are pursued and destroyed until orders to break contact are given by the patrol leader. If the patrol is fired upon from beyond 50 meters, the patrol must break contact as quickly as possible and continue the mission. If it engages the enemy any longer than necessary to break contact, it may put the mission in jeopardy.

Fire and Maneuver

Fire and maneuver is one means to break contact. One portion of the patrol returns the enemy fire while another portion moves by bounds away from the enemy. Each portion of the patrol covers the other by fire until contact is broken by all.

Clock System

The clock system is another means to break contact. Twelve o'clock is the direction of movement of the patrol. The patrol leader shouts a direction and a distance. For example: "TEN O'CLOCK-TWO HUNDRED," means the patrol should move in the direction of ten o'clock for 200 meters. Patrol members keep their same relative positions as they move so the original formation is not disrupted. Subordinate leaders must be alert to ensure that the members of their elements and teams receive the correct order and move as directed.

Counter Ambush

When a patrol is ambushed, the immediate action drill used depends on whether the ambush is a near ambush

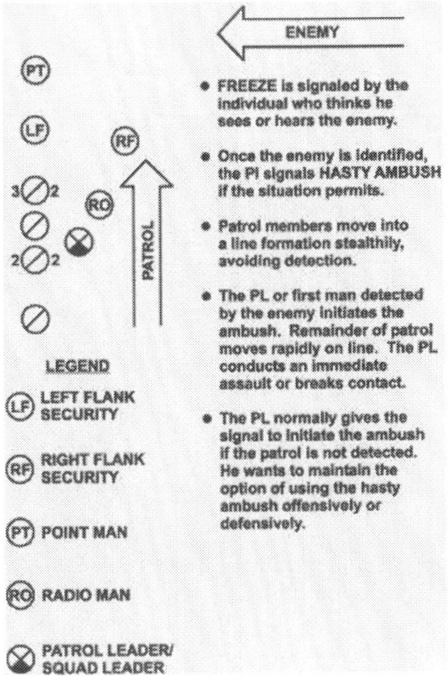

Figure 11-3. Immediate Action, Hasty Ambush.

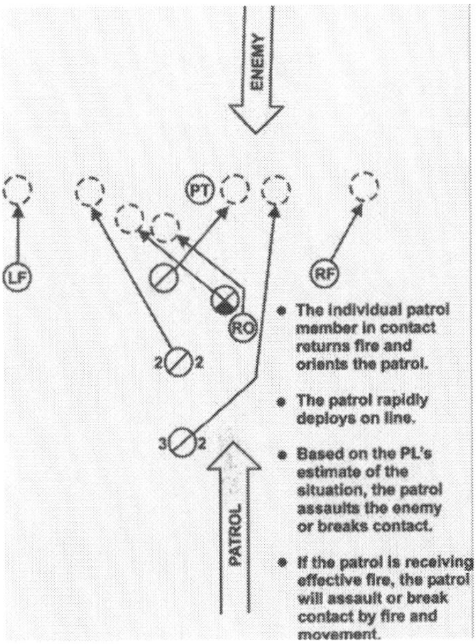

Figure 11-4. Immediate Action, Immediate Assault.

(the enemy is within 50 meters of the patrol) or a far ambush (the enemy is beyond 50 meters of the patrol). Fifty meters is considered the limit from which the ambushed patrol can effectively launch an assault against the enemy.

Near Ambush. In a near ambush, the killing zone is under very heavy, highly concentrated, close range fires. There is little time or space for members to maneu-ver or seek cover. The longer they remain in the killing zone, the greater the chance they will become casualties. Therefore, if members in the kil-ling zone are attacked by a near ambush, they immediately assault without order or signal directly into the ambush position, occupy it, and continue the assault or break contact, as directed. This action moves them out of the killing zone, prevents other elements of the ambush from firing on them without firing on their own members, and provides positions from which other actions may be taken (see fig. 11-5 on page 11-10). Members not in the killing zone maneuver against the enemy as directed. The assault continues until all patrol members are outside of the killing zone.

Far Ambush. In a far ambush, the killing zone is also under very heavy, highly concentrated fires, but from a greater range. This greater range provides members in the killing zone maneuver space and some opportunity to seek cover at a lesser risk of becoming a casualty. If attacked by a far ambush, members in the killing zone, without order or signal, immediately return fire, take the best available positions, and continue firing until directed otherwise. Members not in the killing zone maneuver against the ambush force, as directed (see fig. 11-6 on page 11-10). The assault is continued against the enemy or until the order to break contact is given.

In each situation, the success of the counter ambush employed depends on the members being well trained in quickly recognizing the distance from which an ambush is initiated and well rehearsed in the proper reaction.

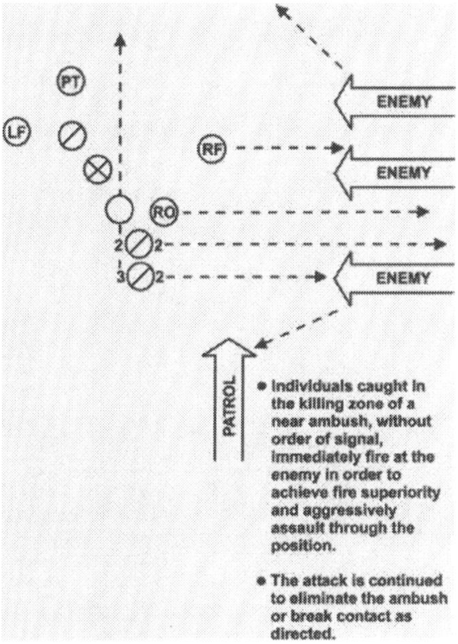

Figure 11-5. Near Ambush.

11007. PATROL LEADER'S ACTION IN A DEVELOPING SITUATION

While good patrolling depends on good planning, the patrol leader's plan must be flexible. Every combat situation develops differently than expected, and the patrol leader must be ready to quickly adapt to the situation as it develops. One of the most difficult tasks in battle is to recognize the correct moment for making a decision. Generally, it is more difficult to determine the moment for making a decision than it is to formulate the decision itself. When the situation demands, decisions must be made promptly without waiting for more information. In a developing situation, the patrol leader should use the following questions as a guide for battlefield decisionmaking:

- How has the situation changed?
- How does the change affect mission accomplishment and the immediate superior's mission?
- Must a decision be made now?
- What are the options?
- Which option best serves the mission and the unit as a whole?
- Which option offers the greatest chance of success?

11008. RETURN FROM OBJECTIVE AREA

After performing actions in the objective area, the patrol reassembles at the objective rally point. This phase of the patrol is perhaps the most difficult and dangerous. Patrol members are experiencing fatigue, emotional letdown, and wounds; they may be low on water and ammunition. Above all, the enemy was likely alerted if not in pursuit. At this point, the patrol leader must move the patrol rapidly but carefully and maintain patrol security at a high level. In returning to friendly lines, the patrol neither uses nor travels near the same route used to get to the objective area. The enemy may have that route covered. For information on reentry of friendly lines, refer to paragraph 11001.

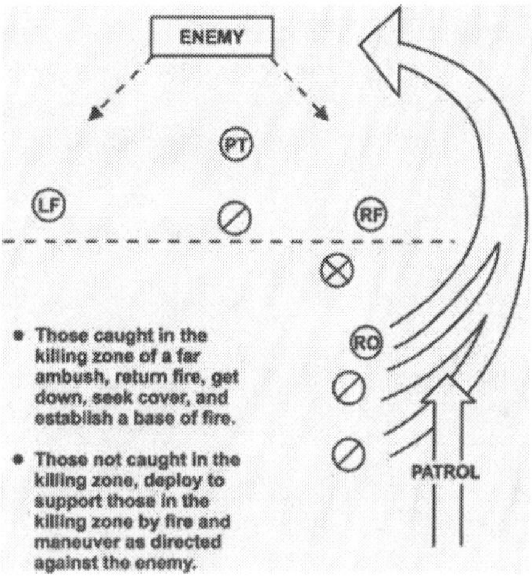

Figure 11-6. Far Ambush.

CHAPTER 12. RECONNAISSANCE PATROLS

The commander needs accurate, timely information about the enemy and the terrain the enemy controls to assist in making tactical decisions. Reconnaissance patrols are one of the most reliable means for obtaining this information. These patrols engage the enemy only when necessary to accomplish their mission or for protection. They generally avoid combat, accomplish the mission by stealth, and do not maintain contact with the unit that sent them out. A reconnaissance patrol is capable of carrying the search for information into the area occupied by enemy forces—usually beyond the range of vision of friendly ground observation posts—and examining objects and events at close range.

12001. GENERAL MISSIONS

General missions for reconnaissance patrols include gathering information about location and characteristics of friendly or hostile positions and installations, terrain (routes, stream crossings), and obstacles.

12002. SPECIFIC MISSIONS

Locate the Enemy

Reconnaissance patrols try to determine the—

- Location of enemy forces, installations, and equipment.
- Identification of enemy units and equipment.
- Strength of enemy forces.
- Disposition of enemy forces.
- Movement of enemy personnel and equipment.
- New or special types of weapons.
- Presence of mechanized units.
- Unusual enemy activity.
- Presence of NBC equipment.

Reconnaissance of Enemy Wire Obstacles

A patrol with the mission of investigating enemy-emplaced wire obstacles employs a formation providing for all-around security and takes precautions against being observed by the enemy. The patrol leader and one Marine inspect each gap and establish its location by means of compass bearings to prominent objects in the rear of enemy or friendly positions or through the use of global positioning satellite receivers.

Reconnaissance of Contaminated Areas

A patrol with the mission of investigating a contaminated area reconnoiters and marks the area's boundaries. Patrol members wear protective clothing and field protective masks. The patrol leader reports the extent of the area, the ttype of agent used, the terrain and vegetation and the method of marking the contaminated area. A sketch of the contaminated area, prepared by the patrol leader, should be included in the patrol report.

Reconnaissance of Enemy Minefields

Patrols assigned to reconnoiter enemy minefields are often composed of specially trained personnel. The type of mine is identified and the possibility of bypassing the area determined.

Reconnaissance of Terrain

The reconnaissance patrol is particularly suited for gathering information about the terrain within their area of operations. (Refer to MCRP 2-15.3B, *Reconnaissance Reports Guide*.) Commanders must know the location and condition of bridges, streams, and roads in order to make proper plans for the movement of troops. The following information should be obtained by a reconnaissance patrol assigned such missions:

- Bridges:
  - Maximum load capacity.
  - Material used in construction (wood, stone, concrete or steel).

- Material used for abutments (dirt, wood, masonry, concrete).
- Type and number of supporting members of the bridge.
- Condition and dimensions of bridge flooring.
- Presence or absence of mines and demolition charges on the bridge and at entrances/exits.
- Terrain crossed (ravine, stream, road).
- Primary use (rail, auto, foot).
- Location and type of an alternate crossing if bridge is unusable.

- Streams and fords:
  - Width and depth of stream.
  - Composition of the bottom (mud, sand, gravel, rock).
  - Speed of current in miles per hour.
  - Composition and gradient of banks.
  - Surrounding terrain and vegetation cover.
- Roads:
  - Width.
  - Composition of surface.
  - Condition.
  - Road blocks.
  - Gradient/degrees of slopes.
  - Curves (location and length).
  - Pitch and culvert locations.
  - Primary use and frequency of use.

12003. TYPES OF RECONNAISSANCE

Route

Route reconnaissance is a directed effort to obtain detailed information of a specified route and all terrain from which the enemy could influence movement along that route. (Refer to MCRP 5-12C.) It is also described as a form of reconnaissance focused along a specific line of communications—such as a road, railway, or waterway—to provide new or updated information on route conditions and activities along the route. (Refer to MCRP 5-12A, *Operational Terms and Graphics*.)

A route reconnaissance normally precedes the movement of forces. It provides detailed information about a specific route and the surrounding terrain that could be used to influence movement along that route. Considerations include trafficablility, danger areas, critical points, vehicle weight and size limitations, and locations for friendly obstacle emplacement.

The objective area is normally defined by a line of departure, a route, and a limit of advance.

Area

An area reconnaissance is a directed effort to obtain detailed information concerning the terrain or enemy activity within a prescribed area such as a town, ridge line woods or other features critical to operations. (Refer to MCRP 5-12C.) An area reconnaissance could also be made of a single point, such as a bridge or installation. (Refer to MCRP 5-12A.)

Emphasis is placed on reaching the area without being detected. Enemy situations encountered en route are developed only enough to allow the reconnoitering unit to report and bypass.

Zone

A zone reconnaissance is a directed effort to obtain detailed information concerning all routes, obstacles (to include chemical or radiological contamination), terrain, and enemy forces within a zone defined by boundaries. A zone reconnaissance is normally assigned when the enemy situation is vague or when information concerning cross-country trafficability is desired. (Refer to MCRP 5-12C.)

The commander specifies specific routes or areas of interest within the zone. The zone to be reconnoitered usually is defined by a line of departure, lateral boundaries, and a limit of advance.

12004. TASK ORGANIZATION

A reconnaissance patrol is organized around the current structure of the Marine rifle squad with one or more of the squad's fire teams assigned as the reconnaissance element to reconnoiter or maintain surveillance over the objective. At least one fire team acts as a security element whose functions are to—

- Secure the objective rally point (see para. 11003).
- Give early warning of enemy approach.
- Protect the reconnaissance unit.

Reinforcing the squad is based on METT-T. If two squads are required, then the mission should be assigned to a platoon.

A company may send a reconnaissance patrol to specifically reconnoiter a given area (a riverbed or a bridge), or to maintain surveillance on a particular streambed for the next four nights. Reconnaissance patrols perform three basic functions en route to and from the objective:

- Provide control.
- Provide security while moving and in the objective area.
- Conduct reconnaissance or surveillance.

Depending on the size of the reconnaissance patrol, an element must be tasked to reconnoiter the area or zone. In the case of a surveillance mission, two teams are recommended to share the same task (so one could lay up and rest, while the other lays low and hides to maintain strict silence and no movement while surveillance is being maintained on the objective). Whatever the case, each unit that has a surveillance mission or reconnaissance mission must be thoroughly briefed as to what essential elements of information are to be collected as well as their location in the objective area.

The patrol should be organized with one or two fire teams to actually conduct the reconnaissance mission and the remaining fire team to provide security. Each fire team should be prepared to assume either mission. A small area reconnaissance patrol needs only one fire team for the assigned mission. A patrol with a wider area reconnaissance mission should use two fire teams to physically conduct the mission and one fire team for cover and/or security.

The security element for a reconnaissance patrol should be organized to cover the likely avenue of approach into the objective area, to protect the units conducting the reconnaissance, and to cover the objective rally point.

The variations of the special organization for reconnaissance patrols illustrate the patrol leader's flexibility in organizing the patrol to meet mission requirements. The patrol leader is not limited to the variations reflected herein but can choose any special organization as long as that task organization accommodates the requirement for command and control, reconnaissance, and security.

12005. SIZE OF RECONNAISSANCE PATROLS

A reconnaissance patrol should be kept to the minimum number of personnel required to accomplish the mission. A mission requiring a patrol to remain away from its unit for a considerable period of time, or one requiring a patrol to send back information by messenger, increases the size of the patrol. Reconnaissance patrols seldom exceed a squad in strength. Unit integrity should be preserved whenever possible. Intelligence personnel, interpreters, and other specialists, such as radio operators or engineers, are assigned to a patrol if the particular mission requires.

12006. RECONNAISSANCE EQUIPMENT

Patrol members are armed and equipped as necessary for accomplishing the mission. The automatic rifle in each fire team provides a degree of sustained firepower in case of enemy contact. The patrol should have at least two: pairs of binoculars, pairs of wire cutters, maps, compasses, and watches. Night observation equipment may be used. Pencils and small notebooks are carried so notes and sketches can be made. A message book with message blanks and overlay paper is mandatory.

12007. RECONNAISSANCE PATROL ACTIONS AT THE OBJECTIVE AREA

Route

The patrol leader halts and conceals the patrol near the objective area in the final preparation position. The patrol leader then conducts leader's reconnaissance to confirm the plan for positioning the security teams and employing units assigned to the reconnaissance mission. The patrol leader returns to the patrol and positions the security to provide early warning of enemy approach and secure the objective rally point. The reconnaissance unit(s) then reconnoiters the objective area (route). The reconnaissance unit may move to several positions, along or adjacent to the specific route, in order to conduct a thorough reconnaissance. After completing the reconnaissance, each reconnaissance team moves to the objective rally point and reports to the patrol leader. The patrol then returns to friendly lines and the leader makes a full report.

Area

The patrol leader halts and conceals the patrol near the objective area in the final preparation position. The patrol leader then conducts leader's reconnaissance to pinpoint the objective and confirm the plan for positioning the security teams and employing units assigned the reconnaissance mission. The patrol leader returns to the patrol and positions security to provide early warning of enemy approach and secure the objective rally point. The reconnaissance unit(s) then reconnoiters the objective area. The reconnaissance unit may move to several positions, perhaps making a circle around the objective area, in order to conduct a thorough reconnaissance. When the reconnaissance is completed, the patrol leader assembles the patrol and tells members what has been observed and heard. Other patrol members contribute observations. The patrol then returns to friendly lines and the patrol leader makes a full report.

Zone

The patrol leader halts the patrol at the final preparation position, confirms the plan, and conducts leader's reconnaissance. The patrol leader positions the security team and sends out the reconnaissance team. When the entire patrol is used to reconnoiter the zone, it provides its own security. After completing the reconnaissance, each reconnaissance team moves to the objective rally point and reports to the patrol leader. The patrol then returns to friendly lines.

CHAPTER 13. COMBAT PATROLS

Combat patrols are assigned missions that usually include engaging the enemy. They are fighting patrols. Every combat patrol has a secondary mission: gaining information about the enemy and terrain. Combat patrols are employed in both offensive and defensive combat operations and they assist the parent unit in accomplishing its mission by inflicting damage on the enemy; establishing and/or maintaining contact with friendly and enemy forces; denying the enemy access to key terrain; and probing enemy positions to determine the nature and extent of enemy presence.

13001. TASK ORGANIZATION

A combat patrol is organized around the current structure of the Marine rifle platoon. A combat patrol leader should use the unit's normal organization (fire team, squad, and platoon) in assigning functions, patrol missions, and chain of command. Combat patrols must be able to perform the following four basic functions en route to and from the objective:

- Provide control.
- Provide security.
- Provide support by fire.
- Attack or assault the objective.

A rifle platoon could task-organize as a combat patrol as follows:

- Platoon headquarters (patrol headquarters).
- First squad (security).
- Second squad (support).
- Third squad (assault).

Every combat patrol must—

- Provide a control mechanism in the form of a headquarters.
- Designate a unit (a fire team or squad) to provide security while moving en route to the objective or while at the objective. At the objective area, this unit isolates the objective area, secures the objective rally point, and covers the withdrawal.
- Designate a unit to act as support. This unit provides the base of fire in the attack or covers withdrawals or advances.
- Designate a unit(s) to conduct the attack or assault. This unit(s) engages the enemy at the objective area by fire and maneuver or movement. It also operates immediately in the objective area (searching, demolition, prisoners of war, etc.). Paragraph 9001 outlines the general organization of combat patrols. As in the case with reconnaissance patrols, the task organization of a combat patrol depends on the specific mission assigned. If any special requirements are generated because of the specific mission, the patrol is task-organized to fit the needs of the mission.

13002. EQUIPMENT

Combat patrols are armed and equipped as necessary for accomplishing the mission. In addition to binoculars, wire cutters, compasses, and other equipment generally common to all patrols, it usually carries a high proportion of automatic weapons and grenades. Communications with higher headquarters is important as success of the mission may depend on being able to call for supporting fires. Also, internal radio communications with the units and teams may be useful. However, the patrol must not be so overburdened with equipment as to impede movement or mission accomplishment.

13003. RAID PATROLS

A raid is a surprise attack on an enemy force or installation with the attacking force withdrawing after accomplishing its mission. Raids destroy or capture enemy personnel or equipment, destroy installations, or free friendly personnel who have been captured by the enemy. Patrolling techniques are used in planning and when moveing to and from the objective. (Refer to MCWP 3-41.2, *Raids*.) Surprise, firepower, and violence of action are the keys to a successful raid. Patrols achieve surprise by attacking—

- When the enemy is least prepared (e.g., during periods of poor visibility such as darkness, rain, fog, or snow).
- From an unexpected direction. (This might be accomplished by approaching through a swamp or other seemingly impassable terrain.)
- With concentration of firepower at critical points within the objective.

Planning

A successful raid requires detailed planning. The leader of a combat patrol engaged in raiding must anticipate probable situations and decide upon definite courses of action to meet them. Rehearsals are imperative.

A raid patrol conducts such missions as destroying an enemy outpost or seizing prisoners from an observation post or lightly defended position.

While preparing for the mission, the patrol leader requests fire support required for the accomplishment of the mission. If practical, artillery and mortars should be employed to isolate the objective to prevent movement of enemy reinforcements into the area.

Execution

The leader's plan must be detailed and complete. All of the considerations outlined in chapter 11 must be covered. Patrol formations must provide for ease of control and all-around security while moving to and from the objective area and provide for rapid and coordinated deployment of the various units once the objective area is reached. The leader's plan usually includes the encirclement of the hostile position—either physically or by fire—in order to isolate it during the assault.

The final simultaneous assault against the objective develops when enemy defensive fires at the objective are suppressed by either friendly fire superiority or surprise. The assault is covered by the fire of the unit assigned the function of support by fire.

Grenades, SMAWs, and demolitions are most effective for clearing bunkers.

Security units are posted to isolate the objective. The patrol leader signals them when the withdrawal begins. As a minimum, security is on each flank and to the rear (at the objective rally point).

Actions at the Objective

The patrol leader halts the patrol near the objective at the final preparation position. Security is established and the leader's reconnaissance is made with appropriate subordinate leaders. When the leaders return to the patrol, they confirm previous plans or announce any changes. Movements are arranged so all units reach their positions simultaneously. This improves the patrol's capability for decisive action, if prematurely detected by the enemy.

The teams of the security element move to positions to secure the objective rally point, give early warning of enemy approach, block avenues of approach into—and prevent enemy escape from—the objective area. As the assault element moves into position, the security element informs the patrol leader of all enemy activity, firing only if detected or on the patrol leader's order. Once the assault element commences action, the security element prevents enemy entry into or escape from the objective area. The security element covers the withdrawal of the assault element (and support element, if employed) to the objective rally point, withdrawing only upon order or prearranged signal.

As the assault element approaches the objective, it deploys early enough to permit immediate assault if detected by the enemy. Each team uses stealth while moving into proper position. On command, or if one or more of the assault element is detected and fired upon by the enemy, the support element opens fire to neutralize the objective, then ceases or shifts fire according to prearranged plans and signals. As supporting fires cease or shift, the assault element assaults the objective. Demolition, search, and other teams are protected by the assault element while they work. On order, the assault element withdraws to the objective rally point.

If a support element is employed, its leader deploys teams to provide fire support for the assault element. Each member of the support element must know the scheme of maneuver to be used by the assault element, specific targets or areas to be neutralized by fire, and the signals that will be employed to commence, shift, and cease fires. The support element withdraws on order of the patrol leader. At the objective rally point, the patrol leader quickly reorganizes the patrol and begins the return movement to friendly lines.

13004. CONTACT PATROLS

General

Contact patrols establish and/or maintain contact to the front, flanks or rear by—

- Establishing contact with an enemy force when the definite location of the force is unknown.
- Maintaining contact with enemy forces through direct and/or indirect fires, or observation.
- Avoiding decisive engagement with the enemy.

Task Organization and Equipment

Task organization and equipment depend on the known enemy situation and anticipated enemy contact. A patrol sent out to establish contact with an enemy force is organized, armed, and equipped to overcome resistance of light screening forces in order to gain contact with the main enemy force. It is not organized and equipped to engage the main enemy forces in combat. Communication is paramount; radios must be reliable over the entire distance covered.

Actions at the Objective

The patrol leader selects a series of objectives. Once an objective is reached, the patrol leader initiates a planned set of actions in order to establish and maintain contact with the enemy. These plans and actions are guided by the missions to establish or maintain contact—not to engage in decisive combat. Contact with the enemy is maintained for the purposes of surveillance, applying pressure, and preventing seizure of the initiative. If the contact patrol becomes decisively engaged with the enemy, many of the tasks originally assigned to the patrol cannot be accomplished, since the enemy has seized the initiative and friendly forces are not forced to react.

13005. AMBUSH PATROLS

General

An ambush is a surprise attack from a concealed position upon a moving or temporarily halted target. It is one of the oldest and most effective types of military actions. Ambush patrols conduct ambushes of enemy patrols, resupply columns, and convoys. The intent of an ambush is to place the enemy in a dilemma where staying in the kill zone or attempting to move out of it prove equally lethal. The ambush may include an assault to close with and decisively engage the enemy, or the attack may be by fire only.

Purpose of Ambushes

Ambushes are executed for the general purpose of reducing the enemy's over-all combat effectiveness and for the specific purpose of destroying its units. The cumulative effect of many small ambushes on enemy units lowers enemy troop morale and harasses the enemy force as a whole.

Destruction is the primary purpose of an ambush because loss of men killed or captured, and loss of equipment and supplies destroyed or captured, reduces the overall combat effectiveness of the enemy.

Harassment, though less apparent than physical damage, is a secondary purpose of ambushes. Frequent ambushes force the enemy to divert men from other missions to guard convoys, troop movements, and carrying parties. When enemy patrols fail to accomplish their mission because they are ambushed, the enemy is deprived of valuable information. A series of successful ambushes causes the enemy to be less aggressive and more defensive minded. His men become apprehensive, overly cautious, reluctant to go on patrols, seek to avoid night operations, are more subject to confusion and panic if ambushed, and in general, decline in effectiveness.

Classification of Ambushes

A *deliberate ambush* is one in which prior information about the enemy permits detailed planning before the patrol departs for the ambush site. Information needed to plan a deliberate ambush includes the size, composition, and organization of the force to be ambushed; how the force operates; and the time it will pass certain points or areas. A deliberate ambush may be planned for such targets as—

- Any force if sufficient prior information is known.
- Enemy patrols that establish patterns by frequent use of the same routes or habitually depart and re-enter their own areas at the same point.
- Logistic columns.
- Troop movements.

An *ambush of opportunity* is conducted when available information does not permit detailed planning before the patrol departs. This is the type of ambush that an infantry unit normally conducts. An ambush of opportunity should not be confused with a hasty ambush. An ambush of opportunity is a planned ambush; a hasty ambush is an immediate action. In planning for an ambush of opportunity, the patrol must be prepared to execute any of several courses of action based on the types of targets that may be ambushed and must rehearse prior to departure. The course of action taken is determined when the opportunity for ambush arises.

The patrol leader may be directed to reconnoiter an area for a suitable ambush site, set up at the site selected, and execute an ambush against the first profitable target that appears.

The patrol may depart just after dark, move to a specific point, observe until a designated time, ambush the first profitable target after that time, and return before daylight.

A *hasty ambush* is an immediate action where the patrol makes visual contact with an enemy force and has time to establish an ambush without being detected. The actions for a hasty ambush must be well rehearsed and accomplished through the use of hand and arm signals given from the patrol leader.

Types of Ambushes

There are two types of ambushes: point and area. The *point ambush* is one where forces are deployed to attack along a single killing zone. The *area ambush* is one where forces are deployed as multiple related point ambushes.

A point ambush, whether independent or part of an area ambush, is positioned along the enemy's expected route of approach. Formation of the forces conducting the ambush is important because, to a great extent, the formation determines whether a point ambush is able to deliver the heavy volume of highly concentrated fire necessary to isolate, trap, and destroy the enemy.

The ambush formation to be used is determined by careful consideration of possible formations and the advantages and disadvantages of each in relation to—

- Terrain.
- Visibility.
- Forces.
- Weapons and equipment.
- Ease or difficulty of control.
- Target to be attacked.
- Combat situation.

For a detailed discussion of ambush formation, see appendix D.

Ambush Operation Terms

The *ambush site* is the location where an ambush is established.

The *killing zone* is that portion of an ambush site where fires are concentrated to trap, isolate, and destroy the target. On little-traveled roads, an obstacle placed in a defile, in the woods, on a bridge or on a steep upgrade can be used effectively to force vehicles to halt, and thus render the occupants vulnerable to attack. Antitank mines may be emplaced and the occupants of the wrecked vehicle killed or captured while still dazed by the explosion.

A *near ambush* is a point ambush where the attacking force is located within reasonable assaulting distance of the killing zone (50 meters is a guide). A near ambush is most often conducted in close terrain, such as a jungle or heavy woods.

A *far ambush* is a point ambush where the attack force is located beyond reasonable assaulting distance of the killing zone (beyond 50 meters is a guide). A far ambush may be more appropriate in open terrain offering good fields of fire or when the target will be attacked by fire only.

Factors for a Successful Ambush Patrol

There are many factors that give the ambush its best chance of success. The ideal situation would be to position the ambush on favorable terrain and have detailed planning completed beforehand.

The patrol leader of an ambush looks for the most favorable terrain in which the enemy is canalized between two obstacles with limited opportunity to attack or escape. Suitable areas include defiles, small clearings, bends in trails, and steep grades. Dense undergrowth adjacent to the ambush site permits observation from concealed positions. The ambush patrol should have maximum cover and concealment, not only for the firing positions, but for the routes of withdrawal. The enemy should be in an area offering as little protection from fire as possible. Favorable fields of fire include stretches of road, trail or open ground of at least 100 meters for machine guns and 15 meters for rifle fire and grenades. The ambush site can be improved by constructing obstacles—such as felled trees, wire, land mines, or booby traps—to impede the enemy.

Planning

A deliberate ambush or an ambush of opportunity requires thorough planning.

A deliberate ambush plan is based on extensive knowledge of the enemy and terrain, and is planned and rehearsed in great detail. A physical reconnaissance of the ambush site is made during the preparation phase and information gained is incorporated into the plan. All likely immediate actions of the enemy when ambushed are examined. Planned counteractions are developed and rehearsed.

In planning an ambush of opportunity, any available information on the enemy and terrain is used. A tentative plan for the ambush that incorporates all anticipated actions is developed and rehearsed. However, the bulk of planning is done concurrently during the patrol leader's reconnaissance of the prospective ambush site. In a rapidly developing situation, hasty ambush immediate action is employed.

The route and ambush site considerations apply to both deliberate ambushes and ambushes of opportunity. A primary route that allows the patrol to enter the ambush site from the rear is planned. Entering the prospective killing zone is avoided. If the killing zone must be entered to place mines or explosives, care is taken to remove any tracks and signs that might alert the enemy and compromise the ambush. If mines or explosives are to be placed on the far side of the ambush site, or if the appearance of the site from the enemy's viewpoint is to be checked, a wide detour is made around the killing zone. Care is taken to remove any tracks that might reveal the ambush. An alternate route from the ambush site to the objective rally point, as in other patrols, is planned.

Maps and aerial photographs are used to carefully analyze the terrain. When possible, an on-the-ground reconnaissance of the ambush site is made prior to occupation. Obvious ambush sites are avoided as the element of surprise is even more difficult to achieve in these areas. An ambush site must provide for—

- Favorable fields of fire.
- Occupation and preparation of concealed positions.
- Canalization of the target into the killing zone. (An ideal killing zone restricts the enemy on all sides, confining him to an area where he can be quickly and completely destroyed. Natural obstacles, such as cliffs, streams, embankments, or steep grades, are used whenever possible to force vehicles to slow down. Man-made obstacles, such as barbed wire, mines, and craters in the roads, are used to supplement natural obstacles.)
- Covered routes of withdrawal that enable the ambush force to break contact.
- Avoidance of enemy pursuit by fire.

Occupation of Ambush Site

The surrounding area is searched for enemy patrols prior to occupation of the ambush site. Ambush formations are used to physically deploy the patrol in a manner to inflict maximum destruction upon the enemy and to provide maximum security to the patrol. Ambush formations are contained in appendix D.

Positions

The patrol is moved into the ambush site from the objective rally point. Security is positioned first to prevent surprise while the ambush is being established. Automatic weapons are then positioned so each can fire along the entire killing zone. If this is not possible, overlapping sectors of fire are provided to cover the entire killing zone. The patrol leader then selects his position, located so he can tell when to initiate the ambush. Riflemen and grenadiers are positioned and sectors of fire are assigned to cover any dead space left by the automatic weapons. The patrol leader sets a time by which positions are to be prepared. Patrol members clear fields of fire and prepare positions in that order, with attention to camouflage for both.

Suitable Objective Rally Point

An easily located objective rally point is selected and made known to all patrol members. The objective rally point is located far enough from the ambush site so that it will not be overrun if the enemy assaults the ambush. Routes of withdrawal to the objective rally point are reconnoitered. Situation permitting, each man walks the route he is to use and picks out checkpoints. When the ambush is to be executed at night, each man must be able to follow his route in the dark. After the ambush has been executed, and the search of the killing zone completed, the patrol is withdrawn quickly but quietly, on signal, to the objective rally point where it reorganizes for the return march. If the ambush was not successful and the patrol is pursued, withdrawal may be by bounds. The last group may arm mines, previously placed along the withdrawal route, to further delay pursuit.

Local Security

Security must be maintained. Security elements do not usually participate in the initial attack, but protect the rear and flanks, and cover the withdrawal.

Patience

The Marines of the ambush force must control themselves so that the ambush is not compromised. Patience and self-discipline are exercised by remaining still and quiet while waiting for the target to appear, particularly if the patrol occupies the ambush site well ahead of the arrival of the enemy. Patience is necessary so as not to alert the enemy to the presence of the ambush.

Surprise

Surprise must be achieved, or the attack is not an ambush. If complete surprise cannot be achieved, it must be so nearly complete that the target is not aware of the ambush until too late for effective reaction. Surprise is achieved by careful planning, preparation, and execution so that targets are attacked when, where, and in a way for which they are least prepared.

Coordinated Fires

Properly timed and delivered fires contribute heavily to the achievement of surprise, as well as to destruction of the enemy. The lifting or shifting of fires must be equally precise; otherwise, the assault is delayed and the enemy has an opportunity to recover and react. All weapons, mines, and demolitions are positioned and all fires, including those of available artillery and mortars, are coordinated to achieve the following results:

- Isolation of the killing zone to prevent the enemy's escape or reinforcement.
- Surprise delivery of a large volume of highly concentrated fires into the killing zone.

Control

Close control of the patrol is maintained during movement to, occupation of, and withdrawal from the ambush site. This is best achieved through rehearsals and establishment and maintenance of good communications. When the enemy approaches, the temptation to open fire before the signal is given is resisted. The patrol leader must effectively control all elements of the ambush force. Control is most critical at the time the enemy approaches the killing zone. Control measures must provide for—

- Early warning of enemy approach.
- Fire control. Withhold fire until the enemy has moved into the killing zone, then open fire at the proper time.
- Initiation of appropriate action, if the ambush is prematurely detected. Individual patrol members must be prepared to react if detected by the enemy prior to the initiation of the ambush.
- Timely and orderly withdrawal of the ambush force from the ambush site and movement to the objective rally point.

It is important to remember that an ambush patrol should have four distinct signals: one to open fire (with an alternate signal to open fire to be used at the same time as the primary); a signal to cease fire or shift fire; a signal to assault or search the killing zone; and a signal to withdraw. The signal to open fire should meet two criteria: first, it should be the firing of a weapon that will kill the enemy; secondly, it should be a weapon reliable in any weather condition. A good primary signal is a Claymore mine, and an alternate signal would be a closed bolt weapon (M16A2). Open bolt weapons (M240G, M249) should not be relied upon to initiate an ambush.

Execution of an Ambush

The manner in which the patrol executes an ambush depends primarily on whether the ambush's purpose is harassment or destruction. To a lesser degree, the execution of the ambush is determined by whether the ambush is deliberate or an ambush of opportunity.

When the primary purpose is harassment, the patrol seals off the area with security teams to prevent enemy reinforcement and escape. Maximum damage is inflicted with demolitions and automatic weapons fire. The patrol delivers a very heavy volume of fire for a short time and withdraws quickly and quietly. The patrol avoids being seen by the enemy.

When the primary purpose of the patrol is destruction, the area is sealed off with security units. Maximum damage is inflicted with demolitions, antitank weapons, and automatic weapons fire from the support team or element. When these fires cease or shift, an assault is launched into the killing zone with heavy fire and violence to complete destruction. The assault unit provides security, while designated teams search and/or capture personnel and destroy vehicles and equipment. On the patrol leader's command, or by prearranged signal, all units withdraw to the objective rally point and move out quickly.

When the patrol's primary purpose is to obtain supplies or capture equipment, security units seal off the area. Demolitions and weapons are used to disable vehicles. The assault unit must use care to ensure its fire does not damage the desired supplies or equipment. Designated teams secure the desired items; other teams then destroy enemy vehicles and equipment.

The most successful ambush is one where the attacker is deployed and concealed in such a way that the enemy will unknowingly be surrounded by fire. The usual method is for the attackers to deploy themselves along a trail or route the enemy will travel. The enemy is permitted to pass by the center of the attacker's force so that the attack can be made from the front, flank, and/or rear. One or two men are posted well forward and to the rear along the route to prevent any enemy from escaping. All fires should be delivered simultaneously on a prearranged signal.

An effective method of luring the enemy is for an ambushing patrol to cut communication or electrical wire. The patrol then deploys and ambushes the line repair crew when it arrives. Since the line crew may be protected by riflemen, the attackers must be careful to engage the entire party.

Vehicles and foot personnel moving on well-established transportation routes can sometimes be captured by altering or moving directional signs so as to divert the enemy into an area where he can be more readily attacked. The attack can best be accomplished at an obstacle, such as a stream or gully, that forces the enemy to stop or slow down.

After the enemy has been ambushed and destroyed, the unit quickly withdraws over a prearranged route to the objective rally point. Speed is very important, since the noise of the ambush could alert other nearby enemy units.

13006. SECURITY PATROLS

General

Security patrols are assigned missions that may or may not require them to engage the enemy. They are used in proximity to defensive positions, on the flanks of advancing units or in rear areas. Purposes of security patrols are to detect infiltration by the enemy, destroy infiltrators, and protect against surprise and ambush.

In any situation where there is a threat of attack, such as a rear area threatened by guerrillas or a facility that is under threat of a terrorist attack, all Marines, not just the infantry, must know how to conduct a security patrol.

In just the offensive operations, infantry units provide security patrols to screen their flanks, areas, and routes. Whereas, in defensive operations, security patrols are used to prevent the enemy from infiltrating an area, detect and destroy infiltrators, and prevent surprise attacks. In rear areas, particularly when there is guerrilla or terrorist threat, the requirement to conduct security patrols increases for all Marine air-ground task force (MAGTF) units ashore, particularly aviation and combat service support units.

Task Organization and Equipment

Generally, a Marine rifle squad or similar sized organization is considered ideal for security patrols. Communications are important to higher headquarters so that they receive information from the patrol; and communications are important to the patrol to request fire support, etc. The radio the patrol carries must have the range necessary for higher headquarters to be able to receive transmissions from anywhere along the patrol route, and the patrol must have a secondary means of reporting (i.e., flare signals upon contact).

Patrol Procedures

All of the procedures presented in previous sections are to be used in security patrols.

Patrol Planning

Security patrol planning includes—

- Rehearsing prior to departing friendly lines.
- Maintaining communications.
- Support by organic weapons.
- Reinforced if necessary.
- Using varied routes and never establishing a routine pattern.
- Staying within proximity of friendly units.

Patrol Techniques

Within rear areas, an irregular pattern of patrol is established and changed daily. Outside of friendly lines it would be prudent to establish a definite preplanned route for the patrol, of which all adjacent units know the route. The parent unit commander establishes frequent checkpoints for control. If checkpoints are designated, the patrol leader treats them as individual objectives to be searched and cleared.

The patrol has a definite plan as to what to do if contact with enemy is made, how to break contact, how to defend itself, and how to call for supporting fires. It is imperative that patrol members know what to do if they become split or separated; i.e., location of rally points and how to be recovered.

13007. URBAN PATROLS

General

As national strategy continues to focus on regional vice global conflicts, the Marine Corps will continue to conduct urban operations in various operational environments. Cities and towns are often the center of economic and political power and are therefore extremely vulnerable to urban insurgent activities and violence. The lessons learned from recent operations in Somalia, as well as experiences gained by British forces in Northern Ireland, provide the foundations of urban patrolling.

Enemy Forces

Enemy forces in urbanized areas range from organized military forces to low intensity engagements with insurgents, such as terrorists or local gangs. For Marine Corps doctrine and tactics, techniques, and procedures (TTP) for dealing with the higher intensity threat in urbanized areas, refer to MCWP 3-35.3, *Military Operations on Urbanized Terrain.*

Common Threat Tactics

Cities provide cover and concealment for both friendly forces and enemy forces. However, enemy forces generally find active support only in certain areas of the town or city. The urban-based insurgent or terrorist usually lives in a friendly community or in one where the people are too frightened to withhold support or inform anyone about the situation. The insurgent or terrorist normally maintains close contact with leaders and others friendly to the cause. The enemy will often have efficient communication and intelligence systems, sometimes involving women and children to provide cover for its activities.

The urban insurgent normally cannot, like his rural counterpart, establish bases and recruit large military units. He is generally an individual or a member of a relatively small group. He relies on the cover afforded by the city's populace and terrorizing them to coerce loyalty or support.

Urbanized areas tend to give the insurgent and/or terrorist many opportunities to initiate action and gain advantage. The normal presence of large numbers of people in cities provides the foe an opportunity to mass crowds quickly and manipulate demonstrations easily. The presence of women and children during mass demonstrations may restrict the courses of action available to friendly forces. Major incidents stemming from overreaction or excessive use of force by friendly forces may provide the insurgent with propaganda material. Publicity is easily gained in an urbanized area because major incidents can't be completely concealed from the local population. Insurgent successes can be exploited to discredit the ability of host nation police, friendly forces, and the civil government, and gain recognition for the insurgents' cause.

The urban insurgent or terrorist can usually be expected to operate more boldly than his rural counterpart. This is reflected in the enemy's tactics. A single sniper or bomber may be the norm in the urbanized area, whereas the rural threat is generally the more conventional ambush. In urbanized areas, explosive devices can be easily emplaced and used effectively against large groups or select individuals. Enemy forces may be expected to employ the following tactics in urbanized areas:

- Using local communications, such as radio and newspapers, for propaganda purposes.
- Disrupting industry and public services through strikes and sabotage.
- Generating widespread disturbances designed to stretch the resources of the security force.
- Creating incidents or massing crowds in order to lure the patrol or reaction force into a trap.
- Provoking security forces in the hope that they may react improperly, therefore discrediting the security force by means of propaganda.
- Sniping at roadblocks, outposts, sentries, and patrols.
- Attacking friendly bases with rockets and mortars.
- Planting explosive devices, either against specific targets or indiscriminately, to cause confusion and destruction, and lower public morale and confidence.
- Using ambush patrols.
- Firing on friendly helicopters.

Principles of Urban Patrolling

Patrolling in an urban environment often presents conditions considerably different and often more complex than those encountered in rural and less inhabited areas. While the principles of patrolling are still relevant in an urban situation, the nature of urban patrolling has led to the development of six specific urban patrolling principles. They are—

1. **Depth.** The restrictive, canalizing nature of urbanized terrain usually limits a patrol's ability to disperse laterally. To prevent the patrol from bunching up, patrols normally maintain dispersion along the length of a patrol formation.

2. **Mutual Support.** The positioning of units in-depth within the patrol enables one unit to cover another unit's movement and facilitates immediate action during various situations. Aircraft, vehicles, and snipers also provide good mutual support.

3. **Deception and Pattern Avoidance.** Deception and pattern avoidance are normally a planning consideration of the headquarters directing the entire patrolling plan. By varying patrol routes, durations, and departure times, hostile actions commonly used against urban patrols—such as enemy ambushes, and roadblocks—are more difficult to plan and may be preempted.

4. **Intra-patrol Communication.** Elements of an urban patrol must have the means to communicate with each other. Ideally, each element will possess a radio enabling it to remain in continuous communication that facilitates rapid response and reporting to higher headquarters, reaction force coordination, and coordination of actions with other patrols or fire support agencies.

5. **Establishment of a Reaction Force.** The nature of urbanized terrain (its compartmentalization) makes urban patrols more vulnerable to a wider range of hostile actions. The requirement for immediate, coordinated reinforcement of a patrol is best satisfied by employment of an established reaction force. The reaction force requires superior mobility (relative to the enemy's) and fire support to be effective for this task.

6. **Three-Dimensional Threat.** Patrolling in an urban environment requires constant attention to its three-dimensional aspect; hostile actions can originate from rooftops, streets, subsurface levels or combinations of all levels at once.

Classification of Urban Patrols

Mission

The vast majority of urban patrols are overt in nature, with their presence readily apparent to the local populace. Most urban patrols are combat vice reconnaissance patrols. The vast majority of urban patrols are security patrols. However, units may be assigned secondary tasks of reconnoitering specific or general areas along the patrol routes.

Raids normally involve a swift penetration of an objective to secure information, confuse the enemy or destroy installations. Raids include a planned withdrawal upon completion of the assigned mission.

Movement

Means of movement are as follows:

- Dismounted: movement on foot.
- Vehicular: movement by motorized, mechanized or armored vehicle.
- Helicopterborne: movement by helicopter, however, helicopterborne patrols will usually involve dismounted or vehicular movement after patrol insertion.
- Combination: movement using a combination of methods.

Task-Organization

Units task-organize to meet the specific requirements of the mission and situation. The Marine rifle squad is ideally suited for urban patrolling and can easily integrate attached specialists required by specific missions. Specialists who may accompany urban patrols include—

- Interrogator-translator team (ITT) and counterintelligence team (CIT) Marines.
- Explosive ordnance disposal (EOD) personnel.
- Members of host nation or allied military forces.
- Interpreters.
- Local community leaders.
- Local law enforcement officers.
- Public affairs personnel escorting media representatives.

Dismounted Patrol Organization

Patrols should maintain unit integrity (fire team and/or squad) in organizing elements for an urban patrol. Similar to reconnaissance patrols, the vulnerability of urban patrols necessitates that all elements must provide for their own security in addition to the combat aspect of the mission. Combined assault and security teams are an effective method to organize for all-around security.

Urban Patrol Base Operations

Urban patrols may operate from an established patrol base that may be located within the unit's assigned area of operations or an area designated for a patrol. The patrol base should be located in a building used exclusively for this purpose. Patrol bases may also be located within a larger site that houses other agencies, such as a higher headquarters' command echelon.

If located within a larger site, the patrol base will be included in the overall facility security plan. If isolated, the patrol base must consider the following security factors:

- External security:
  - Barrier plan.
  - Sentry posts.
  - Local security patrols.
- Internal security:
  - Covered positions for all Marines.
  - Contingency plan for hostile actions against patrol base.

The Reaction Force

The high probability of an urban patrol becoming involved in a hostile or volatile incident requires the establishment of a dedicated reaction force for rapid reinforcement, support or extraction of the patrol. Ideally, the reaction force is—

- Large enough and task-organized in a manner that it can meet and quickly defeat the expected threat. Reaction forces are normally tiered with a lead element (normally one-third of the unit's size) and a main body force (the other remaining two-thirds of the force). Reaction force response times routinely are determined in advance by higher headquarters.
- Ready to respond immediately.
- Motorized or mechanized and supported by close air support and other fire support.
- Familiar with the area of operations.
- Briefed on the patrol's plans and monitors the status of patrols in progress.
- Task-organized to be multimission capable.
- Able to communicate with the higher headquarters, fire support assets, patrol base, and the patrol.
- Controlled by higher headquarters, once employed.

Patrol Preparation

Planning

Higher headquarters will—

- Designate the area for patrol.
- Provide intelligence briefs and updates.
- Ensure liaison with allied forces and the civilian populace.
- Provide special equipment and personnel required for the mission (scout snipers, public affairs officer, interpreters, etc.)
- Provide urban maps, photos, terrain models as required.
- Consider deception and pattern avoidance when issuing mission.
- Prescribe rules of engagement (ROE).

Intelligence Brief

An intelligence brief is conducted by the S-2 officer or representative prior to a patrol conducting its mission. The brief adresses the situation relevant to the specific patrol (e.g., routes, areas, updated enemy situations).

Coordination

Higher headquarters will effect liaison with adjacent and allied forces, as well as civil authorities and other agencies, having a possible effect on the patrol. The patrolling unit generally follows the same procedures as those used during patrol planning and execution in a jungle or forest environment.

The Urban Patrol Order or Warning Order

The Urban Patrol or Warning Orders use the same format and considerations as noted in this publication for patrolling. They rely heavily on a detailed terrain model, photographs, and subterranean construction to ensure complete understanding of the plan.

Rehearsals

The limited size of the patrol base usually precludes the need for full-scale rehearsals. Immediate action drills, such as crossing danger areas, are rehearsed in as much detail as possible, despite the limited available space.

Inspections

Initial and final inspections are conducted in the same manner as other patrols. Attached personnel must be fully integrated into the patrol and familiar with the plan and unit standing operating procedures (SOPs).

Conducting an Urban Patrol

Movement

Individual and unit movement considerations are generally the same as those for other patrols. However, urban environments require consideration of additional factors. Because of these factors, an urban patrol leader should—

- Ensure that each movement within a patrol takes place under the observation or cover of another individual or element of the patrol.
- Know where cover can be taken in the event of a hostile incident or action.
- Be prepared for contact with civilians, especially children, during the patrol and be aware that they may intentionally attempt to distract patrol members.
- Expect the presence of vehicles (both moving and stationary) along the patrol route.
- Expect members of the patrol to be approached by dogs and what action to take if threatened.

Patrol Formations

Squad-sized Patrols. The need for immediate fire power outweighs the dangers of becoming canalized. In contrast to other types of patrols, the headquarters element of an urban patrol will normally locate at the lead of the patrol column. This allows the patrol leader greater flexibility through control of two combined assault and security (A&S) teams. The leaders of these elements tactically stagger their members on each side of the road (see fig. 13-1 on page 13-12).

A&S teams follow in trace of the headquarters element and maintain unit integrity on separate sides of the street. One unit will remain slightly to the rear to create a staggered interval between Marines on either side of the street. This allows A&S teams to take lateral routes in support of headquarters element without having to cross a street to do so.

Platoon-sized Patrols. Squads will generally travel abreast of each other, moving along parallel routes. The interval between squad-sized units and/or teams is situation-dependent, but is usually between 100 and 150 meters (roughly two city blocks; this often prevents visual contact between the units). The intent is to create less of a target to an aggressor, yet still allow the patrol to quickly react to an incident. Individuals within units or teams will move in a staggered column as in a squad-sized patrol. See figure 13-2, which is located on page 13-13).

Night

Night patrols will generally be at least squad-sized and will generally use the same formation as that for day patrols. At night, it may be necessary to close distances between individuals or elements to maintain control. Consideration should be given to the use of night vision devices and thermal weapons sights.

Navigation, Control, and Security Measures

The patrol leader is ultimately responsible for the navigation. The headquarters element normally functions as the base unit during movement. The designated navigator is normally assigned from within the headquarters element. City maps are often inaccurate or outdated; however, when used with aerial photographs and other navigational aids, they can be effective for urban navigation.

Checkpoints and phase lines should be related to major streets (alleys, buildings, bridges) for easy identification. Arrival at checkpoints and crossing of phase

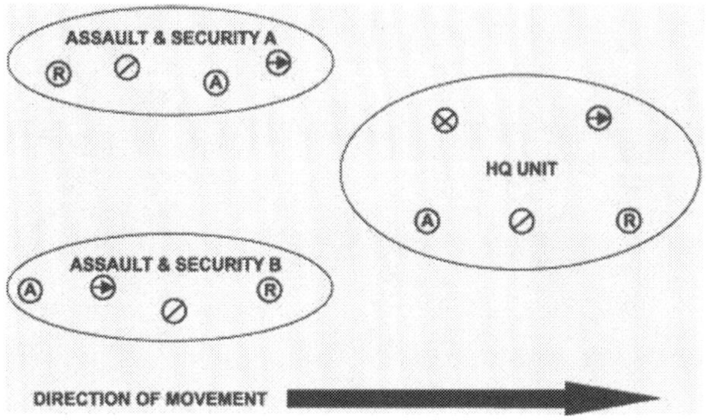

Figure 13-1. Squad-sized Dismounted Urban Patrol Formation.

lines should be relayed to higher headquarters using established brevity or codewords. A detailed patrol overlay is issued to both the reaction force and higher headquarters to keep them advised of routes and control measures used. A terrain model should be maintained at higher headquarters to aid in supervision and control of the patrol.

The use of camouflage should be limited to avoid frightening and confusing the local populace. The often overt nature of urban patrols may negate the need for camouflage. Patrols are sometimes deployed to show force presence and usually move on the urban street in plain view.

During daylight, patrols will routinely vary their rate of movement ranging from short halts to brief periods of double-timing. The British term for this urban patrolling technique is *hard-targeting*, meaning it makes the patrol harder for an enemy to target. Altering the rate of movement is intended to frustrate the enemy's ability to coordinate an attack or ambush against a targeted patrol.

Patrols should use short security halts, with Marines taking up mutually supporting firing positions. Marines must always work in pairs, ensuring mutual support. The last Marine in the element will provide rear security, but stays in his buddy's sight.

Individual Tasks

Individuals may be assigned collateral tasks performed throughout the patrol that may increase the patrol's situational awareness. Individual tasks may include—

- Vehicle spotter: looks for suspicious or known insurgent vehicles.
- Personnel spotter: observes and attempts to recognize previously identified enemy in crowds.
- Talker: attempts to gain information from casual conversations with the local populace. (Talkers are usually subordinate leaders or Marines with foreign language skills.)
- Searcher: conducts physical searches of vehicles and personnel while other patrol members provide cover and security.
- Marksman: engages point targets when the tactical situation does not permit massed or high volume fire. Other patrol members provide security to cover the marksman's engagement.

Departure of Friendly Lines

Urban patrols must vary their departure times to prevent being ambushed while exiting the patrol base.

Individual elements will usually depart exits at staggered times and at different movement rates, especially where sentries cannot provide cover. (This technique is used when the environment adjacent to the patrol base is dangerous.)

Once an element has exited friendly lines, a short halt is conducted in a predetermined, covered initial rally point 50 to 100 meters from the base. This ensures all elements are in position before the patrol continues.

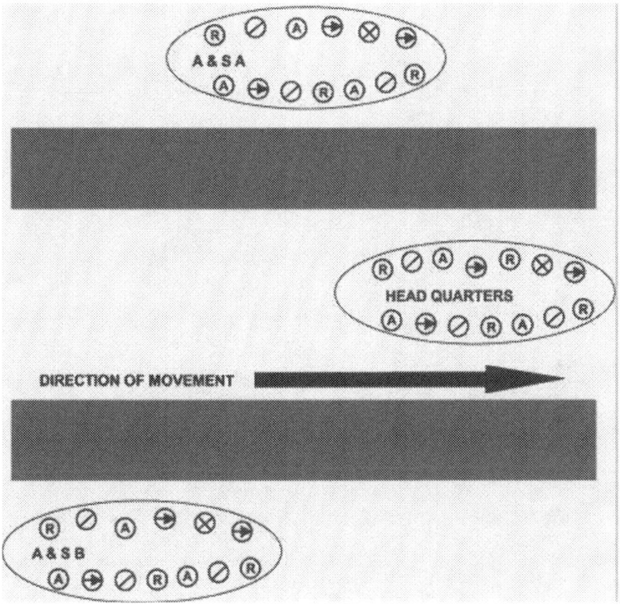

Figure 13-2. Platoon-sized Dismounted Urban Patrol Formations.

Exit points, routes from the base, departure techniques, and locations of IRPs should be varied constantly to avoid setting a pattern. This pattern avoidance may also include using vehicles to insert patrols away from the urban patrol base and employing empty vehicles as part of a deception plan.

Danger Areas

Urban patrols may encounter hundreds of danger areas during a single patrol. The three-dimensional threat requires keen situation awareness by every patrol member. Many danger areas can be dealt with simply by avoidance, while others require an adjustment of patrol formation, movement rate, etc. In the urban environment, places to be treated as danger areas are points that pose a major threat to the patrol, such as local political and religious headquarters, weapons containment areas, roads and routes that canalize movement and direct fire, and any area with a history of repeated contact.

Near and far side rally points are designated and briefed during the issuance of the patrol order. Squad-sized formations may use the A&S teams to provide flank security for the headquarters element and for each other. The headquarters element identifies the danger area and takes up a position on the near side of intersection. Individuals provide all-around security (see fig. 13-3 on page 13-14).

Two Marines (one from each A&S team) are designated to move through the headquarters element and establish respective firing positions on the near side of the danger area covering the patrol's near side flanks. They are followed by a second pair (again, one Marine from each A&S team) that moves across to the far side of the danger area and establishes respective firing positions covering the patrol's far side flanks (see fig. 13-4 on page 13-14).

Once near and far side flank security is established, the headquarters element moves across to the far side of danger area (see fig. 13-5 on page 13-15).

The remaining A&S team members then cross the danger area and join the headquarters element on the far side (see fig. 13-6 on page 13-15).

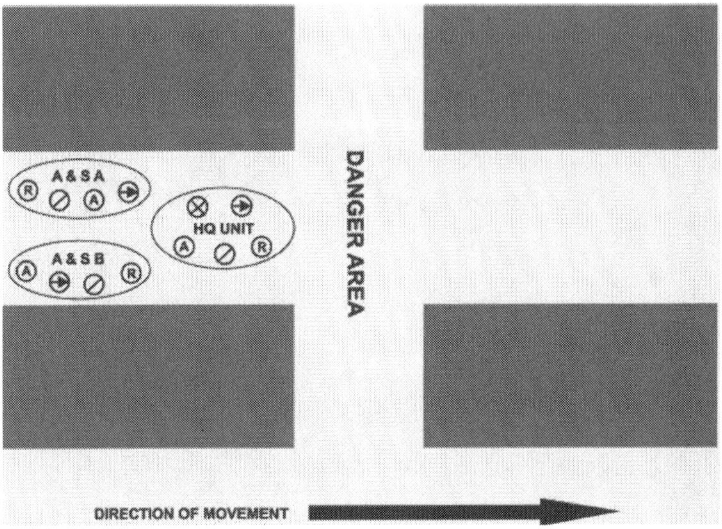

Figure 13-3. Approaching Danger Area.

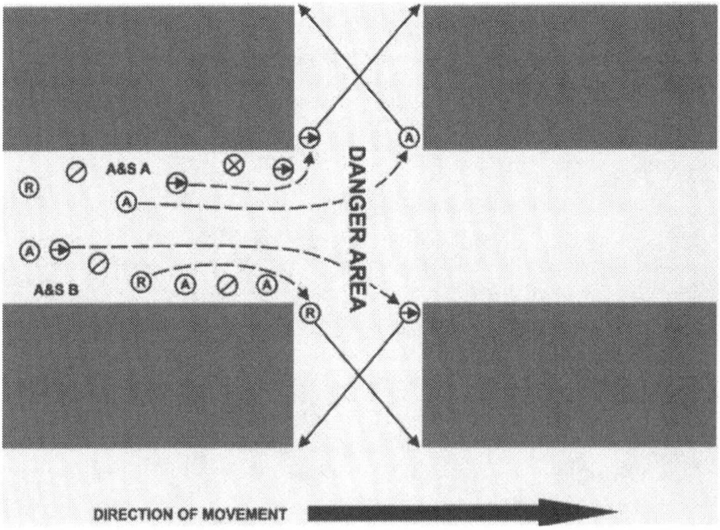

Figure 13-4. Securing Flanks.

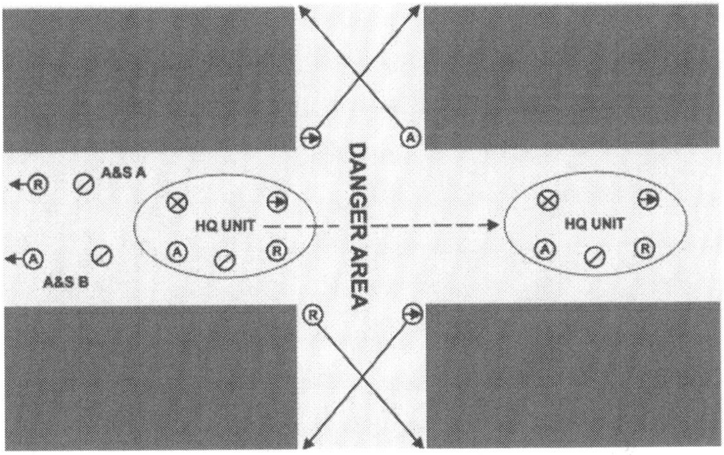

Figure 13-5. HQ Element Moves Through.

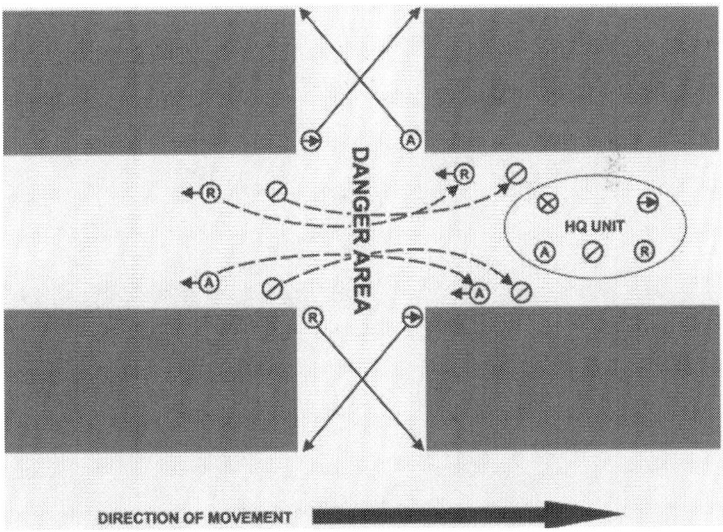

Figure 13-6. A&S Elements Move Through.

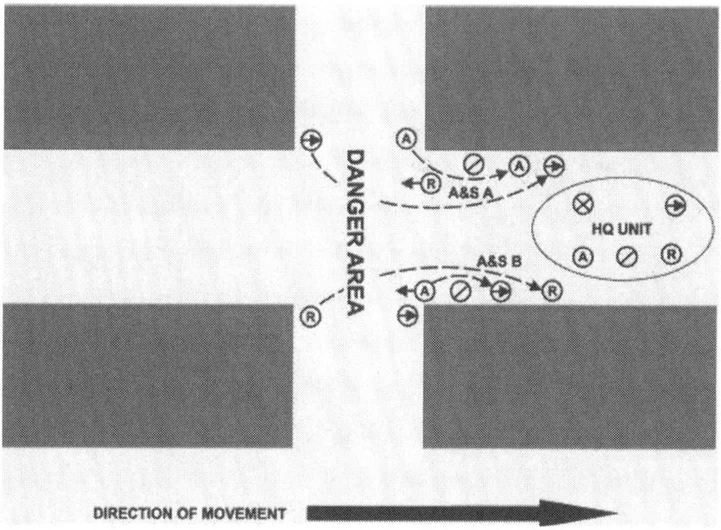

Figure 13-7. Patrol Resumes Movement Along Intended Route.

Once the trail A&S team members cross the danger area, they take up rear security and cover the movements of the near and far side flank security teams as they return to their positions in the patrol formation. The near side security team should collapse back first, followed by their far side counterparts. The patrol then resumes its advance away from the danger area (see fig. 13-7).

Interaction with Local Populace

Urban patrols must interact with the local populace. Patrols are at first a novelty to the civilians but can quickly become an unwelcome intrusion. The movement of the patrol must be fast enough to prevent the enemy from massing their fires upon it, but deliberate enough to ensure adequate security and mutual support. Patrol members must realize that they are usually the only Marines the local populace will encounter and that an inappropriate gesture, comment or act could lead to the deterioration of rapport between U.S. forces and the general population. Marines must remember that the vast majority of the individuals with whom they come in contact will be noncombatants attempting to survive in trying political, economic, and social situations.

Hostile incidents often seem to occur spontaneously, but there are usually indications that can alert Marines to imminent danger. The most obvious are the sudden alteration of normal routines, patterns, and attitudes of the local populace or other unusual activity. Some examples include—

- Observers on rooftops, in windows, etc., who are obviously tracking the patrol.
- The unusual absence of pedestrian traffic and people on porches.
- Stores, markets or street vendors closed suddenly or without explanation.
- Changes in civilian attitude toward patrol members.
- Unknown individuals or vehicles in the patrol area.
- Unfamiliar vehicles parked in the patrol area (possible car bomb).
- Roadblocks.
- Children throwing rocks at patrols to possibly draw the patrol's attention away from a more serious danger, such as a deliberate ambush.
- Vehicles riding unusually low due to overloading (possibly ferrying people, weapons, explosives).
- Agitators trying to provoke an incident with patrol members.

- Absence of the usual stray dogs (dogs are adept at sensing danger and avoiding it).
- Anti-American graffiti suddenly appearing in the patrol area.
- Pictures of enemy leaders and martyrs posted in the patrol area.
- Civilian workers failing to appear at U.S. or friendly bases.
- Normal deliveries and pick-ups conducted late or early without reason.
- Sudden change of civilian sentiment in newspaper articles, radio broadcasts or other media.
- Women and children leaving to live elsewhere.

Immediate Actions Upon Enemy Contact

Reaction to Sniping

Snipings are often executed from a single firing point, but coordinated snipings delievered from multiple points are not uncommon.

Patrol element or team leaders should constantly try to identify likely firing points and anticipate their own reactions to a possible shooting. Normally, snipers in an urban environment have a detailed withdrawal plan. Once a patrol comes into contact with a sniper, the patrol leader must immediately assess the situation and maneuver his patrol accordingly. The patrol's mission, location, size, ROE, and location of the threat often determine whether the patrol will attempt to neutralize the targeted sniper. If the patrol leader decides to kill or capture the sniper, he uses planned and rehearsed immediate actions to maneuver and counter the sniper's assault. The goal is to kill the sniper or cut off his escape and capture him. There are three immediate reactions to neutralizing a sniper: initial contact, immediate follow-up, and subsequent follow-up.

Initial Contact. The initial contact is made when the sniper fires the first shot. The patrol must react immediately and positively to get behind the firing position in order to kill or capture the gunman. The period of contact ends when the gunman is killed or captured, or the patrol element or team leader on the scene ends it. The following technique is the same for both squad- and platoon-sized patrols:

- The element or team in contact attempts to identify the firing position and maneuvers designated marksman into position to return well-aimed and controlled fire. Other members of the patrol take up positions to cover the marksman's engagement.
- The patrol element or team leader in contact sends initial contact report to the patrol leader, who notifies higher headquarters.
- The element or team leader in contact determines appropriate cut-off positions and relays them to flanking elements or teams.
- The patrol element or team leader in contact continues to observe the firing point, but does not enter it due to the possibility of booby traps. Flank elements or teams set up along likely escape routes.
- The incident ends when either the sniper ceases fire or is neutralized.

Immediate Follow-Up. Regardless of the fate of the gunman, isolation of the firing point is necessary to prevent reinforcement and preserve forensic evidence (scent, spent casings, etc.). If not under fire, members of the patrol element or team cordon off the area surrounding the firing point. Flanking elements or teams maintain their positions and prevent civilians from entering the area. The patrol leader moves to link-up with the element or team in contact (if not his own), and makes an estimate of the situation. The patrol leader sends a SPOTREP to higher headquarters.

Subsequent Follow-Up. The aim of the subsequent follow-up is to use follow-on forces to clear the building of remaining resistance or to obtain evidence that can be used to capture the gunman. The patrol leader establishes a position where he can brief arriving units (reaction force commander, S-2 representative, EOD personnel, etc.). Once the arriving units have been briefed, recommendations are made to higher headquarters via radio. No one is allowed into the cordon without the patrol leader's approval.

Reaction to Becoming Decisively Engaged

If a patrol becomes decisively engaged from numerous firing positions, the following immediate action should be taken:

- All patrol members move to available cover and return accurate fire on identified firing points.
- The patrol leader assesses the situation and makes a decision to either request the reaction force or break contact.
- If the reaction force is requested, the patrol will maintain its position until the reaction force arrives. The patrol should use fire and maneuver to gain better tactical positioning and support the arrival of the reaction force. When the reaction force arrives,

its commander may decide to either clear occupied buildings or cover the patrol during its extract.

Reaction to Bomb Threat or Discovery

The use of command-detonated explosive devices is a common ambush tactic employed by a terrorist or insurgent in an urban environment. The appropriate response to a reported threat or an actual discovery is generally involves four steps (known as the four Cs)—

1. The patrol leader **CONFIRMS** the presence of the suspicious item.

2. Without touching or moving anything suspicious, patrol elements or teams **CLEAR** the immediate danger area to a minimum of 100 meters. The area is cleared from the suspected device outward, inform civilians as to the reason for evacuation.

3. A&S teams establish a **CORDON** to secure the cleared area. Avenues of approach are cordoned off to keep people out and to protect EOD or engineer personnel clearing the device. The assistant patrol leader acts as the cordon commander and informs the patrol leader when the cordon is secure. An effective cordon technique is to tape off the area with engineer tape, creating both a physical and psychological boundary.

4. **CONTROL** of the area is maintained throughout the bomb clearing operation by the patrol leader. The patrol leader sends a report to a superior concerning details of the device (if known) and the area affected. The patrol leader coordinates with arriving personnel (EOD, engineers, etc.). The patrol leader maintains communication with the assistant patrol leader and keeps the Marines informed of the progress of the clearing operation.

Reaction to a Bomb Detonation

Bombs may be used by an insurgent as a means of initiating an ambush on mounted or dismounted patrols, in which case the actions for decisive engagement apply. Immediate action in response to an isolated explosion is similar to that used in reaction to a sniping and breaks down into the same three phases:

Initial Contact. The patrol leader attempts to identify the likely initiation point and sends an initial contact report to higher headquarters. If the bomb was command-detonated, the patrol leader sends his A&S teams deep to cut off the bombers' escape routes. Any casualties are moved a minimum of 100 meters from explosion and out of the line of sight to it.

Immediate Follow-Up. The A&S teams may need to maneuver to positions behind the likely initiation point to cut off escape. Once in position, personnel checks are conducted and any suspects are detained. The patrol leader coordinates requests for required support (MEDEVAC, reaction force, etc.)

Subsequent Follow-Up. Due to the possibility of secondary detonations, the four Cs (confirm, clear, cordon, and control) can be conducted as in reaction to a bomb discovery or bomb threat.

Civil Disturbances

Urban patrols must prepare to react to spontaneous aggression by the local populace. In many cases, civil disturbances are organized by the enemy to draw dismounted patrols into a targeted area, or to distract them from enemy activity occurring elsewhere. Civil disturbances are generally divided into two categories: minor aggressive actions, and full-scale rioting.

Minor aggressive actions are activities characterized by rock-throwing or use of devices such as Molotov cocktails and may either be directed at the patrol or take place between different ethnic factions of the population. Minor aggressive actions are normally spontaneous in nature and may have minimal or limited objectives for the insurgents.

Full-scale rioting events are usually in response to another major event or incident that may enflame the populace. Full-scale riots are well-planned and orchestrated, with clear objectives or targets in mind. At times, patrols will need to attempt to maintain control of a civil disturbance situation; however, dismounted and mobile small unit patrols should generally avoid potential flashpoints. Procedures to handle civil disturbances are as follows:

- The patrol leader reports the incident to headquarters and attempts to diffuse the crowd by talking to crowd leaders.
- If the patrol leader determines the size of the disturbance is too large for the force to handle, the patrol should move away from the disturbance to a safer, more remote covered area and occupy positions to observe and report the situation to higher headquarters. To prevent the patrol from being pursued by the crowd, the patrol should move quickly and change direction, often at road junctions, to gain distance from it.

Patrol members should maintain dispersion to create a more difficult target. They should face the crowd at all times to see and avoid any projectiles thrown. Individual self-discipline must be maintained throughout the disturbance. Marines charging into the crowd or throwing objects back at the crowd will only worsen the situation. If pursued or trapped, the patrol leader may consider using riot control agents (combat support, pepper gas, etc.) to disperse or slow a crowd's movement. The use of riot control agents must be authorized under the established ROE. If the patrol leader believes the situation is deteriorating beyond the patrol's ability to control it, the patrol leader should request the reaction force, which may be better equipped to handle a large riot or mob.

Break Contact

As with patrols in rural areas, the patrol leader may be forced to break contact as a result of decisive engagement with the enemy. On the basis of his estimate of the situation, the patrol leader will normally break contact in one of the following ways:

- As a patrol, with elements providing cover for movement as defined by clock direction and distance.
- As individual units/teams taking separate routes out of the area, then linking up at a designated rally point a safe distance away from the engagement.

As in any contact with enemy forces, smoke may be employed to screen movement. Fire support agencies can be utilized to suppress targets; riot control agents can be employed to disrupt enemy movement.

Reentry of Friendly Lines

The reentry of a dismounted patrol into an urban patrol base is no different from that of a patrol conducted in a rural area. The same planning considerations and control methods apply.

Missions Related to Urban Patrolling

House Calls

House calls missions are usually part of a coordinated effort to collect information within the area of operations. They involve obtaining up-to-date information on particular houses and occupants. When possible, local police should accompany patrols to do the actual talking to the occupants. If this is not possible, a technique that may be used by an urban patrol is—

- A&S elements move to provide cover around the target house.
- The headquarters element provides security just outside the house.
- The patrol leader and one Marine for security enter the house, if invited. If not invited, they talk to the occupants on the doorstep. Patience and tact are required in requesting information. An interpreter should be present when language differences exist.

Vehicle Checkpoints

Urban terrorists or insurgents commonly use vehicles to transport personnel, weapons, explosives, and equipment. Civilian vehicles are often used for these purposes, creating the requirement to check as many vehicles as possible. While permanent, fortified checkpoints may be conducted along approaches into an urbanized area, dismounted patrols can be employed to establish hasty vehicle checkpoints to stop vehicles and to keep the enemy off-guard. The two common types of vehicle checkpoints patrols establish are hasty and deliberate.

Hasty checkpoints are deployed anywhere based upon the decision of the patrol leader. Patrols must not set patterns through the frequent use of the same sites.

Deliberate checkpoints are tasked by higher headquarters to achieve a specific purpose. Time and locations are carefully considered to avoid setting patterns.

The general layout for a squad-sized, two-way dismounted checkpoint is depicted in figure 13-8 on page 13-20. The technique is as follows:

- The A&S teams are positioned stealthily in blocking positions on both sides of the road.
- Both the patrol leader and assistant patrol leader act as "talkers" for each direction of traffic (with local police or interrogator or translator Marines acting as interpreters) while a two-man team from the headquarters element physically searches the vehicles.
- Obstacles or parked vehicles may be employed to create a staggered roadblock in center of the checkpoint to slow approaching vehicles.
- The checkpoint location should be sited so that approaching vehicles cannot see it until they have passed a security team, and they have no escape route then available.
- Signs announcing the checkpoint should be displayed a safe distance from the search area for safety to both drivers and Marines.
- Normally higher headquarters will issue criteria that determines which vehicles are searched, but random checks of cars should normally be made as

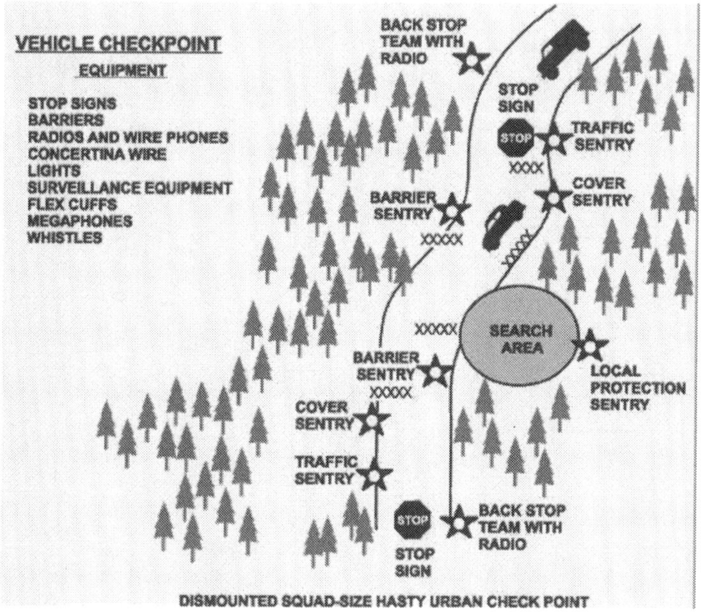

Figure 13-8. Dismounted Squad-sized Hasty Urban Vehicle Checkpoint.

well. While the vehicle is being searched the driver should accompany the searcher around the vehicle.

- Vehicle occupants should be made to exit the vehicle and then searched. Whenever possible, women should be searched by female police or female Marines.
- All patrol members should conduct themselves with courtesy and politeness. If nothing is found, an apology for the inconvenience is recommended.
- A hasty vehicle checkpoint should not be conducted any longer than 30 minutes for security reasons.
- The ROE should dictate whether or not action should be taken against vehicles that fail to stop at the checkpoint. Failure of a vehicle to stop does not automatically give authorization to fire.

Observation Posts

Urban observation posts are established to provide extended security, not only for patrol bases but also for patrols operating within the observation post's sector of observation. Observation posts can be established in conjunction with sniper operations and for providing overwatch for patrols operating within their sector of observation. Observation posts are nor-mally positioned on dominating terrain or in buildings outside the patrol base itself.

Insertion to the observation posts and conduct of observation may be either overt or clandestine in nature. Overt observation posts usually will be hardened positions to increase security. A patrol provides cover while the observation post is being inserted. A clandestine observation post relies on stealth of insertion and occupation for protection. It is normally positioned in abandoned buildings to cover sectors of observation that overt observation posts cannot. Because of their nature, clandestine observation posts are difficult to successfully establish and should not be manned for an extended period of time. Orders establishing observation posts (and patrols) must address the method of extraction as well as actions upon compromise/attack.

Cordon and Search

The cordon and search mission involves isolating a predesignated area by cordoning it off and systematically searching for enemy personnel, weapons, supplies, explosives or communications equipment. While large-scale cordon and search operations are planned and rehearsed in advance and normally entail extensive coordination with local law enforcement agencies, a squad-sized urban patrol may often conduct a cordon and search of a point target—searching one house or building identified by intelligence as a possible weapons cache.

The basic principle of a search of a populated area is to conduct it with limited inconvenience to the population. The populace may be inconvenienced to the point where they will discourage urban guerillas or insurgents from remaining in the area, but not to the point that they will assist the enemy as a result of the search.

Upon receiving intelligence that warrants the searching of a building or a specific tasking from higher headquarters, A&S elements of the patrol move to establish an inner cordon around the target building to seal it off, with the primary intent of preventing movement out of the targeted building.

On order, the designated reaction force deploys to establish an outer cordon, oriented outward some distance from the inner cordon and covering routes leading into the area in order to prevent outside interference/reinforcement. The reaction force maintains a reserve to reinforce either cordon or react to unfolding events (civil disturbance in response to the operation).

Once the cordons have been established, the patrol leader, with the assistance of local police or interpreters, informs the local populace that a building is about to be searched, that a house curfew is in effect (if permitted by higher headquarters), and that all occupants should remain indoors. Occupants of the target house are instructed to gather at a central location to stay out of the way of the search party.

The headquarters element, having linked up with any required assistance (explosive ordnance disposal (EOD), ITT, etc.) now acts as the search party and accompanies local police. A female searcher should be included in the party, if necessary.

Occupants are searched and screened first for possible enemy personnel. Apprehended persons are evacuated as soon as possible.

The head of the household should accompany the search party throughout the operation to be able to counter incriminating evidence and possible accusations of theft and looting against Marines. If possible, a prominent member of the local community should act as a witness.

Buildings are best searched from top to bottom. Ideally, the search is conducted with the assistance of combat engineers using mine detectors to locate hidden arms and ammunition.

If the targeted building is empty or the occupant refuses entry, it may be necessary to forcefully enter the premises to conduct a search of the dwelling. If an unoccupied house containing property is searched, arrangements should be made with the local community to secure it until its occupants return. Unnecessary force and damage to property should be avoided during the search.

Motorized Urban Patrols

The advantages of a motorized urban patrols is their ability to capitalize on the speed, mobility, and protection offered by various vehicles. They may be motorized, mechanized or armored vehicles or a combination. Generally, motorized urban patrols possess greater combat power than dismounted patrols and can cover larger areas faster than dismounted patrols.

The disadvantages of motorized urban patrols is that they are restricted to roads and are vulnerable to ambush by the enemy. They are also restricted in their ability to interact with the local populace.

Motorized patrols are generally organized in the same manner as dismounted patrols (see fig. 13-9). Unit integrity is maintained when assigning personnel to specific vehicles.

The urban patrolling principles apply to motorized patrols in much the same manner as dismounted patrols. Mutual support and depth are achieved by maintaining constant observation between vehicles and coordinating support with any dismounted patrols in the area. All-around security is achieved through the use of constant observation as well as the vehicle's mobility and firepower. Positive communications between units or teams are maintained through vehicle radios. Patrol routes and speeds are varied to promote deception or pattern avoidance.

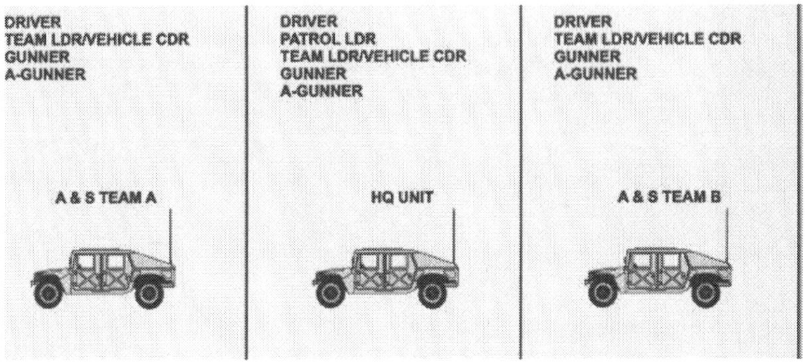

Figure 13-9. Organization of a Squad-sized Motorized Patrol.

The canalizing nature of streets and alleys force vehicular patrols to use a traveling overwatch movement technique to reduce vulnerability to ambushes. All vehicles travel at a moderate rate of speed with the lead vehicle stopping only to investigate potential danger areas. If vehicles must stop in danger areas, designated crew members' will dismount to provide security. The gunner will remain at the ready and in the turret while the driver remains in the driver's seat with the vehicle running.

Vehicles should move at a high rate of speed only when responding to an incident. At all other times, vehicle speed should be between 15 to 20 mph to allow for quick reaction and good observation. Distances between vehicles should be approximately 50 meters (one half to one city block) or such that visual contact and mutual support are ensured. Particular care is taken at major road junctions and other danger areas to ensure individual vehicles do not become isolated.

Vehicles with doors removed generally enhance observation and overall security, yet expose Marines to thrown objects, theft and concealment.

Chapter 14. Information and Reports

It is necessary that patrol leaders and all patrol members be trained in observing and reporting their observations accurately. The leader of a patrol should have all members of the patrol immediately signal or report any information obtained. These reports should not be restricted to information about the enemy, but should also include information about the terrain, such as newly discovered roads, trails, swamps, and streams. The leader includes all information in the report to the officer dispatching the patrol. (Refer to MCRP 2-15.3, *Reconnaissance Reports Guide,* for detailed report formats.)

14001. REPORTING

The officer dispatching the patrol instructs the patrol leader on whether and when messages are to be sent back during the patrol and what communication means to use. Messages may be oral or written. They must be accurate, clear, and complete. Every message should answer the question what, where, and when. For a detailed discussion on reporting, see paragraph 6002.

Verbal Messages

A patrol leader sending a verbal message should make it simple, brief, and avoid using numbers and names. The messenger should accurately repeat the message back to the patrol leader before leaving.

Written Messages

In preparing written messages, the patrol leader must distinguish between fact and opinion. Information about the enemy should include: strength; armament and equipment; actions; location and direction of movement; unit destination, if known; time enemy was observed; and the patrol's location when the observation was made. Use of an overlay or sketch may often simplify the message.

Messengers

A messenger team is given exact instructions as to where to deliver the message and the route to take. Any information obtained along the route should be reported at the time the message is delivered. Messengers must be given all practical assistance. If in danger of capture, the messenger immediately destroys the message.

Radio and Signals

If the patrol is provided with a radio, a definite radio schedule for checking in must be arranged before departure of the patrol. The patrol leader takes every precaution to ensure that codes and copies of messages are not captured by the enemy. If a close reconnaissance of enemy lines is required, the radio should be left in a concealed location at a safe distance from the enemy. Once a report is sent by radio, the patrol should immediately leave the area to avoid the possibility of detection by enemy locating devices. Pyrotechnics (flares, colored smoke, grenades) and air-ground panels may also be used by patrols for reporting information by a prearranged signal.

SALUTE Report

Information must be reported as quickly, accurately, and as completely as possible. An established method to remember how and what to report about the enemy is to use the acronym SALUTE:

Size
Activity
Location
Unit
Time
Equipment

An example of such a report is: "Seven enemy soldiers, unit unknown, traveling SW, crossed road junction on BLACK RIDGE at 211300 August carrying one machine gun and one rocket launcher."

14002. CAPTURED ITEMS

Every patrol should make a practice of searching enemy casualties, prisoners, and installations first for booby traps, then for equipment, papers, maps, messages, orders, diaries, and codes. Search techniques should be practiced often in order to reduce the time exposed to potentially dangerous situations. Items found are collected by the patrol leader and turned in with the patrol report. The items found are marked as to time and place of capture. When possible, captured items should be linked to a specific prisoner who possessed the items or to the place where the items were found. When this is done, the enemy prisoner of war tag and item tag are marked accordingly. The patrol leader must impress upon the members of the patrol the importance of turning in all documents and equipment. Furthermore, the patrol leader must ensure all information gathered by the patrol is quickly disseminated.

14003. PRISONERS

A patrol normally does not capture prisoners unless required by the mission. If prisoners are taken, the "5S and T" rule applies. This memory aid stands for: search, segregate, silence, speed, safeguard, and tag.

Search. Prisoners are body-searched thoroughly for weapons and documents as soon as they have been captured. This search must include the helmet, body armor, and gas mask. These items are left with the prisoner for protection until the patrol is completed. Weapons, equipment, and documents are tagged and immediately sent to the patrol leader.

Segregate. Prisoners are segregated into isolated groups: officers, noncommissioned officers (NCOs), privates, deserters, and civilians. By segregating prisoners, it makes it more difficult for leaders to organize escapes and issue orders to subordinates.

Silence. Silence is essential. Do not allow prisoners to talk to each other.

Speed. Speed is required in getting prisoners to the commander who dispatched the patrol. Timely information secured from prisoners is essential.

Safeguard. Prisoners are safeguarded as they are moved. They are restrained, but not abused. If the patrol will soon reach friendly positions, prisoners are not given cigarettes, food or water until they have been questioned by interrogators. If the patrol will not return to a friendly position for a lengthy time, the prisoners are given food, water and medical aid.

Tag. Equipment and personal effects recovered from prisoners are tagged so that they may later be matched back to specific individuals.

14004. PATROL REPORT

Every patrol leader makes a report when the patrol returns. Unless otherwise directed, the report is made to the person ordering the patrol. If the situation permits, the report is written and supported by overlays and/or sketches. The patrol leader's report should be a complete account of everything of military importance observed or encountered by the patrol while on the assigned mission. It should include the following information:

- Size and composition of patrol.
- Tasks and purpose (mission).
- Time of departure.
- Time of return.
- Routes, out and back (show by sketch, azimuth, trace on map).
- Terrain (general description to include any man-made or natural obstacles and critical terrain features which, if occupied by either enemy or friendly forces, would allow them to control the surrounding area).
- Enemy (size, activity, location, unit, time, equipment).
- Any map corrections (show on map).
- Miscellaneous information not covered elsewhere in report.
- Results of enemy encounters.
- Condition of patrol, including disposition of any dead or wounded.
- Conclusion and recommendations.

14005. PATROL CRITIQUE

After the patrol has rested and eaten, the patrol leader should hold a critique. Constructive criticism is made. It is an excellent time to prepare for future patrols by going over lessons learned as a result of the patrol.

Appendix A. Patrol Warning Order

The warning order is issued as soon as practical with all available information included to assist patrol members in preparation.

1. Situation. Friendly and enemy situation information necessary for initial preparation.

2. Mission. Statement of what the patrol is to accomplish, and the purpose for accomplishing it. When, how, and where will be discussed in the patrol order.

3. Execution

a. Task Organization. General patrol organization and assignment of responsibilities if known; otherwise, promulgate in the patrol order.

b. Tasks. Alert subordinate leaders to patrol tasks (e.g., stream crossing, helicopter rappelling, demolitions) requiring preparation prior to departure.

c. Coordinating Instructions

(1) Time schedule and location for individual preparation, rest, briefings, in-spection, rehearsal, and departure.

(2) Time, place, uniform, and equipment for receiving the patrol order.

(3) Tasks for subordinate leaders to direct and supervise the initial preparation which may include drawing ammunition, rations, and special equipment; conducting immediate action drills or other necessary individual or unit training; meeting and briefing attachment personnel; reconnoitering the area for passage of lines; and coordinating with the necessary unit leaders.

(4) Preliminary guidance to specialists and key individuals regarding their roles and organization within the patrol.

4. Administration and Logistics

a. Individual uniform, equipment, weapons, and prescribed load of rations, water, and ammunition.

b. Crew-served weapons (if required) and guidance regarding distribution of weapons and ammunition during movement.

c. Special equipment requirements (wire cutters, demolitions, radios, flashlights, infrared equipment, mines, binoculars) and their distribution during movement.

d. Restricted or prohibited items.

5. Command and Signal

a. Designation of assistant patrol leader and his role in preparation.

b. Designation of navigators and radio operators (if required).

c. Brief outline of patrol leader's schedule for preparation, and where he can be reached.

Appendix B. Patrol Order

A patrol order follows a warning order. Any of the following subjects that have been addressed in the warning order may be omitted with the exception of the mission statement. A patrol order is more detailed than a 5-paragraph squad order in that a great deal of attention is given to individual duties.

The patrol order also provides orientation information, both enemy and friendly, that impacts the patrol and enemy forces. This information can include, but is not limited to, weather, terrain, visibility, NBC considerations, local population situation, terrain model and/or map orientation, and behavior.

1. Situation

   a. Enemy Forces

      (1) Composition, disposition, and strength are based on size, activity, location, unit, time, and equipment (SALUTE).

      (2) Capabilities and limitations to defend, reinforce, attack, withdraw, and delay (DRAW-D).

      (3) Enemy's most probable course of action.

   b. Friendly Forces

      (1) Mission of next higher unit (task and commander's intent).

      (2) Adjacent unit missions (task and intent). Also identify left, front, right, and rear.

   c. Attachments and Detachments. (date and time effective).

2. Mission. The mission is the task to be accomplished, and its purpose (who, what, where, when, and why). For patrols, specify if the mission or time has priority.

3. Execution

   a. Commander's Intent and Concept of Operations

      (1) Commander's intent.

      (2) The concept of operations tells the where, how, and who and lays out the patrol leader's general scheme of maneuver and fire support plan. It outlines the following:

(a) Task organization of the patrol.

(b) Movement to the objective area, to include navigation method.

(c) Actions in the objective area.

(d) The return movement, to include navigation method.

(e) Use of supporting forces (including illumination, if required).

b. Subordinate Element Missions. Subordinate element missions (task and purpose) are assigned to elements, teams, and individuals, as required.

c. Coordinating Instructions. This paragraph contains instructions common to two or more elements, coordinating details, and control measures applicable to the patrol as a whole. At a minimum, it includes—

(1) Time of assembly in the assembly area.

(2) Time of inspections and rehearsals (if not already conducted).

(3) Time of departure and estimated time of return.

(4) Location of departure and reentry of friendly lines and the actions associated with departure and reentry.

(5) Details on the primary and alternate routes to and from the objective area.

(6) Details on formations and order of movement.

(7) Rally points and actions at rally points.

(8) Final preparation position and actions at this position.

(9) Objective rally point and actions at this point.

(10) Actions at danger areas.

(11) Actions in the event of enemy contact.

(12) Details on actions in the objective area not covered elsewhere.

(13) Estimated time of patrol debriefing upon return.

4. Administration and Logistics

a. Changes/additions to uniform, equipment, and prescribed loads from that given in the warning order.

b. Instructions for handling wounded and prisoners.

5. Command and Signal

a. Command Relationships. Identify key leaders and chain of command.

b. Signal. Challenge and password, arm and hand signals, special signals, and radio frequencies and call signs.

Appendix C. Patrol Evaluation Checklist

This appendix provides a comprehensive checklist of critical patrolling steps, techniques, and procedures to aid unit leaders to critique patrol performance during training. Unit leaders may use the list as they observe the performance of a patrol to provide the patrol leader with a detailed analysis of the performance.

Warning Order	Yes	No
Did the PL ensure all patrol members were present before issuing the warning order?		
Did the PL issue a brief statement on the enemy situation?		
Did the PL issue a brief statement on the friendly situation?		
Did the PL state the mission in a clear, positive manner?		
Did the PL list all members of the patrol including attachments?		
Was the chain of command for the patrol covered fully?		
Were all members of the patrol asigned positions and duties in particular squads and teams?		
Were all the necessary individual duties assigned?		
Did the PL follow established principles in organizing the patrol into squads and teams?		
Was each patrol member assigned a particular weapon to carry on patrol?		
Did the PL list all special equipment needed to accomplish the mission?		
Was required special equipment assigned to the proper element to carry?		
Did the PL select uniform and equiment common to all based on METT-T?		
Coordination with Adjacent Units		
Did the PL coordinate with other patrols operating to the right and left?		
Was the route out and back coordinated?		
Was the time of departure and return cocordinated?		
Were call signs and frequencies coordinated?		
Was a signal for the FPF coordinated so as not to approach friendly lines during this time?		
Coordination With Front Line Units		
Did the coordinator pass the size of the patrol?		
Was the time of departure and return coordinated?		
Did the coordinator give a general area of oeprations for the patrol?		
Did the coordinator ask for information on known or suspected enemy positions and/or obstacles?		
Did the coordinator ask about information on the latest enemy activity?		
Did the coordinator ask for detailed information on friendly fire support available and the unit's barrier plan?		
Was the location of the IRP established and coordinated?		
Did the coordinator ask the forward unit to monitor their patrol frequency?		
Was the current challenge and password confirmed?		
Did the coordinator request that all information coordinated be passed on to any relieving unti?		
Patrol Order		
General		
Did the PL check to ensure all patrol members were present before issuing the patrol order?		

	Yes	No
Did the PL issue the patrol order in a forceful, confident manner?		
Did the PL make maximum use of available visual aids in issuing the patrol order (i.e., terrain model, sand table, map board, chalkboard)?		
Did the PL issue the patrol order in correct sequence?		
Did the PL issue the entire patrol order without allowing interruptions by patrol members?		
Did the PL adequately answer all questions asked by patrol members		
Did paragraph 1a (Enemy Situation) include—		
A weather forecast for the period of operation?		
A description of the terrain over which the patrol was to operate?		
Identification or description of enemy units known to bein the area of operations?		
Known locations of enemy units?		
Recent activity of enemy units?		
Stength of enemy units in the operating area?		
Did paragraph 1b (Friendly Situation) include—		
Mission of the next higher unit?		
Locations, missions, and planned actions of units on right and left?		
Fire support available to support the patrol (as per the coordination)?		
Missions and routes of other patrols operating in the immediate area?		
Did paragraph 1c (Commander's Intent) include—		
All views and ides of what the commander wants to be accomplished?		
The final result desired by the commander?		
Did paragraph 1c (Attachments and Detachments) include—		
All attachments to patrol and effective time of attachment?		
All detachments from patrol and effective time of detachment?		
Did paragraph 2 (Mission) include, at a minimum—		
Who was to conduct the patrol?		
What the patrol was tasked to do (i.e., conduct a point recon patrol)?		
Where the action was to take place?		
Why the action was to take place (i.e., the purpose)?		
Did paragraph 3 a (Commander's Intent and Concept of Operations) include—		
A complete concept of operations?		
A detailed description of the mission of elements?		
A detailed description of the mission of teams?		
A detailed description of the duties of specific individuals (i.e., navigator, compass man, pace m an, corpsman, APL)?		
Time of departure and time of return?		

	Yes	No
Type of formations and order of movement to be used?		
Description of the route and alternate routes(s) to include azimuths and distances between checkpoints as a minimum?		
Techniques to be used in the departure from friendly areas?		
Techniques to be used for the reentry into friendly areas?		
Location of the IRP and a tentative ORP, to include grid coordinates and recognizable terrain features?		
Method for designating and passing rally points?		
Actions to be taken at rally points if their use became necessary?		
Actions to be taken in the event of enemy contact?		
Actions to be taken at danger areas?		
Complete, detailed description of actions to be taken at the objective?		
Times and locations for rehearsals and the order of priority for rehearsals?		
Times and lcoations for inspections and the methods of conduct (to include uniform and equipment to be worn and/or carried)?		
Where, when, and by whom the debriefing is to be conducted?		
Did paragraph 4 (Administration and Logistics) include—		
Rations to be carried or reference to the warning order if there is no change?		
Arms and ammunition to be carried or reference to the warning order if there is no change?		
Uniform and equipment to be worn and carried or reference to the warning order if there is no change?		
Method for handling wounded or dead?		
Method for handling prisoners?		
Did paragraph 5a (Signal) include—		
Complete description of all signals, code words, frequencies, and call signs to be used within the patrol?		
Were signals adequate?		
Radio call signs to be used?		
Primary and alternate frequencies to be used?		
Call signs and frequencies to be used with other units in the area of oeprations (if applicable)?		
Required reports to be transmitted to higher headquarters, along with occasions for reporting?		
Code words and brevity codes to be used between the patrol and higher headquarters?		
Current challenge and password to be used in friendly held areas?		
Current challenge and password to be used forward of FEBA?		
Did paragraph 5b (Command) include—		
Chain of command or reference to the warning order if there is not change?		
Location of the PL during all stages of operation?		

	Yes	No
Location of the APL during all stages of operation?		
Inspection Rehearsal		
General		
Did the P L adequately use the allotted inspection time?		
Was the inspection conducted in an orderly manner?		
During the inspection, did the PL check the following:		
Completeness and correctness of uniform?		
Completeness of all equipment necessary to accomplish the mission?		
Operational condition of equipment?		
Did the PL question members to ensure they knew—		
The mission of the patrol?		
The concept of operations?		
Their individual duties and responsibilities?		
Chain of command and succession?		
Duties and responsibilities of key personnel?		
Did the PL utilize the rehearsal area to conform as much as possible to area of operations?		
Were the following major actions rehearsed:		
Actions at the objective?		
Actions at danger areas?		
Actions on enemy contacts?		
Departure and reentry of friendly lines?		
Did the PL critique each action after it had been rehearsed?		
Were interpatrol communications and control measures checked at rehearsal?		
Did the PL maintain control of the patrol during the rehearsal?		
Movement		
Did the PL use proper formations for movement?		
Did the PL adequately control the point team?		
Did the PL check the compass men?		
Was the PL aware of his position at all times?		
Was light and noise discipline enforced?		
Was the rate of movement appropriate for the mission?		
Did the PL make full use of pace?		
Did the PL make full use of count?		
Was proper security maintained during movement?		
Did the PL select and pass proper rallying points?		
Did the PL ensure that all members knew location of rallying points?		

	Yes	No
Did the PL properly use arm and hand signals?		
Did the PL recognize and halt the patrol a safe distance from a danger area?		
If the PL conducted a map check, did he conduct it properly (light discipline, security, etc.)		
Did designated individuals reconnoiter the far side of a danger area properly?		
Were support teams emplaced properly?		
Did the PL use a proper formation in crossing the danger area?		
Were reports made on enemy contact and at checkpoints?		
Actions on Enemy Contact		
Did the PL halt patrol a safe distance from a tentative ORP?		
Did the PL issue an adequate frag order before departing to look for a tentative ORP?		
Did the PL take appropriate personnel to look for the ORP?		
Did the PL select a suitable ORP?		
Did the PL secure the site adequately?		
Did the patrol occupy the ORP as stated in the patrol order or subsequent frag order?		
Did the patrol occupy the ORP in an orderly manner?		
Did the PL maintain control of the patrol during occupation of the ORP?		
Did the PL issue a satisfactory contingency plan to the APL before departing on a leader's recon?		
Did the PL take appropriate personnel on a leader's recon?		
Did the PL maintain or ensure that the patrol leader had communications with the patrol?		
Did the patrol avoid being detected by the enemy during the leader's recon?		
If contact was made, did the PL take appropriate action?		
Was security maintained during the leader's recon?		
Did the PL ensure the objective was kept under surveillance?		
Were appropriate orders given to surveillance teams?		
If a leader recon proved a tentative ORP unsuitable, did the PL move the ORP?		
Was a satisfactory leader's recon conducted?		
Was the ORP move conducted in a satisfactory and orderly manner?		
On return to the ORP, did the PL issue appropriate frag order and allow enough time for dissemination?		
Actions at Objective		
Did the PL issue a frag order for action at the objective?		
Was security emplaced prior to the teams departing from the ORP?		
Did the PL employ the terrain at the objective to the best advantage?		
Was stealth maintained while moving into the objective?		

	Yes	No
Did the Pl exercise positive control of squads, teams, and individuals at the objective?		
Were actions at the objective in accordance with the details outlined in the patrol order?		
Did the PL make use of supporting arms at the objective?		
Was the action at the objective successful?		
Was withdrawal from the objective accomplished quickly and orderly?		
Did units withdraw according to the details reflected in the patrol order?		
Was reorganization at the ORP completed in an expeditious and orderly manner?		
Did the patrol withdraw from the ORP quickly and quietly?		

Appendix D. Ambush Formations

Formations discussed here are identified by names which correspond to the general pattern formed on the ground by deployment of the assault element.

Line Formation

The assault element is deployed generally parallel to the route of movement of the enemy. This positions the assault element parallel to the long axis of the killing zone and subjects the target to heavy flanking fire. The size of the force which can be trapped in the killing zone is limited by the area the assault element can effectively cover. The enemy is trapped in the killing zone by natural obstacles, mines, demolitions, and direct fires. (See fig. D-l.)

A disadvantage of the line formation is the chance that lateral dispersion of the target may be too great for effective coverage. The line formation is appropriate in close terrain that restricts enemy maneuver, and in open terrain where one flank is protected by natural obstacles or can be protected by mines and demolitions. Similar obstacles can be placed between the attack force and the killing zone to provide protection from possible enemy counter attack. When an ambush is deployed in this manner, access lanes are left so that the enemy can be assaulted. (See fig. D-2.)

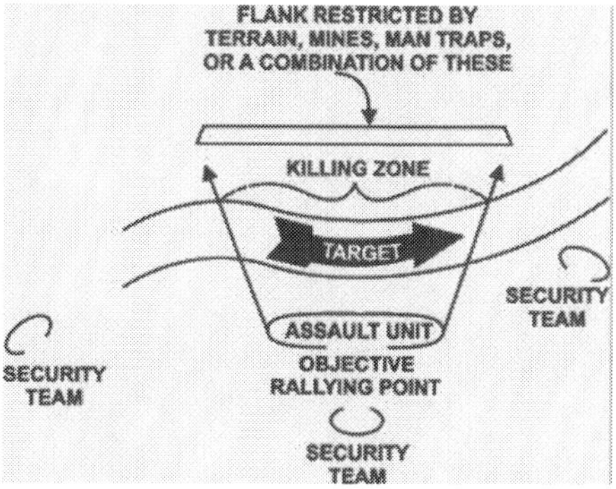

Figure D-1. Line Formation (Harassing or Destruction Ambush).

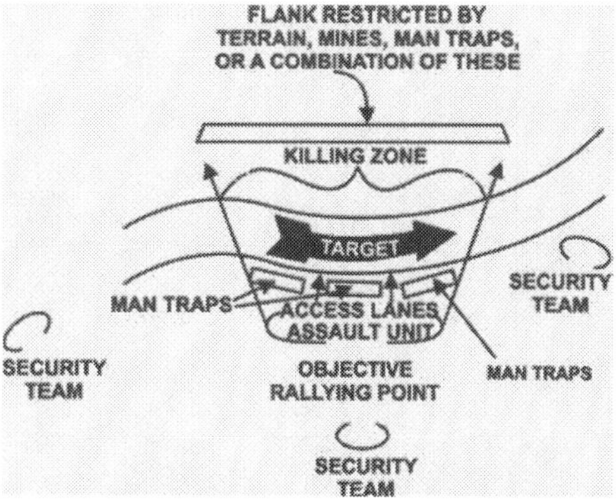

Figure D-2. Line Formation (Access Lanes for Assault of Target).

The main advantage of the line formation is its relative ease of control under all conditions of visibility.

L Formation

The "L" shaped formation is a variation of the line formation. This formation is very flexible because it can be established on a straight stretch of a trail or road (see fig. D-3) or at a sharp bend in a trail or a road (See fig. D-4 on page D-2).The long side of the assault element is parallel to the killing zone and delivers flanking fire. The short side of the attack force is at the

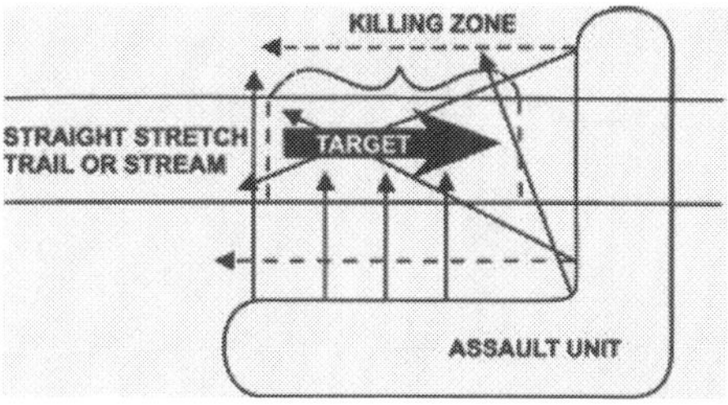

Figure D-3. "L" Formation (Destruction Ambush).

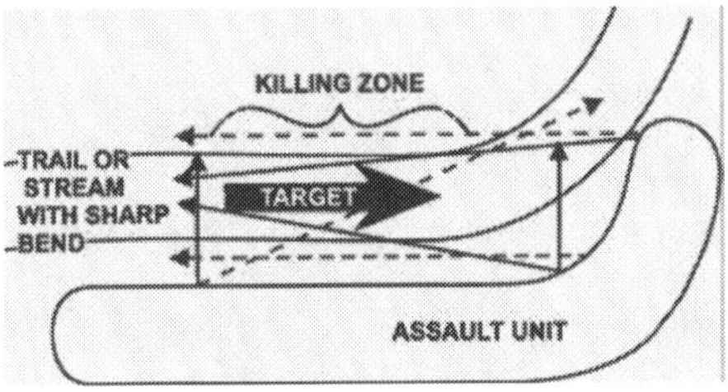

Figure D-4. "L" Formation (Bend of Trail or Stream).

end of, and at right angles to, the killing zone and delivers enfilading fire that interlocks with fire from the long side. When appropriate, fire from the short side can be shifted to parallel the long side if the enemy attempts to assault or escape in the opposite direction. In addition, the short side prevents escape and reinforcement. (See fig. D-5.)

"Z" Formation

The "Z" shaped formation is another variation of the line formation.

The assault element is deployed as in the "L" formation, but with an additional side so that the formation resembles a "Z". (See fig. D-6).

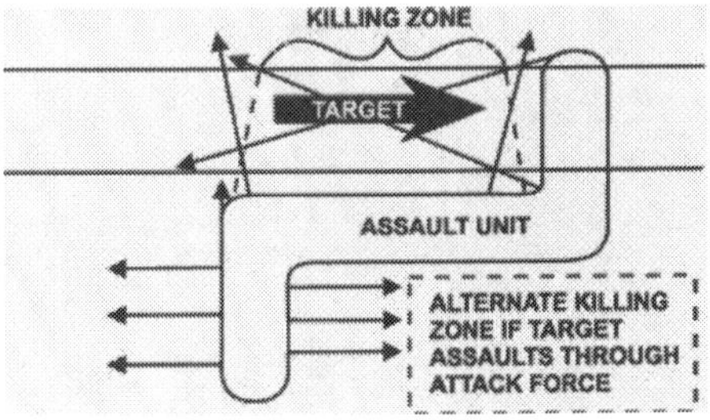

Figure D-6. "Z" Formation.

The additional wing may serve any of the following purposes:

- To engage an enemy force attempting to relieve or reinforce the enemy unit engaged in the kill zone.
- To seal the end of the killing zone.
- To restrict a flank.
- To prevent an envelopment.

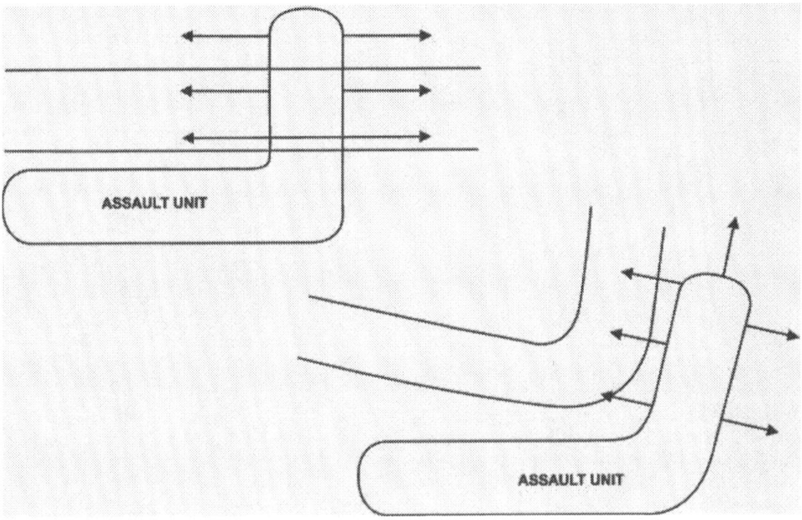

Figure D-5. "L" Formation.

"T" Formation

In the "T" shaped formation, the assault element is deployed across (perpendicular to) the enemy's route of movement so that its position forms the crossing of a "T" at the top. (See fig. D-7.)

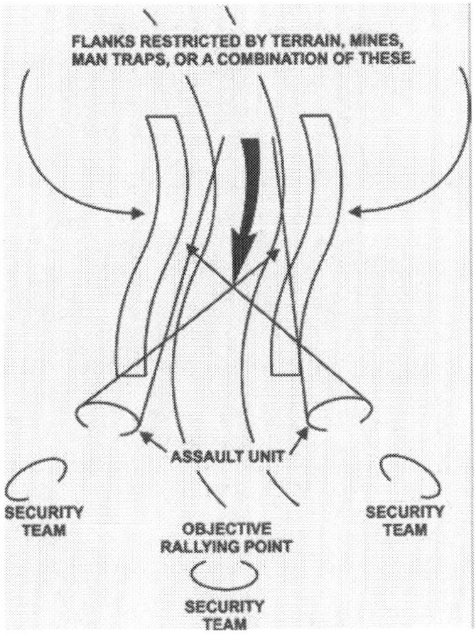

Figure D-7. "T" Formation.

This formation can be used day or night to establish an ambush to interdict movement through open areas that are hard to seal off.

A small force can use the "T" formation to harass, slow, and disorganize a larger force. When the lead elements of the enemy are engaged, they will normally attempt to maneuver right or left to close with the ambush. Mines and other obstacles placed to the flanks of the killing zones slow the enemy's movements and permit the ambush force to deliver heavy fire and withdraw without becoming decisively engaged.

The "T" formation can be used to interdict small groups attempting night movement across open areas. For example, the assault element is deployed along an avenue of approach with every second man facing the opposite direction. The attack of the enemy approaching from either direction requires only that every second man may shift to the opposite side of the formation. Each man fires only to his front and only when the enemy is at a very close range. Attack is by fire only and each man keeps the enemy under fire as long as it remains to his front.

If the enemy attempts to escape in either direction along the killing zone, each Marine takes the enemy under fire as the enemy comes into the Marine's sector of fire. The "T" formation is very effective at halting infiltration. But it has one chief disadvantage; there is a possibility that the ambush will engage a superior force at night while spread out. (See fig. D-8.)

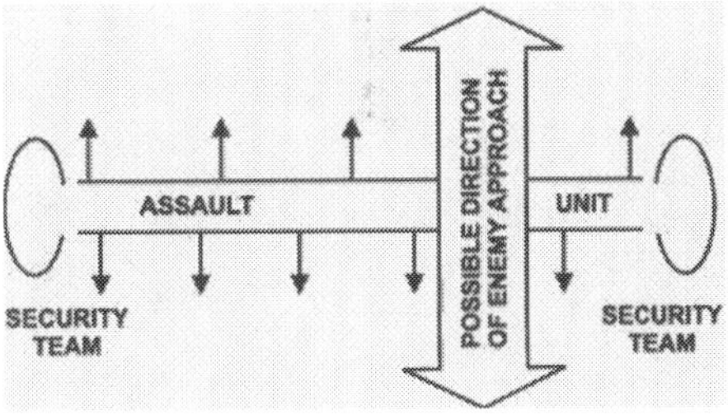

Figure D-8. "T" Formation (Target Approaching from Either Direction).

The "V" Formation

The "V" shaped formation is deployed along both sides of the enemy's route of movement so that it forms a "V"; care is taken to ensure that neither group (within the "V") fires into the other.

This formation subjects the enemy to both enfilading and interlocking fire. The "V" formation is best suited for fairly open terrain but can also be used in close terrain. When established in close terrain, the legs of the "V" close in as the lead element of the enemy force approaches the apex of the "V", and opens fire at a close range.

Here, even more than in open terrain, all movement and fire must be carefully coordinated and controlled to ensure that the fire of one wing does not endanger the other wing. The wider separation of forces makes this formation difficult to control, and there are few sites that favor its use. Its main advantage is that it is difficult for the enemy to detect the ambush until it is well into the killing zone. (See figs. D-9 and D-10.)

Triangle Formation

The triangle is a variation of the "V" formation and can be varied in three ways:

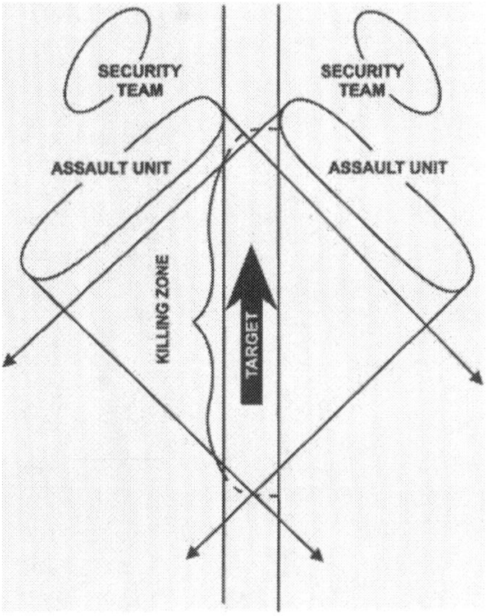

Figure D-9. "V" Formation (Open Terrain).

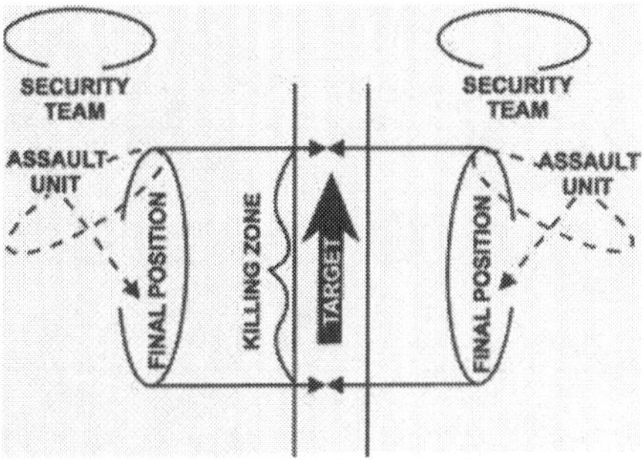

Figure D-10. "V" Formation (Close Terrain).

Closed Triangle Formation

(See fig. D-11.) The assault element is deployed in three teams, positioned so that they form a triangle (or closed "V"). An automatic weapon is placed at each point of the triangle and positioned so that it can be shifted quickly to interlock with either of the others. Men are positioned so that their sectors of fire overlap. Mortars may be positioned inside the triangle.

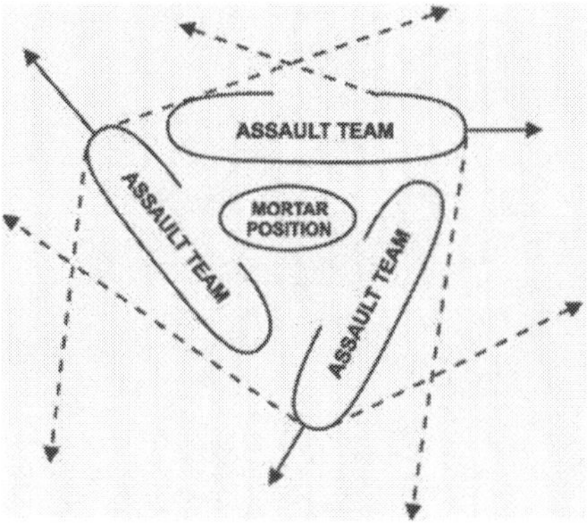

Figure D-11. Closed Triangle Formation (Night Ambush).

When deployed in this manner, the triangle ambush is used to interdict night movement through open areas. When enemy approach is likely to be from any direction, this formation provides all-around security, and security forces are deployed only when they can be positioned so that if detected by an approaching enemy, they will not compromise the ambush. Attack is by fire only, and the enemy is allowed to approach within close range before fire is initiated.

The advantages of the closed triangle formation are ease of control, all-around security, and the enemy can be brought under the fire of at least two automatic weapons, regardless of the direction they approach.

Disadvantages are that it requires a force of platoon size or larger to reduce the danger of being overrun by an unexpectedly large force and that one or more sides of the triangle may come under enfilade fire. The lack of dispersion, particularly at the points, increases the danger from enemy mortar fire.

Open Triangle Harassing Formation

This variation of the triangle ambush enables a small force to harass, slow, and inflict heavy casualties upon a large force without itself being decisively engaged. The assault element is deployed in three teams, positioned so that each team becomes a corner of a triangle containing the killing zone. (See fig. D-12.)

When the enemy enters the killing zone, the team to the enemy's front opens fire on the leading element. When the enemy counterattacks, the group withdraws and the team on the enemy's flank opens fire. When this team is attacked, the team to the opposite flank opens fire. This process is repeated until the enemy is pulled apart. Each team reoccupies its position, if possible, and continues to inflict the maximum damage possible without becoming decisively engaged.

Open Triangle Destruction Formation

The assault element is again deployed in three teams, positioned so that each team is a point of the triangle, 200 to 300 meters apart. The killing zone is the area within the triangle. The enemy is allowed to enter the killing zone; the nearest team attacks by fire. As the enemy attempts to maneuver or withdraw, the other teams open fire. One or more teams, as directed, assault or maneuver to envelop or destroy the enemy. (See fig. D-13 on page D-6.)

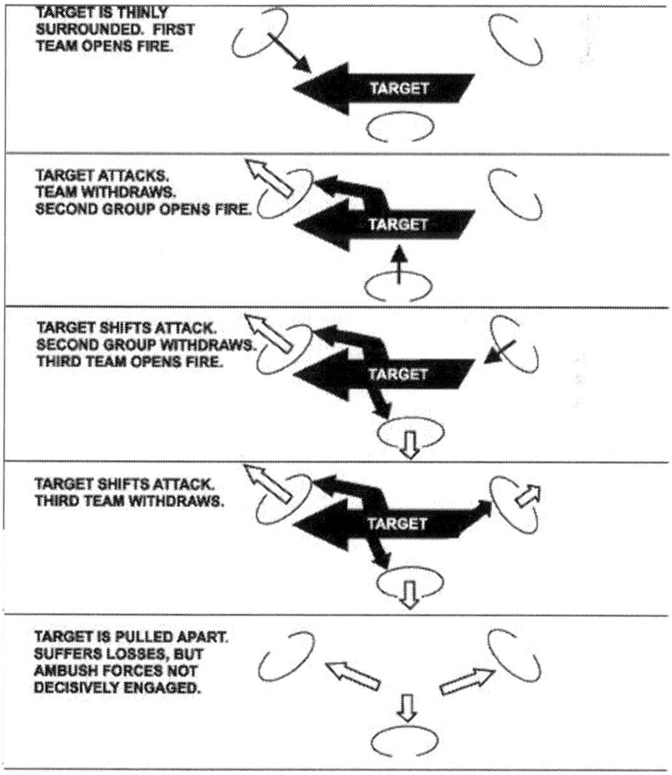

Figure D-12. Open Triangle Formation.

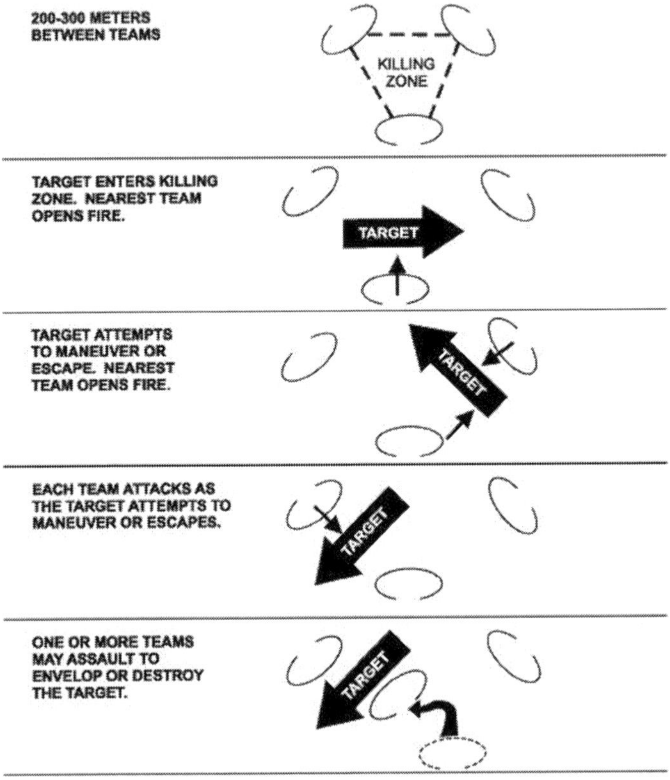

Figure D-13. Open Triangle Formation (Destruction Ambush).

This formation is suitable for platoon-size or larger ambush forces. A smaller force would be in too great a danger of being overrun. Another disadvantage is that control, in assaulting or maneuvering, is very difficult. Very close coordination and control are necessary to ensure that assaulting or maneuvering teams are not fired on by another team. The ambush site must be a fairly level open area that provides concealment around its border for the ambush force.

The Box Formation

The "box" formation is similar in purpose to the open triangle ambush. The assault element is deployed in four teams, positioned so that each team becomes a corner of a square or rectangle containing the killing zone. It can be used for a harassing or destruction ambush in the same manner as the variations of the open triangle formation. (See figs D-14 on page D-7 and D-15 on page D-8.)

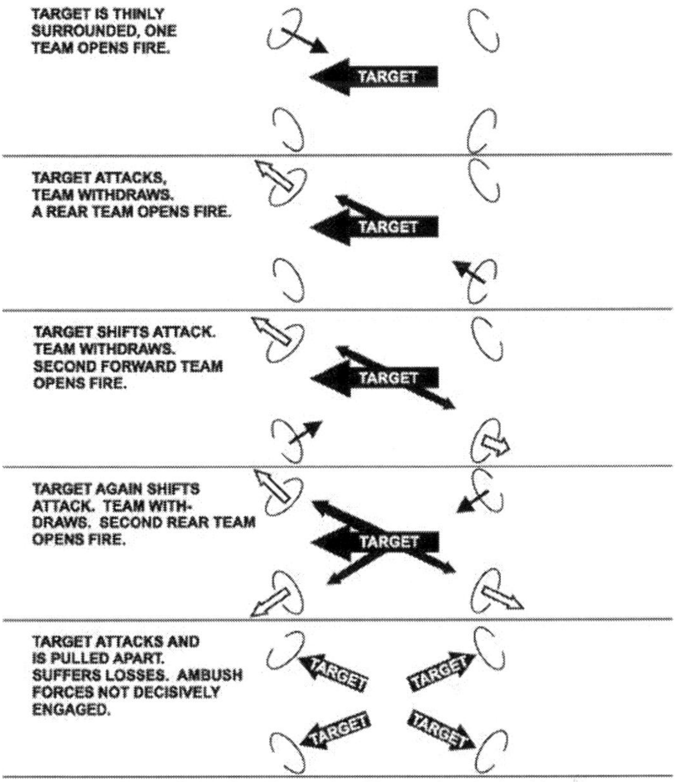

Figure D-14. Box Formation (Harrassing Ambush).

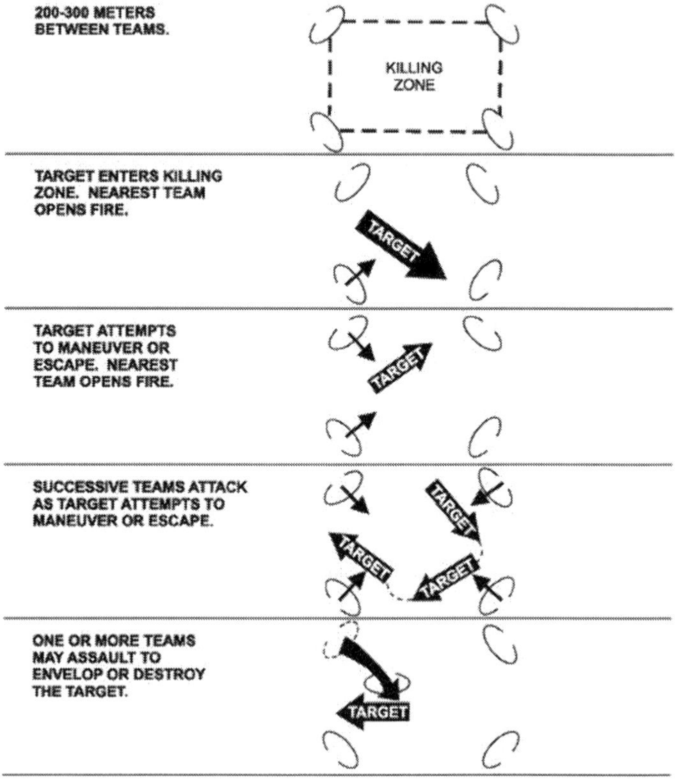

Figure D-15. Box Formation (Destruction Ambush).

Appendix E. Acronyms

A&S . . . assault and security

CIT . . . counterintelligence team

EOD . . . explosive ordnance disposal

ITT . . . interrogator-translator team

MAGTF . . . Marine air-ground task force
MEDEVAC . . . medical evacuation
METT-T . . . mission, enemy, terrain and weather, troops and support available

NBC . . . nuclear, biological, chemical
NCO . . . noncommissioned officer

OTL . . . overserver-target line

RCA . . . riot control agents
ROE . . . rules of engagement

SMAW . . . shoulder-launched multipurpose assault weapon
SMEAC . . . situation, mission, execution, administration and logistics, and command and signal
SOP . . . standing operating procedures
SPOTREP . . . spot report

TTP . . . tactics, techniques, and procedures

Appendix F. References

Joint Publication (JP)

1-02 DOD Dictionary of Military and Associated Terms

Fleet Marine Force Manual (FMFM)

6-5 Marine Rifle Squad (proposed MCWP 3-11.2)

Marine Corps Warfighting Publications (MCWP)

3-15.3 Scout Sniping (under development)

3-35.3 Military Operations on Urbanized Terrain (MOUT)

3-41.2 Raids (under development)

Marine Corps Reference Publications (MCRP)

2-15.3A Reconnaissance Patrol Leader's Planning Handbook (under development)

2-15.3B Reconnaissance Reports Guide

3-02C Water Survival Handbook (under development, currently exists as Fleet Marine Force Manual 0-13)

5-12A Operational Terms and Graphics

5-12C Marine Corps Supplement to the DOD Dictionary of Military and Associated Terms

Made in United States
Orlando, FL
04 November 2026

90203576R10208